BIOCHEMISTRY

The LABFAX series

Series Editors:

B.D. HAMES Department of Biochemistry and Molecular Biology, University of Leeds, Leeds LS2 9JT, UK

D. RICKWOOD Department of Biology, University of Essex, Wivenhoe Park, Colchester CO4 3SQ, UK

MOLECULAR BIOLOGY LABFAX

CELL BIOLOGY LABFAX

CELL CULTURE LABFAX

BIOCHEMISTRY LABFAX

Forthcoming titles

VIROLOGY LABFAX

PLANT MOLECULAR BIOLOGY LABFAX

IMMUNOCHEMISTRY LABFAX

BIOCHEMISTRY

EDITED BY
J.A.A. CHAMBERS
1487 West 5th Avenue, Apt. 311,
Columbus, OH 43212, USA

and
D. RICKWOOD
Department of Biology, University of Essex,
Wivenhoe Park, Colchester CO4 3SQ, UK

β IOS
SCIENTIFIC
PUBLISHERS

ACADEMIC PRESS

©BIOS Scientific Publishers Limited, 1993

All rights reserved by the publisher. No part of this book may be reproduced or transmitted, in any form or by any means, without permission in writing from the publisher.

First published in the United Kingdom 1993 by
BIOS Scientific Publishers Limited
St Thomas House, Becket Street, Oxford OX1 1SJ, UK

ISBN 0 12 167340 5

A CIP catalogue entry for this book is available from the British Library.

This Edition published jointly in the United States of America by Academic Press, Inc. and BIOS Scientific Publishers Limited.

Distributed in the United States, its territories and dependencies, and Canada exclusively by Academic Press, Inc., 1250 Sixth Avenue, San Diego, California 92101 pursuant to agreement with BIOS Scientific Publishers Limited, St Thomas House, Becket Street, Oxford OX1 1SJ, UK.

Typeset by Unicus Graphics Ltd, Horsham, UK.
Printed by Information Press Ltd, Oxford, UK.

The information contained within this book was obtained by BIOS Scientific Publishers Limited from sources believed to be reliable. However, while every effort has been made to ensure its accuracy, no responsibility for loss or injury occasioned to any person acting or refraining from action as a result of the information contained herein can be accepted by the publishers, authors or editors.

PREFACE

Biochemistry continues to be one of the most important areas of research in the life sciences. Indeed, with the need to characterize the diverse range of gene products generated as a result of genetic engineering techniques, there has been a resurgence of interest in biochemistry as molecular biologists examine the changing properties of modified proteins.

There have been handbooks on biochemistry, both encyclopedic and selective. However, the former type of book tests physical fitness and patience and the selection of topics for the latter often does not reflect the current biochemical practices and techniques because the books were published some time ago. We, as editors, have attempted to assemble the most important facts in a way that reflects current biochemical techniques to provide data that biochemists will need to use in their work on an almost daily basis. The length restriction of this book has required us to be very selective in terms of which topics to include and which to exclude. We have tried to emphasize those aspects of the subject that have not been included in previous books. We hope that this will prove to be a useful data book for everyone working in the area of biochemistry.

J.A.A. Chambers

D. Rickwood

HAZARD WARNING

Some of the chemicals and procedures described in this book may be associated with chemical and biological hazards. In addition, the reader should be aware of the hazards associated with the handling of animal tissue samples. While efforts have been made to indicate the hazards associated with the different reagents and procedures covered in this book, it is the ultimate responsibility of the reader to ensure that safe working practises are used.

In several chapters Chemical Abstracts Service Registry numbers have been supplied for the reagents discussed. These numbers, found in square brackets and of the general formula XXXXX-YY-Z, are assigned to a unique chemical structure regardless of name and are of particular use in the literature searches and in the recovery of information from a number of chemical and biological databases. We decided to include these numbers when it became obvious that the number of names for some of the substances listed were in the hundreds and that the numbers provided another way to help the researcher find further information.

CONTENTS

Contributors   xv
Abbreviations   xvii

1. BUFFERS, CHELATING AGENTS AND DENATURANTS   1

Buffers   1
   Introduction   1
   Goods' buffers   1
   Preparation of buffers   1
   Counter ions   1
   Effects of temperature and concentration   1
   Biological compatibility and chemical reactivity   2
      pH ranges of non-zwitterionic buffers (Figure 1)   2
      pH ranges of zwitterionic buffers (Figure 2)   3
      Components of non-zwitterionic buffers (Table 1)   5
      Zwitterionic (Goods') buffers (Table 2)   8
   Preparation of phosphate buffers   9
      Preparation of sodium phosphate buffer (Table 3)   9
   Preparation of acetate buffers   9
      Preparation of acetate buffer (Table 4)   10
   Buffered salines, balanced salts and osmotic supports   10
      Balanced salt solutions (Table 5)   11
Chelating agents   11
      Chelating agents: solubility and pK_a values (Table 6)   12
      Chelating agents: stability constants (Table 7)   13
Denaturing agents for proteins   13
   Mechanisms of denaturing agents   13
   Mixing denaturants   14
   Characteristics of denaturants and precipitants   14
      Solubilizing denaturants (Table 8)   15
      Precipitating denaturants (Table 9)   16
Thiol reagents   17
      Thiol reagents (Table 10)   17
Detergents   17
      Detergents and surfactants (Table 11)   19
Properties of acids, bases, salts and organic solvents   22
      Properties of acids, bases, salts and organic solvents (Table 12)   23
      Acid–base indicators (Table 13)   29
Amino acids   22
      L-Amino acids: physical and chemical data (Table 14)   31
      Amino acid classification by solution properties of side chains (Table 15)   33
Ammonium sulfate precipitation chart for proteins   33
      Chart for ammonium sulfate precipitation of proteins (Table 16)   34
      Saturated ammonium sulfate solutions at various temperatures (Table 17)   35
References   35

2. RADIOISOTOPES IN BIOCHEMISTRY   37

Definitions   37
   Units of radioactivity   37
   Specific activity   37
   Electron volt (eV, or MeV = 10^6 eV)   37
   Gray   37
   Dose equivalent man (sievert)   37

Radioisotopes used in biochemistry	37
Radioactive isotopes used in biochemical studies (Table 1)	38
Radioactive decay correction	39
Half-life activity corrections for selected radioisotopes (Table 2)	40
Autoradiography	39
Sensitivities of film detection methods for commonly used radioisotopes (Table 3)	41
Scintillation counting	41
Cerenkov counting	41
Counting efficiency using Cerenkov counting (Table 4)	41
Counting of finely dispersed or solvent-soluble substances	41
Structures of compounds used in scintillants (Figure 1)	42
Counting of aqueous samples	42
Gamma counting	43
Use of radioisotopes as tracers	43
Radiological protection	45
Shielding for β-particles and γ-rays	46
Shielding for β-particle emitters (Table 5)	46
Shielding for γ-ray emitters (Table 6)	46
Methods for decontaminating laboratories	46
Decontamination methods (Table 7)	47
References	48
Further reading	48

3. CHROMATOGRAPHIC FRACTIONATION MEDIA 49

Introduction	49
Media for size exclusion	50
Fractionation ranges of commercially available gel filtration matrices (Figure 1)	51
Carbohydrate-based column support materials for separation by size exclusion (Table 1)	52
Silica-based column support materials for size exclusion HPLC of proteins and peptides (Table 2)	55
Polymer-based column support materials for size exclusion HPLC of proteins and peptides (Table 3)	56
Controlled pore glass for permeation chromatography (Table 4)	57
Media for ion-exchange separations	50
Ion-exchange cellulose media: physical and chemical properties of Whatman cellulose media (Table 5)	58
Ion exchange: anion exchangers on polystyrene (Table 6)	59
Ion exchange: cation exchangers on polystyrene (Table 7)	61
Some commercially available column support materials for ion-exchange HPLC of proteins (Table 8)	62
Media for reversed-phase separations	50
List of silica-based reversed-phase column support materials (Table 9)	64
List of non-silica-based reversed-phase column support materials (Table 10)	66
Media for affinity chromatography	50
Some characteristics of commonly used matrices for affinity chromatography (Table 11)	67
Manufacturers and suppliers of chromatography column support materials	66
References	68
Further reading	68

4. ELECTROPHORESIS OF PROTEINS AND NUCLEIC ACIDS 69

Proteins	69
Separations on denaturing gels	69
Recipe for gel preparation using the SDS-PAGE discontinuous buffer system (Table 1)	70
Separations on non-denaturing gels	69
Buffers for non-denaturing discontinuous systems (Table 2)	71
Recipe for gel preparation using non-denaturing continuous buffer systems (Table 3)	72
Separations of proteins by isoelectric focusing	69
Commercially available carrier ampholytes (Figure 1)	72
Recipes for isoelectric focusing gels (Table 4)	73
Two-dimensional gel electrophoresis	71

Marker proteins	74
Standard marker proteins (Table 5)	74
Staining protein gels	74
Staining procedures for proteins separated on polyacrylamide gels (Table 6)	77
Nucleic acids	74
Gels for separating nucleic acids and nucleoproteins	74
Recipes for preparation of polyacrylamide gels for the electrophoresis of nucleic acids (Table 7)	78
Recipes of gels used for the electrophoresis of polysomes and ribosomes (Table 8)	79
Running buffers for the electrophoresis of nucleic acids (Table 9)	80
Denaturants used in denaturing gels for separating nucleic acids (Table 10)	80
Markers for nucleic acids	74
Molecular weight markers for gel electrophoresis of RNA (Table 11)	81
Sizes of the restriction fragments of pBR322 (Table 12)	82
Sizes of the restriction fragments of phage λ cl *ts* 857 (Table 13)	83
Staining nucleic acid gels	74
Visualization of nucleic acids in gels (Table 14)	84
Methods manuals on electrophoresis	84

5. GENERAL CENTRIFUGATION DATA   85

Calculation of centrifugal force	85
Applications of centrifuge rotors	85
Applications of centrifuge rotors (Table 1)	85
Calculation of k-factors of rotors and pelleting times	85
Derating rotors for use with dense solutions	86
Properties of centrifuge tubes and bottles	86
Centrifuge tube and bottle materials (Table 2)	87
Centrifuge tube and bottle care and use (Table 3)	88
Chemical resistance chart (Table 4)	89
Sterilization and disinfection procedures	96
Sterilization techniques	96
Biological disinfection	96
Sterilization and disinfection procedures (Table 5)	97
Equations relating the refractive index to the density of solutions	98
Ionic gradient media (Table 6)	98
Non-ionic gradient media (Table 7)	99
Properties of sucrose solutions	99
Dilution of stock solutions of sucrose (Table 8)	99

6. ENZYMOLOGY   101

Introduction	101
Enzyme kinetics	101
Enzyme kinetics of single substrate reactions	101
Components of the Michelis-Menten reaction (Figure 1)	102
Initiation velocity versus substrate concentration (Figure 2)	103
Lineweaver-Burke plot (Figure 3)	105
Hane-Woolf plot (Figure 4)	105
Eadie-Hofstee plot (Figure 5)	106
Direct linear plot (Figure 6)	107
Alternative form of the direct linear plot (Figure 7)	108
Sigmoidal kinetics and allosteric enzymes	107
Hill plot (Figure 8)	109
Positive and negative homotropic cooperativity (Figure 9)	111
Enzyme kinetics of bisubstrate reactions	110
Types of Bi Bi reaction kinetics (Table 1)	112
Product inhibition for sequential bisubstrate mechanisms (Table 2)	115
Enzyme inhibitors	115
Reversible inhibition	115
Types of reversible inhibition (Table 3)	117
Irreversible inhibition	115
Inhibitors of specific enzymes (Table 4)	126
Diagnostic plot for irreversible inhibition (Figure 10)	116

Effects of pH on enzymes	130
Amino acid side-chain pK_a values (Table 5)	130
Optimal pH of some important enzymes (Table 6)	131
Coenzymes — structure and functions	130
Structure and functions of some important coenzymes (Table 7)	132
Enzyme assays	130
Summary of main enzyme assay methods (Table 8)	140
Handling and storage of enzymes and coenzymes	141
References	142
Further reading	143

7. HYDROLYTIC ENZYMES 145

Introduction	145
Properties of selected nucleases (Table 1)	146
Inhibition of nucleases and proteases	145
Deoxyribonucleases	145
Selected nuclease inhibitors (Table 2)	147
Ribonucleases	145
Proteases	147
Broadly specific or nonspecific proteases (Table 3)	150
Endoproteases (Table 4)	151
Aminopeptidases (Table 5)	154
Carboxypeptidases (Table 6)	155
Inhibitors of proteases (Table 7)	156
Carbohydrases	147
Carbohydrate-degrading enzymes (Table 8)	161
Naturally occurring or physiological inhibitors	147
References	164

8. CHARACTERISTICS OF SELECTED PROTEINS 167

Properties of apolipoproteins (Table 1)	169
Structure of an IgG molecule (Figure 1)	179
Properties of human immunoglobulins (Table 2)	180
Procaryotic and eucaryotic protein synthesis factors (Table 3)	186
References	190

9. GLYCOPROTEINS AND PROTEIN GLYCOSYLATION 193

Introduction	193
Occurrence of glycosylated proteins in nature	193
Range and types of glycosylated proteins	193
Occurrence of proteins in membranes (Figure 1)	194
Examples of glycosylated proteins in nature (Table 1)	194
Typical glycosylated proteins in the animal glycocalyx (Table 2)	196
Typical structural glycosylated proteins in animal membranes (Table 3)	197
Structural features of glycosylated proteins	196
Linkage of carbohydrate to proteins	196
Carbohydrate–protein linkages (Table 4)	198
Linkage and core structures in proteoglycans (Table 5)	199
Oligosaccharide structures	200
Linkages between monosaccharides in eucaryotic protein oligosaccharides (Table 6)	200
Examples of N-linked oligosaccharide structure (Figure 2)	201
Core structures in O-linked oligosaccharides (Table 7)	201
Repeating backbone carbohydrate structures in glycoproteins (Table 8)	202
Common repeating carbohydrate units in proteoglycans (Table 9)	203
Peripheral structures in protein-linked oligosaccharides (Table 10)	204
Post-translational modifications to glycosylated proteins (Table 11)	205
Enzymic glycosylation — synthesis of sequence	203
Monosaccharide transport, interconversion and activation	203
Carbohydrate transport systems in animal cell membranes (Table 12)	205
Metabolic routes in glycoconjugate biochemistry (Figure 3)	206

Glycosyltransferases	206
Glycosyltransferases — examples of transfer (Table 13)	207
Biosynthesis of an oligosaccharide (Figure 4)	209
Structure of a dolichol-phosphate-linked oligosaccharide (Figure 5)	210
Biosynthesis of a dolichol-linked oligosaccharide (Figure 6)	211
Processing pathways for N-linked oligosaccharides (Figure 7)	212
Examples of functional roles for protein glycosylation	211
Functions of some glycosylated proteins (Table 14)	213
References	214

10. CHEMICAL AND POST-TRANSLATIONAL MODIFICATION OF PROTEINS — 215

Chemical modification of proteins	215
Reagents for selective chemical modification of proteins (Table 1)	216
The specificity of reagents used to chemically modify proteins (Table 2)	221
Enzyme-catalyzed covalent modification reactions	215
Introduction	215
Methods for identification of modified amino acids (Table 3)	233
Influence of modification	234
Characteristics of modification reactions	234
Recognition sequences and donors for protein modification (Table 4)	235
Physiological role of the modification	235
Inhibitors of post-translational modifications (Table 5)	237
Reversibility of reactions	238
Examples of reversible post-translational modifications (Table 6)	238
Summary	239
References	239

11. NUCLEIC ACIDS AND THEIR COMPONENTS — 247

Nucleosides and nucleotides	247
Basic structures	247
Bases and sugars of nucleic acids (Figure 1)	248
Unusual bases of nucleic acids (Figure 2)	249
Base pairing in DNA (Figure 3)	250
Nucleotides as acids	247
Optical density of bases, nucleosides and nucleotides	247
Physical properties (Table 1)	250
Nucleotide-derived compounds	247
Nucleotide-derived compounds (Table 2)	252
Selected nucleotide analogs	247
A selection of nucleotide analogs (Table 3)	255
DNA of selected organisms	251
Bacteria (Table 4)	256
Protozoa, algae, fungi, echinoderms, arthropods and Insecta (Table 5)	257
Chordata (Table 6)	258
Animal viruses (Table 7)	260
Plants (Table 8)	260
The genetic code	251
The genetic code (Table 9)	261
Abbreviations of amino acids	261
Amino acid abbreviations (Table 10)	261
Assays for nucleic acids	262
Assays for DNA and RNA (Table 11)	262
References	262

12. LIPIDS — 267

Introduction	267
Structure and characteristics of acyl lipids	267
Fatty acids	267
Selected saturated and monoenoic fatty acids (Table 1)	268
Selected naturally occurring polyunsaturated fatty acids (Table 2)	269

Neutral acyl lipids — wax esters, acylglycerols and glycerol ethers	270
Structures of common wax esters and acylglycerols (Table 3)	271
Glycerophospholipids	272
Structure and distribution of membrane glycerophospholipids (Table 4)	273
Glycerophospholipids containing structural variations of phosphatidyl moiety (Table 5)	275
Glyceroglycolipids	275
Structures of major glycosylglycerolipids of higher plants and bacteria (Figure 1)	276
Sphingolipids	277
Sphingolipid structures (Figure 2)	277
Structures of some major gangliosides (Figure 3)	278
Structure and distribution of terpenoid constituents of membranes	278
Sterols	279
Structures of major membrane sterols (Figure 4)	279
Chlorophylls and carotenoids	279
Structures of plant and bacterial chlorophylls (Figure 5)	280
Structures of carotenoids of plants, algae and bacteria (Figure 6)	281
Composition and distribution of lipids in membranes	282
Composition of membrane lipids	282
Fatty acid composition of membrane lipids	282
Rat liver preparations (Table 6)	283
Sphingolipids of rat liver cells (Table 7)	283
Representative higher plants, algae and cyanobacteria (Table 8)	284
Chloroplast lipids (Table 9)	286
Selected fungi (Table 10)	287
Selected bacterial membrane systems (Table 11)	288
Structure and properties of bioactive lipids	288
Eicosanoids	288
Structures of the prostanoids (Figure 7)	289
Pathways of eicosanoid formation (Figure 8)	291
Structures of leukotrienes and hydroxyeicosatetraenoic acids (Figure 9)	292
Biological effects of leukotrienes and HETEs (Table 12)	293
Platelet activating factor	294
Biological effects of PAF (Table 13)	294
Diacylglycerol	294
Steroid hormones	297
Structures of steroid hormones and analogs (Figure 10)	296
Properties of steroid hormone systems (Table 14)	298
Structure and composition of bile acids and bile salts	300
Structure of primary and secondary bile acids and bile salts (Figure 11)	300
Chemical Abstracts Registry Numbers of lipids cited	301
References	301

13. CARBOHYDRATES AND SUGARS   305

Introduction	305
Structures and characteristics of monomeric carbohydrates	305
Classification of monosaccharides	305
Classification of monosaccharides (Table 1)	306
Distribution and properties of some monosaccharides	306
Origin and properties of some monosaccharides (Table 2)	307
Stereoisomerism	309
Formation of the hemiacetal forms of D-glucose (Figure 1)	310
Oligosaccharides	310
Classification	310
Classification of simple oligosaccharides (Table 3)	311
Distribution and properties	310
Structures of some common disaccharides (Figure 2)	311
Structures and characteristics of polysaccharides	312
Classification	312
Classes of common polysaccharides (Table 4)	312
Plant polysaccharides	313
Examples of common polysaccharides (Figure 3)	314
Bacterial polysaccharides	313

Animal polysaccharides	313
DNA, RNA, nucleosides and nucleotides	313
Other saccharide derivatives	315
Further reading	315

14. SAFETY AND THE DISPOSAL OF TOXIC AND INFECTIOUS MATERIALS — 317

Protection from chemical hazards	317
Risk and safety classification systems	317
Hazard symbols (Table 1)	317
UN Chemical Hazard Classification (Table 2)	318
European Commission Risk and Safety Phrases (Table 3)	319
Disposal of toxic and infectious materials	322
Disposal of toxic materials	322
Toxic materials: effects and methods of disposal (Table 4)	323
Disposal of biohazardous materials	329
Pressure-temperature relationships for autoclaves (Table 5)	329
Effectiveness of disinfectants against infectious agents (Table 6)	329
References	330

15. SOURCES OF FURTHER BIOCHEMICAL DATA — 331

Reference books	331
Databases	331
Nucleic acid sequence databases	342
Contact addresses	343

16. ATOMIC WEIGHTS AND MATHEMATICAL FORMULAE — 345

Atomic weights	345
Atomic weights (Table 1)	345
Mathematical formulae	347
Lengths, areas and volumes in some common geometric figures	347
Inter-relations of sides and angles in a plane triangle	349
Trigonometrical data	350
Mathematical series	352
Mathematical constants	353
Standard equations	353
Differentials and integrals	355

INDEX — 357

CONTRIBUTORS

A.S. BALL
Department of Biology, University of Essex, Wivenhoe Park, Colchester CO4 3SQ, UK

D. BILLINGTON
School of Biomolecular Sciences, Liverpool John Moores University, Byrom Street, Liverpool L3 3AF, UK

T.A. BROWN
Department of Biochemistry and Applied Molecular Biology, UMIST, Manchester M60 1QD, UK

J.A.A. CHAMBERS
1487 West 5th Avenue, Apt. 311, Columbus, OH 43212, USA

A.P. CORFIELD
Department of Medicine Laboratories, Bristol Royal Infirmary, Bristol BS2 8HW, UK

R.E. FEENEY
Department of Food Science and Technology, University of California, Davis, CA, USA

J.M. GRAHAM
Merseyside Innovation Centre, 131 Mount Pleasant, Liverpool L3 5TF, UK

B.L. MARTIN
Department of Biochemistry and Biophysics, Iowa State University of Science and Technology, 1210 Molecular Biology Building, Ames, IA 50011, USA

G.E. MEANS
Department of Biochemistry, 484 West 12th Avenue, Ohio State University, Columbus, OH 43210-1292, USA

D. PATEL
Department of Biology, University of Essex, Wivenhoe Park, Colchester CO4 3SQ, UK

J. QIU
Beatson CRC Laboratories, Garscube Estate, Bearsden, Glasgow G61 1BD, UK

D. RICKWOOD
Department of Biology, University of Essex, Wivenhoe Park, Colchester CO4 3SQ, UK

T.J. WALTON
Biochemistry Research Group, School of Biological Sciences, University College of Swansea, Singleton Park, Swansea SA2 8PP, UK

ABBREVIATIONS

AA	amino acid
Aces	N-(2-acetamido)-2-aminoethanesulfonic acid
ADP	adenosine diphosphate
AMP	adenosine monophosphate
APMSF	amidino-phenylmethylsulfonyl fluoride
APS	adenosine phosphosulfate
Asn	asparagine
ATP	adenosine triphosphate
ATPase	adenosine triphosphatase
Bes	N,N-bis(2-hydroxyethyl)-2-aminoethanesulfonic acid
Bicine	N,N-bis(2-hydroxyethyl)glycine
Bis-Tris	bis(2-hydroxyethyl)imino-tris(hydroxymethyl)methane
butyl PBD	2-(4'-t-butylphenyl)-5-(4''-biphenylyl)-1,3,4-oxadiazole
cAMP	cyclic adenosine 3',5'-monophosphate
Caps	3-(cyclohexylamino)-1-propanesulfonic acid
CAT	chloramphenicol acetyl transferase
CDP	cytosine diphosphate
Cer-gal	monogalactosylcerebroside
Cer-glu	glucosyl cerebroside
CHAPS	3-[(3-cholamidopropyl)-dimethylammonio]-1-propanesulfonate
CHAPSO	3-[(3-cholamidopropyl)-dimethylammonio]-2-hydroxy-1-propanesulfonate
Ches	2-(cyclohexylamino)-ethanesulfonic acid
CMC	critical micellar concentration
CMP	cytosine monophosphate
CPB	cetylpyridinium bromide
CPI	carboxypeptidase inhibitor
CTAB	cetyltrimethylammonium bromide
CTP	cytosine triphosphate
DABA	diaminobenzoic acid
DAGs	diacylglycerols
DAPI	4,6,diamidino-2-phenylindole
DGDG	digalactosyldiacylglycerol
DHA	docosahexaenoic acid
DHFA	dihydrofolate
DMAPN	3,dimethylaminopropionitrile
DMF	dimethylformamide
DMSO	dimethylsulfoxide
DNA	deoxyribonucleic acid
Dol	dolichol
DOPA	3,4-dihydroxyphenylalanine
DPG	diphosphatidylglycerol (cardiolipin)
DTNB	5,5'-dithiobis(2-nitrobenzoic acid)

DTT	dithiothreitol
ε_{max}	extinction coefficient (molar)
EACA	ε-amino caproic acid
EDTA	ethylenediaminetetraacetic acid
EFA	essential fatty acid
EGF	epidermal growth factor
EGTA	ethyleneglycol-bis(β-aminoethylether)N,N,N',N'-tetraacetic acid
EPA	eicosapentaenoic acid
EtOH	ethanol
FAD	flavin adenine dinucleotide
FMN	flavin mononucleotide
FMPI	N_2-(N-phosphono-L-phenylalanyl)-L-arginine
Fuc	fructose
GABA	γ-aminobutyric acid
Gal	D-galactose
GalNAc	N-acetyl-D-galactosamine
GDP	guanosine diphosphate
GLA	γ-linolenic acid
Glc	D-glucose
GlcN	D-glucosamine
GlcNAc	N-acetyl-D-glucosamine
GlcUA	D-glucuronic acid
GM-CSF	granulocyte–macrophage colony-stimulating factor
GMP	guanosine monophosphate
GPI	glycosyl phosphatidylinositol
GSH	glutathione
GTP	guanosine triphosphate
HDL	high-density lipoprotein
Hepes	4-(2-hydroxyethyl)piperazine-1-ethanesulfonic acid
Hepps	4-(2-hydroxyethyl)piperazine-1-propanesulfonic acid
HETEs	hydroxyeicosatetraenoic acids
HiPIP	high-potential iron–sulfur protein
HIV	human immunodeficiency virus
HLB	hydrophile-lipophile balance
HMG	high-mobility group proteins
HMG-CoA	hydroxymethylglutarate-CoA
HPLC	high-pressure liquid chromatography
Hyl	hydroxy-L-lysine
ICAMs	intercellular adhesion molecules
IDL	intermediate-density lipoproteins
IdUA	L-iduronic acid
IEF	isoelectric focusing
Il	interleukin
Ins	1D-myo-inositol
λ_{max}	maximum wavelength
LDL	low-density lipoprotein
LiDS	lithium dodecyl sulfate
LT	leukotriene
LXA	5,6,15-L-trihydroxy 7,9,11,13-eicosateraenoic acid
MAGs	monoacylglycerols
Man	D-mannose
ManNAc	N-acetyl-D-mannosamine

MAP	microtubule-associated protein
MeOH	methanol
Mes	2-morpholinoethanesulfonic acid monohydrate
MGDG	monogalactosyldiacylglycerol
MGP	matrix γ-carboxyl glutamic acid protein
Mops	3-morpholinopropanesulfonic acid
NAD	nicotinamide adenine dinucleotide
NADP	nicotinamide adenine dinucleotide phosphate
NANA	N-acetylneuraminic acid (NeuAc)
NDP	nucleotide diphosphate
NEFA	non-esterified fatty acid
NeuAc	N-acetylneuraminic acid (NANA)
Neu5Gc	N-glycolylneuraminic acid
NHS	N-hydroxysuccinimidyl
P	phosphate
PA	phosphatidic acid
PAF	platelet activating factor
PAGE	polyacrylamide gel electrophoresis
PAPS	adenosine-3′-phospho-5′-phosphosulfate
PAS	p-aminosalicylic acid
PBD	2-phenyl-5-(4-biphenylyl)-1,3,4-oxadiazole
PC	phosphatidylcholine
PCA	perchloric acid
PCMB	p-chloromercuribenzoate
PCNA	proliferating cell nuclear antigen
PE	phosphatidylethanolamine
PG	phosphatidylglycerol
PGs	prostaglandins
PI	phosphatidylinositol
Pipes	piperazine-1,4-bis(2-ethanesulfonic acid)
PKC	protein kinase C
PKI	protein kinase inhibitor
PMA	phorbol myristate acetate
PMSF	phenylmethylsulfonylfluoride
POPOP	1,4-di-(2-(5-phenyloxazolyl))-benzene
PPO	2,5-diphenyloxazole
PS	phosphatidylserine
psi	lb/in^2
PtdIns	phosphatidylinositol
PtdIns(4)P	phosphatidyl-myo-inositol 4-phosphate
PtdIns(4,5)P$_2$	phosphatidyl-myo-inositol 4,5-bisphosphate
PUFA	polyunsaturated fatty acid
RNA	ribonucleic acid
SAM	S-adenosylmethionine
SDS	sodium dodecyl sulfate
Ser	L-serine
SLS	sodium lauryl sulfate (SDS)
SM	sphingomyelin
SQDG	sulfoquinovosyldiacylglycerol
SRS-A	slow release substances of anaphylaxis
TAGs	triacylglycerols
Taps	N-[tris(hydroxymethyl)methyl]-3-aminopropanesulfonic acid

TCA	trichloroacetic acid
TDP	thymidine diphosphate
TEMED	N,N,N',N'-tetramethylethylenediamine
Tes	N-[tris(hydroxymethyl)methyl]-2-aminomethanesulfonic acid
TGF	transforming growth factor
THFA	tetrahydrofolate
Thr	L-threonine
TLCK	$N\alpha$-p-tosyl-L-lysine chloromethyl ketone
TMP	thymidine monophosphate
TNBS	2,4,6-trinitrobenzenesulfonate
TNF	tumor necrosis factor
TPCK	N-tosyl-phenylalanine chloromethyl ketone
Tricine	N-[tris(hydroxymethyl)methyl]glycine
Tris	Tris(hydroxymethyl)aminomethane
tRNA	transfer RNA
TTP	thymidine triphosphate
UDP	uridine diphosphate
UMP	uridine monophosphate
UTP	uridine triphosphate
UV	ultraviolet
VLDL	very-low-density lipoprotein
Xyl	D-xylose

CHAPTER 1
BUFFERS, CHELATING AGENTS AND DENATURANTS

J. A. A. Chambers

1. BUFFERS

1.1. Introduction
Buffers maintain pH within a limited range (*Figures 1* and *2*) despite the generation or loss of hydrogen ions by the reactions going on within the system. Buffers are weak acids or bases that are not fully dissociated in solution. Within a limited pH range the equilibrium between the dissociated and undissociated forms of the buffer remains constant. Many such substances have been used but comparatively few remain widely used. Some unusual buffers remain in use for specific systems. *Table 1* summarizes the properties of non-zwitterionic buffers.

1.2. Goods' buffers
Goods' buffers (1) are a group of zwitterionic organic compounds, primarily sulfonates and tertiary amines, that show excellent buffering properties, solubility and biological and chemical inertness. As a group these buffers show excellent compatibility with biological systems. The properties of these are summarized in *Table 2*.

1.3. Preparation of buffers
Buffers are prepared by two methods:

(i) bringing a dissolved solution of the buffer to the correct pH by titrating with a strong acid or base; or
(ii) mixing of stock solutions; this is used frequently for phosphate and acetate buffers.

When the mixing of stock solutions is used confusion can sometimes arise over what the molarity refers to. In the case of phosphate and acetate buffers the molarity refers to all phosphate or acetate ions in the medium (see Sections 1.8 and 1.9 for the preparation of these buffers). On the other hand, for most buffers the concentration refers to the buffer substance alone and the concentration of the titrating acid or base is ignored. *Tables 3* and *4* give the composition of phosphate and acetate buffers, respectively.

1.4. Counter ions
Buffers with an acid pK_a are brought to their working pH with a strong base (e.g. NaOH, KOH), and those with a basic pK_a use a strong acid (e.g. HCl). The counter ion used is not always important and may be chosen by criteria such as convenience and compatibility with the system. Sometimes compound buffers such as Tris–acetate or Tris–phosphate may be used.

1.5. Effects of temperature and concentration
All buffers are sensitive to temperature to some extent. This is indicated in tables by the value $\Delta t°C$. Some buffers, such as Tris are comparatively temperature sensitive and should be brought to the required pH at the temperature to be used for the experimental work.

Similarly, some buffers are very sensitive to dilution of the stock solution (e.g. phosphate). After a concentrated stock has been prepared an aliquot should be diluted to the working concentration to check the final pH.

1.6. Biological compatibility and chemical reactivity

Most buffers, in their optimum pH ranges, are compatible with most biological systems; however, there can be exceptions. When used outside their optimum ranges some buffers can become reactive, insoluble, or may complex metal ions in addition to acting as a poor buffer. Some of the classical buffers have deficiencies that led to the development of zwitterionic buffers. These include reactivity, metabolism in some systems (glycine, citrate and aconitate) and limited buffer capacity. Tris, for example, inhibits phospholipase C, the Hill reaction of chloroplasts, complexes metal ions and is toxic to cells. The amino group of Tris or glycine reacts with aldehydes (e.g. formaldehyde) and even enzyme-bound aldehyde intermediates (2–6). Carboxyl and hydroxyl groups of aconitate, citrate, glycine and maleate are also reactive and these compounds have extensive chemistries, as does imidazole. Cacodylate reacts with thiol groups and is moderately toxic (7). Phosphate and borate buffers at high ionic strengths can chelate inorganic cations, and form insoluble salts (especially phosphate). In addition, borate complexes 1,2-*vic*-diols including nucleosides (8).

Figure 1. pH ranges of non-zwitterionic buffers.

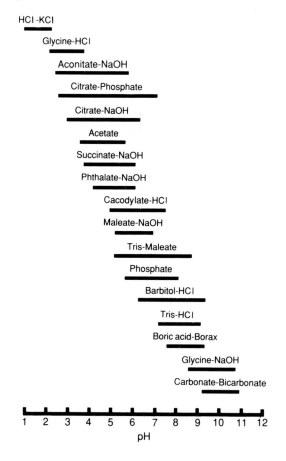

Despite their otherwise excellent behavior, the following reactivity and incompatibility problems have been found for zwitterionic buffers:

Compatibility
Zwitterionic buffers, including Hepes, are not compatible with all tissue cultures (e.g. chondrocytes (9). Hepes also acts as an antagonist of γ-aminobutyric acid (inhibits action of the receptor (10)). Pipes interferes with tubulin polymerization into microtubules but Mes does not (11).

Reactivity
Zwitterionic buffers with ethanolamine groups interfere with the Lowry protein assay (12). Bicine is reactive in the acylation of tertiary amino alcohols with active esters (e.g. *N*-hydroxysuccinimide (13)); similarly, Tes reacts with succinic anhydride (14). Bistris interferes with the diaminobenzoic acid (DABA) fluorimetric assay for DNA and is reactive with the superoxide anion in heme-containing systems (15, 16). Caps, Ches, Taps and Mes are reactive, or catalytically active in some proton transfer reactions (17, 18). If zwitterionic buffers have a weakness, it is transition metals. Hepes is auto-oxidized in the presence of some aromatic compounds (19) and by copper (II) (20). Cyclohexyl groups of some of these buffers are prone to chemiluminescent oxidation in the presence of certain metal ions, for example Ce(IV) (21). Those with piperazine rings can form long-lived radicals in the presence of Fe(III) (22).

Figure 2. pH ranges of zwitterionic buffers.

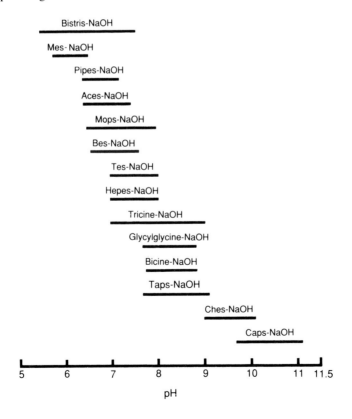

It must be pointed out that these problems are minimal in comparison to the advantages they offer over non-zwitterionic buffers.

1.7. Terms used in the tables

pK_a
Log_{10} apparent dissociation constant for the acid component of the buffer substance. A buffer is usually most effective at about 1 pH unit above or below the value of pK_a. pH is related to pK_a and the concentration of the buffer by the well-known Henderson–Hasselbach equation.

$\Delta t°C$
Temperature coefficient of change of pK_a. A large value, e.g. for Tris, implies that the buffer is temperature sensitive and should be brought to the correct pH at the working temperature.

pH range
The pH values over which the buffer is effective. Outside this range it will not only be ineffective but may also become highly reactive.

Counter ion
The preferred acid or base for bringing the buffer to the correct pH.

Risk and Safety
The numbers given in these columns refer to the European Commission Risk and Safety Phrases (see *Chapter 14, Table 3*).

Table 1. Components of non-zwitterionic buffers

Name	Formula	Mol. wt	Solubility (20°C) (g/100 ml)	pK_a (20°C) 1	2	3	Titrate with	Risk	Safety
Boric acid [10043-35-3]	H_3BO_3	61.8	6.4	9.14	12.74	13.8	Tris, NaOH	—	—
Citric acid [77-92-9]	$C_6H_8O_7$	192.4	146.0	3.14	4.77	6.39 (18°C)	Sodium phosphate, NaOH, KOH	—	—
Imidazole [288-32-4]	$C_3H_4N_2$	68.1	v.sol.	6.95 (25°C)	—	—	NaOH, KOH	R21/R22	S25
Potassium dihydrogen phosphate [7778-77-0]	KH_2PO_4	136.1	33	2.12	7.21	12.61 (25°C)	K_2HPO_4	—	—
di-Potassium hydrogen phosphate [7758-11-4]	K_2HPO_4	174.2	167	2.12	7.21	12.61 (25°C)	KH_2PO_4	—	—
Potassium dihydrogen phthalate [877-24-7]	$C_8H_6KO_4$	204.2	10	2.89	5.51	—	KOH	—	—
Sodium acetate anhydrous trihydrate [1227-09-3]	CH_3COONa $CH_3COONa·3H_2O$	82.0 136.1	119.0 (0°C) 76.2 (0°C)	4.75			Acetic acid	—	—

Table 1. Continued

Name	Formula	Mol. wt	Solubility (20°C) (g/100 ml)	pKa (20°C) 1	pKa (20°C) 2	pKa (20°C) 3	Titrate with	Risk	Safety
Sodium cacodylate [124-65-2]	$(CH_3)_2AsO_2Na \cdot 3H_2O$	214.0	200	6.21	—	—	NaOH	R23/R25–R33	S1/S2–S7 S20–S21 S28–S44
Sodium carbonate anhydrous decahydrate [497-19-8]	Na_2CO_3 $Na_2CO_3 \cdot 10H_2O$	106.0 286.1	7.1 (0°C) 21.5 (0°C)	6.37	10.25	—	CO_2, carbonates	—	—
tri-Sodium citrate [68-04-2]	$Na_3C_6H_5O_7 \cdot 2H_2O$	294.1	72.0	3.09	4.75	5.41 (18°C)	Phosphates, hydroxides	—	—
Sodium dihydrogen phosphate [7558-80-7]	$NaH_2PO_4 \cdot 2H_2O$	156.0	v.sol.	2.12	7.21	12.67	Na_2HPO_4	—	—
Sodium hydrogen carbonate [144-55-8]	$NaHCO_3$	84.0	9.6	6.37	10.25	—	CO_2, carbonates	—	—
di-Sodium hydrogen phosphate anhydrous dihydrate decahydrate [7558-79-4]	Na_2HPO_4 $Na_2HPO_4 \cdot 2H_2O$ $Na_2HPO_4 \cdot 10H_2O$	142.0 178.0 358.1	60.0 4.2	2.12	7.21	12.67	NaH_2PO_4	—	—

Table 1. Continued

Name	Formula	Mol. wt	Solubility (20°C) (g/100 ml)	pK_a (20°C) 1	2	3	Titrate with	Risk	Safety
Sodium tetraborate anhydrous decahydrate [1330-43-4]	$Na_2B_4O_7$ $Na_2B_4O_7 \cdot 10H_2O$	201.2 381.4	1.1 (0°C) 2.0 (0°C)	9.14	12.74	13.80	Boric acid	—	—
Triethanolamine [102-71-6]	$C_6H_{15}NO_3$	149.2	Liq. freely misc. with water		—	—			
Tris[a] [77-86-1]	$C_4H_{11}NO_5$	121.1	39.6	8.3	—	—	HCl, acetic, phosphoric acid	—	—

[a] 2-amino-2-(hydroxymethyl)-1,3-propanediol.
Abbreviations: v.sol., very soluble; liq., liquid; misc., miscible.

Table 2. Zwitterionic (Good's) buffers

Name	Mol. wt	Solubility (0°C) (mol l^{-1})	pK_a at 20°C	Δt °C	Useful pH range
Aces [7365-82-4]	182.2	0.22	6.88	−0.02	6.4–7.4
Bes [10191-18-1]	213.3	3.2	7.17	−0.016	6.6–7.6
Bicine [150-25-4]	163.2	1.1	8.35	−0.018	7.8–8.8
Bis-Tris [6976-37-0]	209.2	—	6.5	—	5.5–7.5
Caps [1135-40-6]	221.3	0.47	10.4	0.032	9.7–11.1
Ches [103-47-9]	207.1	1.15	9.55	−0.011	9.0–10.1
Glycylglycine [556-50-3]	132.1	1.01 (25°C)	8.25	−0.026	7.7–8.8
Hepes [7365-45-9]	238.3	2.25	7.55	−0.014	7.0–8.0
Hepps [16052-06-5]	252.3	1.6	8.0	−0.011	7.6–8.6
Mes [4432-31-9]	213.3	0.65	6.15	−0.011	5.8–6.5
Mops [1132-61-2]	209.3	3.1	7.2	−0.011	6.5–7.9
Pipes [5625-37-6]	302.4	1.4	6.82	0.009	6.4–7.2
Taps [29915-38-6]	243.3	v.sol.	8.4 (25°C)	0.018	7.7–9.1
Tes [7365-44-8]	229.3	2.6	7.5	−0.02	7.0–8.0
Tricine [5704-04-1]	179.2	0.8	8.15	−0.028	7.0–9.0

Abbreviations: Aces, N-(2-acetamido)-2-aminoethanesulfonic acid; Bes, N,N-bis(2-hydroxyethyl)-2-aminoethanesulfonic acid; Bicine, N,N-bis(2-hydroxyethyl)glycine; Bis-Tris, bis(2-hydroxyethyl)imino-tris(hydroxymethyl)methane; Caps, 3-(cyclohexylamino)-1-propanesulfonic acid; Ches, 2-(cyclohexylamino)-ethanesulfonic acid; Hepes, 4-(2-hydroxyethyl)piperazine-1-ethanesulfonic acid; Hepps, 4-(2-hydroxyethyl)piperazine-1-propanesulfonic acid; Mes, 2-morpholinoethanesulfonic acid monohydrate; Mops, 3-morpholinopropanesulfonic acid; Pipes, piperazine-1,4-bis(2-ethanesulfonic acid); Taps, N-[tris(hydroxymethyl)methyl]-3-aminopropanesulfonic acid; Tes, N-[tris(hydroxymethyl)-methyl]-2-aminomethanesulfonic acid; Tricine, N-[tris(hydroxymethyl)methyl]glycine; v. sol., very soluble.

1.8. Preparation of phosphate buffers

The following table is used for the preparation of 200 ml of 0.1 M sodium phosphate buffer. The buffer may be made as a five- or tenfold concentrate simply by dissolving the appropriate masses of the phosphates in a final volume of 200 ml. Phosphate buffers show concentration-dependent changes in pH, so the pH of the concentrate should be checked by diluting an aliquot to the final concentration.

Table 3. Preparation of 200 ml of 0.1 M sodium phosphate buffer (adapted from ref. 23, with permission from CRC Press, Inc.)

Solution A: 0.2 M NaH_2PO_4[a] (27.8 g in 1000 ml water).

Solution B: 0.2 M Na_2HPO_4[a] (53.65 g $Na_2HPO_4 \cdot 7H_2O$ or 71.7 g $Na_2HPO_4 \cdot 12H_2O$ in 1000 ml water).

The appropriate volumes of solutions A and B are mixed, and water is added to a final volume of 200 ml:

Vol. A (ml)	Vol. B (ml)	pH
93.5	6.5	5.7
92.0	8.0	5.8
90.0	10.0	5.9
87.7	12.3	6.0
85.0	15.0	6.1
81.5	18.5	6.2
77.5	22.5	6.3
73.5	26.5	6.4
68.5	31.5	6.5
62.5	37.5	6.6
56.5	43.5	6.7
51.0	49.0	6.8
45.0	55.0	6.9
39.0	61.0	7.0
33.0	67.0	7.1
28.0	72.0	7.2
23.0	77.0	7.3
19.0	81.0	7.4
16.0	84.0	7.5
13.0	87.0	7.6
10.5	90.5	7.7
8.5	91.5	7.8
7.0	93.0	7.9
5.3	94.7	8.0

[a] The corresponding potassium salts may also be used.

1.9. Preparation of acetate buffers

Acetate buffers are prepared in the same way as phosphate buffers. The following recipe is for 100 ml of a 0.1 M acetate buffer.

Table 4. Preparation of 100 ml of 0.1 M acetate buffer (adapted from ref. 23, with permission from CRC Press, Inc.)

Solution A: 0.2 M acetic acid (11.55 ml glacial acetic acid in 1000 ml water).

Solution B: 0.2 M sodium acetate (16.4 g $C_2H_3O_2Na$ or 27.2 g $C_2H_3O_2Na \cdot 3H_2O$ in 1000 ml water).

The appropriate volumes of solutions A and B are mixed, and water is added to a final volume of 100 ml:

Vol. A (ml)	Vol. B (ml)	pH
46.3	3.7	3.6
44.0	6.0	3.8
41.0	9.0	4.0
36.8	13.2	4.2
30.5	19.5	4.4
25.5	24.5	4.6
20.0	30.0	4.8
14.8	35.2	5.0
10.5	39.5	5.2
8.8	41.2	5.4
4.8	45.2	5.6

1.10. Buffered salines, balanced salts and osmotic supports

Salines and buffered salines
Used to keep animal tissue in good condition before analysis, also in immunochemistry, as a perfusant, as an osmotic support, and as a diluent.

Physiological saline. 0.9% w/v NaCl.

Phosphate-buffered saline.

	NaCl	7.2 g l^{-1}
	Na_2HPO_4 (anhydrous)	1.48 g l^{-1}
	KH_2PO_4 (anhydrous)	0.43 g l^{-1}

Tris-buffered saline. Physiological saline containing 10–50 mM Tris–HCl (pH 7.2). However, Tris is toxic to some cells, so with cells it is better to use Tricine as a buffer. *Table 5* lists the composition of selected balanced salt solutions.

Sugars and sugar alcohols as osmotic supports
Non-reactive sugars and sugar alcohols are used as osmotic supports during the formation of spheroplasts of plant, fungal or microbial cells, because of the very high internal osmotic pressures in these systems. Usually they are used at a hypertonic concentration in the range 0.3–0.8 M. Those used most frequently are glucose, mannitol and sorbitol (mol. wt 182.18) and sucrose. All are readily soluble in water and solutions can be sterilized by autoclaving but sucrose is toxic to some cells.

Table 5. Balanced salt solutions

Compound (mg l^{-1})	Dulbecco's	Earle's	Hank's I	Hank's II	Ringer's mammalian	Ringer's amphibian	Tyrode's
$CaCl_2$	100	200	200	140	250	120	200
$MgCl_2 \cdot 6H_2O$	100	—	—	—	—	—	100
$MgSO_4$	—	100	—	—	—	—	—
$MgSO_4 \cdot 7H_2O$	—	—	200	200	—	—	—
KCl	200	400	400	400	420	140	200
KH_2PO_4	200	—	100	60	—	—	—
$NaHCO_3$	—	2200	1273	350	—	200	1000
NaCl	8000	6800	8000	8000	9000	6500	8000
Na_2HPO_4	1150	—	—	—	—	—	—
$Na_2HPO_4 \cdot 2H_2O$	—	—	100	60	—	—	—
$NaH_2PO_4 \cdot H_2O$	—	140	—	—	—	—	50
Glucose	—	1000	2000	1000	—	1000	—

Adapted from the *Biology Data Book, Vol. 1 (2nd edn)*, with permission from the Federation of American Societies for Experimental Biology.

2. CHELATING AGENTS

Chelating agents are used to control metal ion concentrations in reactions or to remove metal ions essential for structural or functional integrity. The affinity of a chelating agent for a metal is indicated by the stability constant. This is the $\log_{10}$ of the dissociation constant for the complex. Although it is generally thought that there is a simple binary reaction:

$$\text{metal} + \text{chelating agent} \rightleftharpoons \text{complex}$$

involved in the formation of a complex, in practice a series of minor complexes may also be formed. In such cases it is the stability constant for the most stable complex that is reported. For some ligands (e.g. EDTA) the complex contains one chelating agent molecule and one metal ion, but for others there may be more than one chelating agent molecule with a series of intermediates; in these cases the stability constant for the most stable form is indicated in *Table 7* and the presence of intermediates is marked.

The ratio of stability constants for a chelating agent with a pair of metal ions shows how effectively the two would be bound in a mixture. For EDTA, the stability constants for Ca^{2+} and Mg^{2+} are 10.61 and 8.83. The difference between these two is ~1.8, which is $\log_{10}$ of ~60. Accordingly, the ratio of complexed Ca^{2+} to Mg^{2+} in an equimolar mixture is about 60:1. For EGTA, with a lower stability constant for Mg^{2+}, this ratio is about 4×10^5.

Tables 6 and *7* include the two most widely used chelating agents, EDTA and EGTA, and two others that have some specialized uses. Citrate is used as a weak chelating agent in some systems and phosphate is included for comparison.

Table 6. Chelating agents: solubility and pK_a values

	EDTA [60-00-4]	EGTA [67-42-5]	1,10-Phenanthroline [66-71-7]	2,2'-Bipyridine [366-18-7]	Citric acid [77-92-9]	Phosphoric acid [7664-38-2]
Formula	$C_{10}H_{16}N_2O_8$	$C_{14}H_{24}N_2O_{10}$	$C_{12}H_8N_2$	$C_{10}H_8N_2$	$C_6H_8O_7$	H_3PO_4
Mol. wt	372.2 (2Na·2H$_2$O)	380.4				
Solubility (20°C, g/100 ml H$_2$O)	11.1	v.sol.				
pK_a (25°C)						
pK_a1	10.17	9.40	4.93	4.42	5.69	11.74
pK_a2	6.11	8.78	1.9	—	4.35	6.57
pK_a3	2.68	2.66[a]	—	—	2.87	1.72
pK_a4	2.0	2.0[a]	—	—	—	—

[a] At 20°C.

Abbreviations: EDTA, ethylenediamine tetraacetic acid; EGTA, ethylene glycol-bis(β-aminoethyl ether)N,N,N',N'-tetraacetic acid; v. sol, very soluble.

Table 7. Chelating agents: stability constants (ref. 51)

	EDTA	EGTA	1,10-Phenan-throline	2,2'-Bipyridyl	Citric acid	Phosphoric acid
Common physiological cations						
Na^+	1.64	—	—	—	0.7	0.60
K^+	0.8	—	—	—	0.59	0.49
Ca^{2+}	10.61	10.86	0.7[a]	—	3.5	3.4[a]
Mg^{2+}	8.83	5.28	1.2	—	3.37	1.5
Other physiological cations						
Co^{2+}	16.26	12.35	19.8[b]	15.9[b]	5.0	2.18
Co^{3+}	41.4	—	—	—	—	—
Cu^+	—	—	15.8[a]	12.95	—	—
Cu^{2+}	18.7	17.57	10.69	6.33	5.9	3.2
Fe^{2+}	14.27	11.8	21.0[b]	17.2[b]	4.4	3.6
Fe^{3+}	25.0	20.5	14.1[b]	16.29	11.5	8.3
Mn^{2+}	13.81	12.18	10.3[b]	2.62	4.15	—
Ni^{2+}	18.52	13.5	24.3	20.16[b]	5.4	2.08
Zn^{2+}	16.44	12.6	—	13.2[b]	5.9[b]	2.4
Other cations						
Ag^+	7.32[a]	6.88[a]	12.6[b]	6.67[b]	—	—
Al^{3+}	16.5	13.9	—	—	—	—
Cd^{2+}	16.36	16.5	14.6[b]	10.3[b]	4.54[b]	—
Hg^{2+}	21.5	22.9	23.35[b]	19.5[b]	10.9	—
Pb^{2+}	17.88	14.5	4.65	2.9	6.1	—

[a] At 20°C.
[b] Most stable complex.
Abbreviations: EDTA, ethylenediaminetetraacetic acid; EGTA, ethylene glycol-bis(β-aminoethyl ether)N,N,N',N'-tetraacetic acid.

3. DENATURING AGENTS FOR PROTEINS

3.1. Mechanisms of denaturing agents

Protein denaturing agents are chaotropic agents that disrupt the higher-order structure (secondary and higher) of proteins. Most often this is by disrupting the hydrogen bonds and other interactions that maintain these higher-order structures. Denaturing detergents act by a slightly different mechanism (24).

Denaturants can be either solubilizing (*Table 8*) or precipitating (*Table 9*). Urea, sodium dodecyl sulfate (SDS or SLS) and guanidinium salts are effective solubilizing denaturants, whereas trichloroacetic acid (TCA), ethanol, methanol and chloroform/isoamyl alcohol are effective precipitants.

Generally, denaturants are used in a large excess but care should be taken to ensure that this is the case. Agents that disrupt hydrogen bonding should be used at a ratio of at least two moles per mole of amino acids. SDS binds proteins in a well-defined complex of 1.4 g SDS/g protein. Many denaturing solutions, especially solubilizing ones, contain a high concentration of a thiol reagent. This helps to reduce disulfide bonds to thiols to complete the loss of tertiary structure.

3.2. Mixing denaturants

Quantitative studies show that the binding of two denaturants to proteins when presented as a mixture results in reduced binding of both denaturants; that is, they act antagonistically (similar to detergents). The best choice of denaturant is the most powerful single component system that is compatible with your requirements.

3.3. Characteristics of denaturants and precipitants

SDS, Sarkosyl, lithium dodecyl sulfate

SDS is soluble up to 25% w/v but is then very sensitive to reductions in temperature and dust particles in solution. A stock of 20% w/v is most convenient. Sarkosyl is usually available as a 30% stock solution. LiDS is more effective at cold-room temperatures than either SDS or Sarkosyl. Potassium dodecyl sulfate is insoluble so it is important to avoid introducing potassium salts into solutions containing SDS.

Urea

Urea is soluble up to 10 M but above 8 M care must be taken with temperature to prevent precipitation. Urea slowly decomposes in water, with the formation of ammonia and the highly reactive cyanate ion. When using urea solutions, pretreat them with a mixed-bed ion-exchange resin or include a low concentration of an amine-containing compound (e.g. Tris buffer, lysine) to react with the cyanate as it is formed and avoid high temperatures.

Guanidinium salts

The chloride is readily available at high purity at a reasonable price. It is generally useful for all except some of the most recalcitrant proteins, in which case the thiocyanate is to be preferred (25, 26). Even the purest grades of the thiocyanate contain significant insoluble material and solutions should be filtered before use. The thiocyanate reacts with Cleland's reagent, so 2-mercaptoethanol is the preferred thiol reagent (27).

Ethanolic perchlorate

A convenient reagent for the preparation of nucleic acids. It precipitates proteins and solubilizes DNA and RNA (28). This solution is a very powerful oxidizing agent and all glassware, etc. should be thoroughly rinsed with water immediately after use.

Phenol

Phenol will take up to 12% (v/v) of water before forming a two-phase system, with proteins preferentially entering the phenol phase. Phenol is a mild denaturant and so preliminary use of SDS to denature proteins is recommended. A 1:1 mixture of phenol and chloroform/isoamyl alcohol is a more effective deproteinization reagent. The solubility of nucleic acids, carbohydrates and proteins is pH-dependent, so the phenol should be neutralized with concentrated Tris buffer before use. Phenol also enters the aqueous phase of the system and should be extracted with ether or chloroform/isoamyl alcohol. 8-Hydroxyquinoline is usually added as an antioxidant and to strip Mg^{2+} from the aqueous phase (the complex is more soluble in the phenol phase). Sometimes *p*-aminosalicylic acid (PAS) is also added as a chelating agent and tri-isopropyl naphthalene sulfonate is added as an emulsifying agent (29, 30).

Trichloroacetic acid (TCA) and perchloric acid (PCA)

These are both extremely effective precipitants and will precipitate both nucleic acids and proteins. TCA is more widely used because of the reactivity of perchloric acid as an oxidizing agent. TCA is opaque in the ultraviolet, whereas PCA is transparent. This allows direct measurement at 260 nm of nucleotide concentration after hot acid hydrolysis of nucleic acids precipitated with PCA (see *Chapter 11*).

Acetone
The addition of 6 volumes of acetone will effectively precipitate most proteins, but it can be used as a selective precipitant for some classes of proteins at low temperatures.

Chloroform/isoamyl alcohol (24:1)
Proteins are effectively denatured by emulsifying with this reagent. The solubility of proteins is limited, and when the emulsion is broken down by centrifugation most of the protein forms a precipitate at the interface. Deproteinization by extraction with this reagent is more effective when in combination with phenol.

Ethanol (EtOH) and methanol (MeOH)
Proteins are denatured and precipitated by the addition of two volumes of EtOH. At higher ratios the proteins may resolubilize. Methanol appears to be more effective. Either may be used in combination with about 10% acetic acid to ensure complete and irreversible precipitation.

Table 8. Solubilizing denaturants

Name	Formula	Mol. wt	Solubility (g/100 ml 20°C)	Working concn	Comments
Sodium dodecyl sulfate [151-21-2]	$C_{12}H_{26}O_4S \cdot Na$	273.3	25[a]	0.1–2%	Cationic detergent, salt-sensitive, cold-sensitive
Sodium dodecyl sarcosinate [137-16-6]	$C_{15}H_{29}NO_3 \cdot Na$	294.4	30[b]	0.5–2.5%	Cationic detergent, less sensitive to salt & cold than SDS but also less effective
Lithium dodecyl sulfate [2044-56-6]	$C_{12}H_{26}O_4S \cdot Li$	257.3	30	0.1–2%	Less sensitive to salt and cold than SDS or Sarkosyl
Urea [57-13-6]	CH_4N_2O	60.06	48–60	4–10 M	Non-ionic, chaotropic agent; does not disrupt ionic bonds
Guanidinium chloride [50-01-1]	CH_6N_3Cl	95.53	76	4–6 M	Ionic chaotropic agent; disrupts ionic bonds
Guanidinium thiocyanate [593-84-0]	$C_2H_5N_4S$	117.1	47	3–4 M	Similar to the chloride but more effective

[a] At 25% SDS is very sensitive to temperature and salt particles, 20% is a useful working concentration.
[b] Sold as a 30% solution.

Table 9. Precipitating denaturants

Name	Formula	Mol. wt	Solubility	Working concn	Comments
Acetic acid	$C_2H_4O_2$	60.05	Freely miscible	5–10 vol.%	Also ppts nucleic acids
Acetone	C_3H_6O	58.08	Freely miscible	2–6 vols	Can be used for selective pptn; also ppts nucleic acids
Chloroform/isoamyl alcohol (24:1 by vol.)	$CHCl_3$/ $C_5H_{12}O$		Immiscible with aq. solns	Equal vol.	Effective denaturant
Ethanol	C_2H_6O	46.1	Freely miscible[a]	1–2 vols	Also ppts nucleic acids
Ethanolic perchlorate	$C_2H_6O/HClO_4$		Miscible with water	5–10 vols	Ppts protein, solubilizes nucleic acids
Methanol	C_2H_4O	32.04	Freely miscible with water[a]	1–2 vols	

[a] A salt (preferably chloride or acetate) must be present at concentrations greater than 0.1 M for effective precipitation.
Abbreviations: aq. solns, aqueous solutions; pptn, precipitation; ppts, precipitates.

4. THIOL REAGENTS

Thiol reagents have two uses in biochemistry: to prevent the uncontrolled oxidation of thiol groups in proteins or elsewhere (e.g. that of coenzyme A) to disulfides, and to maintain a reducing environment for some enzyme reactions. In all cases these reagents work by mass action. Several reagents have been used and some still are preferred for specific reactions. The most widely used are the dithiol butanols (e.g. dithioerythritol and dithiothreitol, DTT) and 2-mercaptoethanol (*Table 10*). Glutathione is frequently used because it is a physiological reducing agent and can be regenerated *in situ* with glutathione reductase. There is no obvious reason for choosing one of these reagents over the other although dithiols are active at lower concentrations. However, chemical reactivity or stereochemistry may be a problem: for example, dithiothreitol appears to be incompatible with guanidine thiocyanate. The redox potential of the —SH group is the same (-330 mV at pH 7) in all these compounds. All thiol reagents have a smell that is more or less offensive.

Table 10. Thiol reagents

Name	Formula	Mol. wt	Solubility	Refs
Cysteine [52-90-4]	C_3H_7NS	121.16	V.sol.	
Dithiothreitol [3843-12-3]	$C_4H_{10}O_2S_2$	154.25	Sol.[a]	31
Dithioerythritol [6892-68-8]	$C_4H_{10}O_2S_2$	154.25	Sol.[a]	31
2-Mercaptoethanol [60-24-2]	C_2H_6OS	78.1	Liquid freely miscible with water[b]	
Glutathione [70-18-8]	$C_{10}H_{17}N_3O_6S$	307.3	Sol.	
2,3-Dimercaptopropanol [59-52-9]	$C_3H_8OS_2$	124.2	8.7 g/100 ml H_2O[c]	

[a] Typical working concentrations for these compounds are 1–5 mM and they can be prepared as stocks of > 100 mM.
[b] The typical working concentration is 0.1–2% by volume. Molarity of the pure liquid is 14.3 M.
[c] Much more soluble in vegetable oils. Commercially sold as a stock solution in 10% peanut oil.

5. DETERGENTS

The primary uses of detergents in biochemistry are:

(i) solubilization of membranes and membrane lipids;
(ii) solubilization and stabilization of proteins, especially membrane proteins and others with significant hydrophobic surfaces;
(iii) as protein denaturants;
(iv) as wetting agents; and
(v) as emulsifying agents.

Structurally, detergents all have a hydrophilic 'head' region and a long hydrophobic 'tail'. This causes them to form aggregates (micelles) with a hydrophobic core into which a hydrophobic molecule may partition from an aqueous medium, with the resulting stabilization of the hydrophilic molecule in the aqueous system (24, 32, 33). Many chemical classes of detergent are recognized, but generally they are discussed in terms of their ionic properties; there are four classes in this respect: non-ionic, zwitterionic, cationic and anionic (*Table 11*). Generally speaking, non-ionic and zwitterionic detergents are non-denaturing and are useful in the solubilization of proteins. Ionic detergents are often denaturing, especially SDS and Sarkosyl; CTAB is widely used as a germicide. Detergents may be quite effective in solubilizing a specific protein without loss of biological activity but without necessarily being effective in membrane solubilization, and vice versa (34, 35).

By definition, all detergents are soluble in water to some extent, and a stock solution of 5–10% detergent can usually be prepared. When the solution is saturated with detergent it separates into two phases. The phase behavior of detergent solutions is extremely complex. Solubility is salt-dependent and so care should be taken when preparing a concentrated stock as the presence of salts may lead to phase separation. In cases where this is likely to happen, unless the detergent is a waxy solid that dissolves only slowly in water (e.g. the Brij series), it is best to add the detergent directly to the diluted reagent immediately before use. Solubility is also temperature-dependent. Although detergents dissolve more readily upon heating, there is an upper limit called the cloud point. At this temperature the detergent reversibly separates from the aqueous phase to form a separate phase as a cloudy suspension. For most detergents it acts as an upper temperature limit for its use and usually lies at temperatures $>60°C$ but Triton X-114 has a cloud point of 22°C. This property of Triton X-114 has been found to be useful in the purification of membrane proteins (36).

5.1. Terms used in the tables

Aggregation number
The average number of molecules per micelle.

CMC: critical micellar concentration
The minimal concentration at which the detergent forms micelles. This is an important guideline in solubilization because the formation of micelles is necessary for protein stabilization and lipid solubilization. A high CMC also indicates that the detergent is more readily dialyzable. Compare octyl pyranoside, which is dialyzable, with Triton X-100, which is not.

HLB: hydrophile-lipophile balance
A quantitative measure of the hydrophilicity and lipophilicity of a detergent, and therefore of its solubilization properties. There is a wide range of HLB values covering many applications. For solubilization of biological membranes and lipids an HLB value of ≈ 15 is most effective. An HLB value between 12 and 14.5 is recommended for solubilization of membranes, and detergents with an HLB value between 15 and 20 are useful for solubilization of extrinsic proteins. Detergents with HLB values outside this range are useful as wetting agents and permeabilizing agents.

Risk and Safety
The numbers given in these columns refer to the European Commission Risk and Safety Phrases (see *Chapter 14, Table 3*).

Table 11. Detergents and surfactants

Name[a]	Mol. wt	HLB[b]	CMC (μM)	Aggregation no. (20–25°C, 0–0.1 M Na$^+$)	Risk	Safety	Refs
Anionic detergents							
Sarkosyl[a] [137-16-6]	295.4						
SDS (SLS) [151-2-3]	288.4	≈40	8200	620	R22–R36/R38		37–39
Cationic detergents							
CPB [140-72-7]	402.5		1000		R21/R22	S24/S25	
CTAB [57-09-0]	364.5		1000	170	R21/R22	S26	
Zwitterionic detergents							
CHAPS [75621-03-3]	614.9		6–10 × 10^3	10	⟨irritant⟩		40–42
CHAPSO [82473-24-3]	630.9		8000	11	⟨irritant⟩		43
Zwittergent 3.08 [15178-76-4]	280		33 × 10^4				44
Zwittergent 3.10 [15163-36-7]	308		25–40 × 10^3	41			
Zwittergent 3.12 [14933-09-6]	336		2–4 × 10^3	55			

Table 11. Continued

Name[a]	Mol. wt	HLB[b]	CMC (μM)	Aggregation no. (20–25°C, 0–0.1 M Na$^+$)	Risk	Safety	Refs
Zwitterionic detergents (continued)							
Zwittergent 3.14 [14933-09-6]	364		100–400	83			
Zwittergent 3.16 [2281-11-0]	392		10–60	155			
Non-ionic detergents							
Brij 35 [3055-98-9]	≈1200	16.9	75	40	⟨irritant⟩		45, 46
Brij 56 [9004-95-9]	682	12.9	2	40	⟨irritant⟩		
Brij 58 [9004-95-9]	1122	15.7	77	40	⟨irritant⟩		
Octyl-β-D-glucopyranoside [29836-26-8]	292.4		25×10^3	84			42, 47, 48
Triton X-100 [9002-93-1]	625	13.5	250	100–155			45, 46
Triton X-114 [9036-19-5]	537	12.4					
Nonidet P40 [9036-19-5]	625	13.1	250	100–155			
Tween 20 [9005-64-5]	1228	10.7	59				47, 48

Table 11. Continued

Name[a]	Mol. wt	HLB[b]	CMC (μM)	Aggregation no. (20–25°C, 0–0.1 M Na$^+$)	Risk	Safety	Refs
Tween 60 [9005-67-8]		14.9					
Tween 80 [9005-65-6]	1310	15.0	12	58			

[a] Names: Sarkosyl, N-laurylsarcosine, sodium salt; SDS, sodium dodecyl(lauryl)sulfate; CPB, cetylpyridinium bromide; CTAB, cetyltrimethylammonium bromide; CHAPS, 3-[(3-cholamidopropyl)-dimethylammonio]-1-propanesulfonate; CHAPSO, 3-[(3-cholamidopropyl)-dimethylammonio]-2-hydroxy-1-propanesulfonate; Zwittergent 3.08, N-octylsulfobetaine; Zwittergent 3.10, N-decylsulfobetaine; Zwittergent 3.12, N-dodecylsulfobetaine; Zwittergent 3.14, N-tetradecylsulfobetaine; Zwittergent 3.16, N-hexadecylsulfobetaine; Brij 35, polyoxyethylene (23) lauryl ether; Brij 56, polyoxyethylene (10) cetyl ether; Brij 58, polyoxyethylene (20) cetyl ether; Triton X-100, nonaethylene glycol octylphenol ether; Triton X-114, heptaethylene glycol octylphenyl ether; Nonidet P40, polyoxyethylene (9)-p-t-octylphenol; Tween 20, polyoxyethylene sorbitan monolaurate; Tween 60, polyoxyethylene sorbitan monostearate; Tween 80, polyoxyethylene sorbitan monooleate.
[b] HLB values are from refs 49 or 50.

6. PROPERTIES OF ACIDS, BASES, SALTS AND ORGANIC SOLVENTS

Table 12 lists selected acids, bases salts and organic solvents used in biochemistry. Details of acid–base indicators are given in *Table 13*.

7. AMINO ACIDS

Table 14 lists the abbreviations and physical and chemical data of L-amino acids and *Table 15* provides a classification of amino acids.

Table 12. Properties of acids, bases, salts and organic solvents

Name	Formula	Mol. wt	m.p. (°C)	b.p. (°C)	Density (20°C)	Molarity of the concd solution	ml l^{-1} to give a 1 M soln	Risk[a]	Safety[a]
Acids									
Acetic acid [64-19-7]	CH$_3$COOH	60.05	17	118	1.05	17.4	57.5	R10–R35	S2–S23–S26
Formic acid [64-18-6]	HCOOH	46.03	8	100	1.22	23.6 (90%) 25.9 (98%)	42.4 (90%) 38.5 (98%)	R35	S2–S3–S23–S36
Hydrochloric acid [7647-01-0]	HCl	36.46	—	—	1.18	11.6	85.9	R34–R37	S2–S26
Nitric acid [7697-37-2]	HNO$_3$	63.01	−42	86	1.42	15.7	63.7	R35	S2–S23–S26–S27
Orthophosphoric acid [7664-38-2]	H$_3$PO$_4$	98.00	22	261	1.7	14.7	67.8	R34	S26
Sulfuric acid [7664-93-9]	H$_2$SO$_4$	98.07	10	330	1.84	18.3	54.5	R35	S2–S26–S30
Bases									
Ammonium hydroxide [1336-21-6]	NH$_4$OH	17.03	—	—	(25%) 0.91 (35%) 0.88	13.3 18.1	75.1 55.2	R34–R36 R37/R38	S2–S3–S7–S26
Potassium hydroxide [1310-58-3]	KOH	56.11	360	1320	2.0	Solid		R35	S2–S26–S37/S39
Sodium hydroxide [1310-73-2]	NaOH	40.0	318	1390	2.12	Solid		R35	S2–S26–S37/S39

Table 12. Continued

Name	Formula	Mol. wt	Solubility (g/100 ml H$_2$O)		Risk[a]	Safety[a]
			Cold (°C)	Hot (°C)		
Selected Salts						
Ammonium acetate [631-61-8]	CH$_3$COONH$_4$	77.08	148(4)	Decomposes	—	—
Ammonium chloride [12125-02-9]	NH$_4$Cl	53.49	29.7(0)	75.8(100)	—	—
Ammonium nitrate [6484-52-2]	NH$_4$NO$_3$	80.04	118.3(0)	871(100)	R8-R9	S24-S25
Ammonium sulfate [7783-20-2]	(NH$_4$)$_2$SO$_4$	132.13	70.6(0)	103.8(100)	—	—
Calcium chloride dihydrate	CaCl$_2$·2H$_2$O	147.02	97.7(0)	326(60)	—	—
Calcium chloride hexahydrate [10043-52-4]	CaCl$_2$·6H$_2$O	219.08	279(0)	536(20)	—	—
Calcium hypochlorite [7778-54-3]	Ca(HOCl$_2$)	142.99	Sol.	Sol.	R8-R31-R34	S2-S26-S43
Lithium chloride [7447-41-8]	LiCl	42.39	63.7(0)	130(95)	R22	S24/S25
Magnesium acetate tetrahydrate [142-72-3]	(CH$_3$COO)$_2$Mg·4H$_2$O	214.4	120(15)	V. sol.	—	—
Magnesium chloride hexahydrate [7786-30-3]	MgCl$_2$·6H$_2$O	203.3	167(0)	367(100)	—	—

Table 12. Continued

Name	Formula	Mol. wt	Solubility (g/100 ml H_2O)		Risk[a]	Safety[a]
			Cold (°C)	Hot (°C)		
Magnesium nitrate hexahydrate [10377-60-3]	$Mg(NO_3)_2 \cdot 6H_2O$	256.41	125(0)	V. sol.	—	—
Magnesium sulfate heptahydrate [7487-88-9]	$MgSO_4 \cdot 7H_2O$	246.47	71(20)	91(40)	—	—
Manganous chloride tetrahydrate [7773-01-5]	$MnCl_2 \cdot 4H_2O$	197.9	151(8)	656(100)	—	—
Manganous sulfate heptahydrate [7785-87-7]	$MnSO_4 \cdot 7H_2O$	223.06	172	—	—	—
Potassium acetate [127-08-2]	CH_3COOK	98.14	253(20)	492(62)	—	—
Potassium chloride [7447-40-7]	KCl	74.55	34.7(20)	56.7(100)	—	—
Potassium iodide [7681-11-0]	KI	166.00	127.5(0)	208(100)	—	—
Potassium nitrate [7757-79-1]	KNO_3	101.1	13.3(0)	247(100)	R8-R12	—
Potassium permanganate [7722-64-7]	$KMnO_4$	158.03	6.4(20)	25(65)	R8-R22	S2-S24/S25-S27
Potassium sodium tartrate tetrahydrate [304-59-6]	$KOCOCH(OH)CH(OH)COONa \cdot 4H_2O$	282.22	26(0)	66(25)	—	—

Table 12. Continued

Name	Formula	Mol. wt	Solubility (g/100 ml H$_2$O)		Risk[a]	Safety[a]
			Cold (°C)	Hot (°C)		
Potassium sulfate [7778-80-5]	K$_2$SO$_4$	174.25	12(25)	24.1(100)	—	—
Sodium chloride [7647-14-5]	NaCl	58.44	35.7(0)	39.1(100)	—	—
Sodium metabisulfite [7681-57-4]	Na$_2$S$_2$O$_5$	190.1	54(20)	81.7(100)	R31	—
Sodium nitrate [7631-99-4]	NaNO$_3$	84.99	92.1(25)	180(100)	R8-R22	—
Sodium nitrite [7632-00-0]	NaNO$_2$	69.0	81.5(15)	163(100)	R8-R25-R31	S44
Sodium salicylate [54-21-7]	C$_6$H$_4$(OH)·COONa	160.1	81.5(15)	163(100)	—	—
Sodium succinate hexahydrate [150-90-3]	(CH$_3$COONa)·6H$_2$O	270.14	21.5(0)	86.6(75)	—	—
Sodium sulfate [7757-82-6]	Na$_2$SO$_4$	142.04	4.7(0)	42.7(100)	—	—
Sodium sulfate decahydrate [7757-82-6]	Na$_2$SO$_4$·10H$_2$O	322.19	11(0)	92.7(30)	—	—
Zinc chloride [7646-85-7]	ZnCl$_2$	136.29	423(25)	615(100)	R34	S7/S8-S28
Zinc sulfate heptahydrate [7733-02-0]	ZnSO$_4$·7H$_2$O	287.54	96.5(20)	663.6(100)	—	S22-S24/S25

Table 12. Continued

Name	Mol. wt	m.p. (°C)	b.p. (°C)	Flash point (°C)	Density	Miscible with water	Risk[a]	Safety[a]
Organic solvents								
Acetone [67-64-1]	58.08	−94	56.5	−18	0.788	Yes		
Acetonitrile [75-05-8]	53.06	−83.5	77.3	0	0.806	Limited		
Benzene [71-43-2]	78.11	5.5	80.1	−11	0.878	No		
n-Butanol [71-36-3]	74.1	−89	118	29	0.81	No	R10-R18-R20	S16
Chloroform [67-66-3]	119.4	−63	61	—	1.48	No	R20	S2-S24/S25
N,N-dimethyl formamide [68-12-2]	73.1	−61	153	57	0.95	Yes	R20/R21-R36	S26-S28-S36
Dimethyl sulfoxide (DMSO) [67-68-5]	78.1	18	190	95	1.10	Yes	R22	S24/S25
Ethanol [64-17-5]	46.1	−117	78	13	0.8	Yes	R11	S7-S16
Ethyl acetate [141-78-6]	88.1	−83	77	7.2	0.898	Yes		
Formaldehyde [50-00-0]	30.0	—	96	49	1.08	Yes	R10-R23/R24 R25–R43	S2-S26-S28
Formamide [75-12-7]	45.1	2.5	210	154	1.13	Yes	R23/R24/R25	S23-S24/S25

Table 12. Continued

Name	Mol. wt	m.p. (°C)	b.p. (°C)	Flash point (°C)	Density	Miscible with water	Risk[a]	Safety[a]
Glycerol [56-81-5]	92.1	18	290	160	1.26	Yes	—	—
Isoamyl alcohol [123-51-3]	88.2	−117	130	45	0.81	No	Irritant	
Isopropanol [67-63-0]	60.1	−89.5	82.4	22	0.79	Yes	R2-R36/R38	S23
Methanol [67-56-1]	32.04	−97.8	64	12	0.81	Yes		
Phenol [108-95-2]	94.1	41	182	80	1.07	Yes	R24/R25-R34	S2-S28-S44
Siliconizing solution [75-78-5][b]	129.06	—	—		1.31	No	R20/R22	S2-S24/S25
Toluene [108-88-3]	92.13	−95	110.6	4.4	0.866	No		
Xylene[c] [1330-20-7]	106.16[c]		137–140[c]	29	0.86	No		

[a] Risk and safety factors are given in *Table 3* of *Chapter 14*.
[b] Dimethyldichlorosilane, sold as a 2% solution in 1,1,1-trichloroethane.
[c] Xylene used as a solvent is a mixture of the ortho-, meta-, and para-isomers. The melting point is poorly defined, as is the boiling point.

Table 13. Acid–base indicators

Indicator (common range)	Preparation 0.1 g in 250 ml of	Acid color	Basic color	pH range
Cresol red (acid range)	Water containing 2.62 ml 0.1 M NaOH	Red	Yellow	0.2–1.8
m-Cresol purple (acid range)	Water containing 2.72 ml 0.1 M NaOH	Red	Yellow	1.0–2.6
Thymol blue (acid range)	Water containing 2.15 ml 0.1 M NaOH	Red	Yellow	1.2–2.8
Methyl yellow	90% EtOH	Red	Yellow	2.9–4.0
Bromophenol blue	Water containing 1.49 ml 0.1 M NaOH	Yellow	Purple	3.0–4.6
Tetrabromophenol blue	Water containing 1.0 ml 0.1 M NaOH	Yellow	Blue	3.0–4.6
Congo red	Water or 80% EtOH	Violet	Reddish-orange	3.0–5.0
Methyl orange	Free acid: water Na salt: water containing 3 ml 0.1 M HCl	Red	Orange-yellow	3.1–4.4
Bromocresol green (blue)	Water containing 1.43 ml 0.1 M NaOH	Yellow	Blue	3.6–5.2
Methyl red	Na salt: water free acid: 60% EtOH	Red	Yellow	4.2–6.3
Chlorophenol red	Water containing 2.36 ml 0.1 M NaOH	Yellow	Violet-red	4.8–6.4
Bromocresol purple	Water containing 1.85 ml 0.1 M NaOH	Yellow	Violet	5.2–6.8
Litmus	Water	Red	Blue	5.0–8.0
Bromothymol blue	Water containing 1.6 ml 0.1 M NaOH	Yellow	Blue	6.0–7.6

Table 13. Continued

Indicator (common range)	Preparation 0.1 g in 250 ml of	Acid color	Basic color	pH range
Phenol red	Water containing 2.82 ml 0.1 M NaOH	Yellow	Red	6.8–8.4
Neutral red	70% EtOH	Red	Orange-brown	6.8–8.0
Cresol red (basic range)	Water containing 2.62 ml 0.1 M NaOH	Yellow	Red	7.2–8.8
m-Cresol purple (basic range)	Water containing 2.62 ml 0.1 M NaOH	Yellow	Purple	7.6–9.2
Thymol blue (basic range)	Water containing 2.15 ml 0.1 M NaOH	Yellow	Blue	8.0–9.6
Phenolphthalein	70% EtOH (60% cellosolve)	Colorless	Pink	8.3–10.0
Thymolphthalein	90% EtOH	Colorless	Blue	9.3–10.5
Alizarin yellow	EtOH	Yellow	Red	10.1–12.0
Tropaeolin O	Water	Yellow	Orange	11.1–12.7

Table 14. L-Amino acids: physical and chemical data

Amino acid	Abbreviations		Mol. wt	Solubility (25°C) (g/100 ml H₂O)	pK_a			pI (25°C)
					–COOH	–NH₂	R-group	
Alanine [56-41-7]	Ala	A	89.1	16.65	2.35	9.87	—	6.11
Arginine [74-79-3]	Arg	R	174.2	15.00[a]	1.82	8.99	12.48	10.76
Asparagine [70-47-3]	Asn	N	132.12	2.99	2.14	8.72	—	5.43
Aspartic acid [56-84-8]	Asp	D	133.11	0.5	1.99	9.90	3.90	2.98
Cysteine [52-90-4]	Cys	C	121.16	V. sol[b]	1.92	10.7	8.37	5.15
Glutamic acid [56-86-0]	Glu	E	147.13	0.86	2.10	9.47	4.07	3.08
Glutamine [56-85-9]	Gln	Q	146.15	3.6[a]	2.17	9.13	—	5.65
Glycine [56-40-6]	Gly	G	75.07	25	2.35	9.78	—	6.06
Histidine [71-00-1]	His	H	155.16	4.16	1.80	9.33	6.04	7.64
Isoleucine [73-32-5]	Ile	I	131.18	4.12	2.32	9.76	—	6.04
Leucine [61-90-5]	Leu	L	131.18	2.43	2.33	9.74	—	6.04
Lysine [56-87-1]	Lys	K	146.19	66.6[a]	2.16	9.06	10.54	9.47
Methionine [63-68-3]	Met	M	149.21	5.14[a]	2.13	9.28	—	5.71

Table 14. Continued

Amino acid	Abbreviations		Mol. wt	Solubility (25°C) (g/100 ml H$_2$O)	pK_a			pI (25°C)
					–COOH	–NH$_2$	R-group	
Phenylalanine [63-91-2]	Phe	F	165.19	2.96	2.2	9.31	—	5.76
Proline [147-58-3]	Pro	P	115.13	162.3	1.95	10.64	—	6.30
Serine [56-45-1]	Ser	S	105.09	25.0[a]	2.19	9.21	—	5.70
Threonine [72-19-5]	Thr	T	119.12	V.sol	2.09	9.10	—	5.60
Tryptophan [73-22-3]	Trp	W	204.23	1.14	2.46	9.41	—	5.88
Tyrosine [60-18-4]	Tyr	Y	181.19	0.045	2.20	9.21	10.46	5.63
Valine [72-18-4]	Val	V	117.15	8.85	2.29	9.74	—	6.02

[a] Temperatures: Arg, 21°C; Gln, 18°C; Lys, 20°C; Met, 20°C; Ser, 20°C.
[b] As hydrochloride. Arginine and lysine are also frequently used as the hydrochloride. In such cases take care with pHs of solutions.

Table 15. Amino acid classification by solution properties of side chains

Hydrophobic (non-polar) R-groups[a]
Alkyl R-groups: alanine, glycine, isoleucine, leucine, methionine[b], valine, glycine
Cycloalkyl R-group: proline
Aromatic R-group: phenylalanine, tryptophan

Uncharged polar R-groups[a]
Hydroxylalkyl: serine, threonine
Amides: asparagine, glutamine
Alkylthiol: cysteine

Positively charged polar R-groups[a]
Alkyl amine: lysine
Alkylguanidino: arginine
Imidazole: histidine

Negatively charged polar R-groups[a]
Carboxylic acids: aspartic acid, glutamic acid

[a] At pH 7.
[b] Strictly an alkyl thioether R-group.

8. AMMONIUM SULFATE PRECIPITATION CHART FOR PROTEINS

Ammonium sulfate precipitation remains as a useful method for the fractionation of proteins prior to further separation by various chromatographic methods. This chart (*Table 16*) shows the amount of solid ammonium sulfate required to bring a solution of known initial saturation to the desired final saturation at 0°C. For higher temperatures more ammonium sulfate is required for saturation. An indication of the different amounts required is given in *Table 17*.

Table 16. Chart for ammonium sulfate precipitation of proteins

Initial concn of ammonium sulfate, % saturation at 0°C	Final concn of ammonium sulfate, % saturation at 0°C																
	20	25	30	35	40	45	50	55	60	65	70	75	80	85	90	95	100
	g solid ammonium sulfate to add to 100 ml of solution																
0	10.7	13.6	16.6	19.7	22.9	26.2	29.5	33.1	36.6	40.4	44.2	48.3	52.3	56.7	61.1	65.9	70.7
5	8.0	10.9	13.9	16.8	20.0	23.2	26.6	30.0	33.6	37.3	41.1	45.0	49.1	53.3	57.8	62.4	67.1
10	5.4	8.2	11.1	14.1	17.1	20.3	23.6	27.0	30.5	34.2	37.9	41.8	45.8	50.0	54.4	58.9	63.6
15	2.6	5.5	8.3	11.3	14.3	17.4	20.7	24.0	27.5	31.0	34.8	38.6	42.6	46.6	51.0	55.5	60.0
20	0	2.7	5.6	8.4	11.5	14.5	17.7	21.0	24.4	28.0	31.6	35.4	39.2	43.3	47.5	51.9	56.5
25		0	2.7	5.7	8.5	11.7	14.8	18.2	21.4	24.8	28.4	32.1	36.0	40.1	44.2	48.5	52.9
30			0	2.8	5.7	8.7	11.9	15.0	18.4	21.7	25.3	28.9	32.8	36.7	40.8	45.1	49.5
35				0	2.8	5.8	8.8	12.0	15.3	18.7	22.1	25.8	29.5	33.4	37.4	41.6	45.9
40					0	2.9	5.9	9.0	12.2	15.5	19.0	22.5	26.2	30.0	34.0	38.1	42.4
45						0	2.9	6.0	9.1	12.5	15.8	19.3	22.9	26.7	30.6	34.7	38.8
50							0	3.0	6.1	9.3	12.7	16.1	19.7	23.3	27.2	31.2	35.3
55								0	3.0	6.2	9.4	12.9	16.3	20.0	23.8	27.7	31.7
60									0	3.1	6.3	9.6	13.1	16.6	20.4	24.2	28.3
65										0	3.1	6.4	9.8	13.4	17.0	20.8	24.7
70											0	3.2	6.6	10.0	13.6	17.3	21.2
75												0	3.2	6.7	10.2	13.9	17.6
80													0	3.3	6.8	10.4	14.1
85														0	3.4	6.9	10.6
90															0	3.4	7.1
95																0	3.5
100																	0

Table 17. Saturated ammonium sulfate solutions at various temperatures

	Temperature (°C)				
	0	10	20	25	30
Percentage w/w	41.42	42.22	43.09	43.47	43.85
Gram required to saturate 1 liter	706.8	730.5	755.8	766.8	777.5
Gram in a liter of saturated solution	514.7	525.1	536.1	541.2	545.9
Molarity of saturated solution	3.90	3.97	4.06	4.10	4.13

9. REFERENCES

1. Good, X. and Izawa, X. (1972) *Methods Enzymol.*, **24B**, 53.

2. Kamberov, E. and Ivanov, A. (1990) *J. Chromatog.*, **525**, 307.

3. Yamashita, T. and Butler, W.L. (1969) *Plant Physiol.*, **44**, 435.

4. Ogilvie, J.W. and Whitaker, S.C. (1976) *Biochim. Biophys. Acta*, **445**, 525.

5. Eriksson, C.G., Nordstroem, L. and Eneroth, P. (1983) *J. Steroid Biochem.*, **19**, 1199.

6. Fischer, B.E., *et al.* (1979) *Eur. J. Biochem.*, **94**, 523.

7. Weakley, B.S. (1977) *J. Microsci.*, **109**, 249.

8. Wesser, U. (1967) *Z. Naturforsch.*, **B22**, 4513.

9. Poole, C.A., Reilly, H.C. and Flint, M.H. (1982) *In Vitro* **18**, 755.

10. Tunnicliff, G. and Smith, J.A. (1981) *J. Neurochem.*, **36**, 1122.

11. Waxman, P.G., *et al.* (1981) *Eur. J. Biochem.*, **120**, 129.

12. Lleu, P.L. and Rebel, G. (1991) *Anal. Biochem.*, **192**, 215.

13. Werber, M.M. and Shalitin, S. (1973) *Bioorg. Chem.*, **2**, 202.

14. Eterovic, V.A., *et al.* (1989) *J. Recept. Res.*, **9**, 107.

15. Kosaka, H. and Tyuma, I. (1982) *Biochim. Biophys. Acta*, **709**, 187.

16. Schy, W.E. and Plewa, M.J. (1989) *Mutat. Res.*, **226**, 263.

17. Sayer, J.M., *et al.* (1990) *J. Am. Chem. Soc.*, **112**, 1177.

18. Schepartz, A. and Breslow, R. (1987) *J. Am. Chem. Soc.*, **109**, 1814.

19. Swader, J.A. and Kumamoto, J. (1975) *Photochem. Photobiol.*, **21**, 313.

20. Hegetschweiler, K. and Saltman, P. (1986) *Inorg. Chem.*, **25**, 107.

21. Koukli, I.I., Sarantonis, E.G. and Calokerinos, A.C. (1990) *Anal. Lett.*, **23**, 1167.

22. Grady, J.K., Chasteen, N.D. and Harris, D.C. (1988) *Anal. Biochem.*, **173**, 111.

23. *CRC Handbook of Biochemistry and Molecular Biology, Physical and Chemical Data*, 3rd edn (1975). CRC Press, Boca Raton, FL, Vol. 1, p. 374.

24. Helenius, A. and Simons, K. (1975) *Biochim. Biophys. Acta*, **415**, 29.

25. Nozaki, Y. and Tanford, C. (1970) *J. Biol. Chem.*, **245**, 1648.

26. Gordon, J.A. (1972) *Biochemistry*, **11**, 1862.

27. Chirgwin, J.M., *et al.* (1979) *Biochemistry*, **18**, 5294.

28. Wilcoxson, J. (1975) *Anal. Biochem.*, **66**, 64.

29. Parish, J.H. (1972) *Principles and Practice of Experiments with Nucleic Acids*. J.H. Longmans, London.

30. Parish, J.H. (1972) in *Subcellular Components, Preparation and Fractionation* (G.D. Birnie, ed.). Butterworths, London.

31. Cleland, W.W. (1964) *Biochemistry*, **3**, 480. (See also the booklet *Cleland's Reagent* published by Calbiochem.)

32. Tanford, C. and Reynolds, J.A. (1976) *Biochim. Biophys. Acta*, **457**, 133.

33. Neugebauer, J. (1988) *A Guide to the Properties and Uses of Detergents*. Calbiochem, La Jolla, California.

34. Hjelmeland, L.M. and Chrambach, A. (1984) *Methods Enzymol.*, **104**, 305.

35. Helenius, A., *et al.* (1979) *Methods Enzymol.*, **56**, 734.

36. Bordier, C. (1981) *J. Biol. Chem.*, **256**, 1604.

37. Waehneldt, T.V. (1975) *Biosystems*, **6**, 176.

38. Weber, K. and Osborn, M. (1969) *J. Biol. Chem.*, **244**, 4406.

39. Simons, K. and Helenius, A. (1970) *FEBS Lett.*, **7**, 59.

40. Hjelmeland, L.M. (1980) *Proc. Natl. Acad. Sci. USA*, **77**, 6368.

41. Bitonti, A.J., *et al.* (1982) *Biochemistry*, **21**, 3650.

42. Gould, R.J., Ginsberg, B.H. and Spector, A.A. (1981) *Biochemistry*, **20**, 6776.

43. Hjelmeland, L.M., Nebert, D.W. and Osborne, J.C. (1983) *Anal. Biochem.*, **130**, 72.

44. Gonenne, A. and Ernst, R. (1978) *Anal. Biochem.*, **87**, 72.

45. Ashani, Y. and Catravas, G.N. (1980) *Anal. Biochem.*, **109**, 55.

46. Lever, M. (1977) *Anal. Biochem.*, **83**, 274.

47. Baron, C. and Thompson, T.E. (1975) *Biochim. Biophys. Acta*, **382**, 276.

48. Stubbs, G.W. and Litman, B.J. (1978) *Biochemistry*, **17**, 215.

49. *McCutcheon's, 1991, Detergents and Emulsifiers*, International Edition. M.C. Publishing, Ridgewood, NJ.

50. Becher, P. (1967) in *Non-ionic Surfactants* (M.J. Schick, ed.). Marcel Dekker, New York, p. 604.

51. Martell, A.E. and Smith, R.M. (1974-1976) *Critical Stability Constants*. Plenum, New York, Vols. 1-6.

CHAPTER 2
RADIOISOTOPES IN BIOCHEMISTRY
D. Rickwood, D. Patel and D. Billington

1. DEFINITIONS

1.1. Units of radioactivity

Becquerel
The SI unit of radioactivity, the becquerel (Bq) is 1 disintegration per second; this is equivalent to 2.70×10^{-11} curies (Ci).

Curie (milli-, micro-)
This is equivalent to the amount of an isotope undergoing 3.7×10^{10} nuclear disintegrations per second (2.22×10^{12} min^{-1}). 1 curie equals 3.7×10^{10} becquerels.

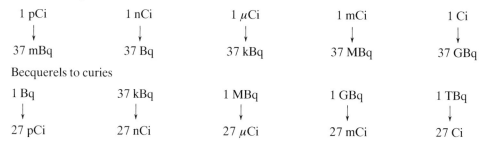

1.2. Specific activity
The radioactivity of an element or compound per unit mass of element.

1.3. Electron volt (eV, or MeV = 10^6 eV)
Measure of energy of radioactive emissions; energy acquired by an electron when accelerating along a potential gradient of 1 electron volt = 1.6×10^{-12} erg.

1.4. Gray
This is an absorbed dose of 1 joule kg^{-1} of tissue. 1 gray equals 100 rads.

1.5. Dose equivalent man (sievert)
This is defined as: (dose in Sv) = (dose in gray) × (quality factor, Q). The value of Q is unity for β particles, X-rays and γ-rays, and 10 for α-particles, neutrons and protons. 100 rem is equal to 1 sievert (Sv).

2. RADIOISOTOPES USED IN BIOCHEMISTRY

The characteristics of the isotopes used most frequently in biochemistry are listed in *Table 1*.

Table 1. Radioactive isotopes used in biochemical studies

Element (mass no.)	Radiation	Half-life	Energy of radiation (MeV)		Detection method
			Particles	γ-rays	
^{3}H	β^-	12.3 yr	0.0185	—	LS
^{14}C	β^-	5736 yr	0.156	—	LS
^{22}Na	β^+, γ	2.6 yr	0.58 (90%)	1.3	LS, SS
^{24}Na	β^-, γ	14.8 h	1.39	1.38, 2.76	LS, C, SS
^{28}Mg	β^-	21.4 h	0.459	—	LS
^{32}P	β^-	14.3 days	1.71	—	LS, C
^{33}P	β^-	25.4 days	0.249	—	LS
^{35}S	β^-	87.1 days	0.169	—	LS
^{36}Cl	β^+, K, β^-	3.1×10^5 yr (β^-)	0.64	—	LS
^{42}K	β^-, γ	12.4 h	2.04 (25%) 3.58 (75%)	1.4, 2.1	LS, C, SS
^{45}Ca	β^-	165 days	0.260	—	LS
^{51}Cr	K, γ	28 days	—	0.323, 0.237	LS, SS
^{52}Mn	β^+ (35%) K (65%)	5.8 days	0.58	1.0, 0.73 0.94, 1.46	LS
^{54}Mn	K, γ	310 days	—	0.835	LS, SS
^{55}Fe	K only	2.94 yr	—	—	LS
^{57}Co	γ	270 days	—	0.136 (10%) 0.122 (88%)	LS, SS
^{58}Co	β^+ (14.5%), γ	72 days	0.472	0.805	LS, SS
^{59}Fe	β^-, γ	46.3 days	0.46 (50%) 0.26 (50%)	1.3 (50%) 1.1 (50%)	LS, SS
^{60}Co	β^-, γ	5.3 yr	0.31	1.16, 1.32	LS, SS
^{64}Cu	K (54%) β^- (31%) β^+ (15%) $\gamma + K$ (1.5%)	12.8 h	0.57 (β^-) 0.66 (β^+)	1.35 (2.5%)	LS, SS
^{65}Zn	β^+ (1.3%) K (98.7%)	250 days	0.32	1.14 (46% of K)	LS, SS
^{75}Se	K, γ	125 days	—	0.076–0.405	SS
^{76}As	β^-, γ	26.8 h	3.04 (60%) 2.49 (25%) 1.29 (15%)	1.705 1.20 0.55	LS, SS
^{82}Br	β^-, γ	34 h	0.465	0.547, 0.787 1.35	LS, SS
^{86}Rb	β^-, γ	18.7 days	1.822 (80%) 0.716 (20%)	1.081	LS, C, SS
^{89}Sr	β^-	53 days	1.46	—	LS, C
^{90}Sr	β^-	28.1 yr	0.61	—	LS
^{99}Mo	β^-, γ	68 h	1.3	0.75, 0.24	LS
^{125}Sb	β^-, γ	2.7 yr	0.3 (65%) 0.7 (35%)	0.55	LS
^{125}I	γ	60 days	—	0.035	LS, SS
^{131}I	β^-, γ	8.1 days	0.605 (86%) 0.25 (14%)	0.637, 0.363 0.282, 0.08	LS, SS

Abbreviations: LS, liquid scintillation counting; SS, gamma counter (solid scintillant); C, Cerenkov; K, electron capture.

3. RADIOACTIVE DECAY CORRECTION

Some radioisotopes decay so quickly that it is necessary to apply corrections to the original activities. *Table 2* gives an indication of the decay of various isotopes. Alternatively, this can be done by calculation:

$$\log_{10}(N_0/N_t) = 0.3010(t/h).$$

Percentage of isotope remaining $= 100/\text{antilog}[0.3010(t/h)]$, or $100 \times e^{-0.693(t/h)}$,

where $t =$ time of decay;
  $h =$ half-life (time taken for radioactivity to decay to 50% of original activity);
  $N_0 =$ radioactivity at zero time;
  $N_t =$ radioactivity at time t.

After H half-lives the percentage of original isotope remaining $= 2^{-(H)} \times 100$. Thus, when (H) $= 1$, the percentage of original isotope remaining $= 50\%$; (H) $= 2$, 25%; (H) $= 3$, 12.5%; (H) $= 4$, 6.25%; (H) $= 5$, 3.13%; and so on.

4. AUTORADIOGRAPHY

Autoradiography has now assumed a very prominent role in both the quantitative and semi-quantitative measurement of radioisotopes in biochemical analyses. Autoradiography can be divided into three different techniques, namely direct autoradiography, fluorography and indirect autoradiography. In direct autoradiography, labeled components are detected simply by placing the sample (e.g. a dried gel, filter, etc.) in contact with one of the several commercial brands of 'direct' X-ray film (e.g. Kodak direct exposure film and Amersham Hyperfilm β-max) and then exposing at room temperature. However, with ^{3}H-labeled components in gels, the low-energy β-particles are unable to penetrate the gel matrix and reach the X-ray film. In this case, detection can be achieved by impregnating the gel with a scintillator, such as PPO (2,5-diphenyloxazole; see Section 5.2). The β-particles then interact with the scintillator to cause the emission of light which exposes blue-sensitive X-ray film (so called 'screen-type' X-ray film) placed next to the gel, forming a detectable image. Many brands of suitable screen-type film are available (e.g. Fuji RX, Kodak XAR-5 and Amersham Hyperfilm-MP). This procedure, fluorography, is also widely used for the detection of ^{14}C and ^{35}S in gels. Exposure is at $-70°$C to stabilize latent image formation during long exposures to the light generated by fluorography; this stabilization can increase the sensitivity of detection at least fourfold. Unfortunately, when untreated X-ray film is used, the absorbance of the fluorographic image is not proportional to the amount of radioactivity in the sample; small amounts of radioactivity produce disproportionately faint images. This is overcome by exposing the film to a flash of light (<1 ms) prior to fluorography.

The detection of ^{32}P and ^{125}I with X-ray films involves a quite different problem. Here the energy of many of the β-particles is such that the emissions pass completely through the film. However, they can be trapped by placing a calcium tungstate intensifying screen on the other side of the X-ray film. Emissions reaching this screen produce multiple flashes of light which now cause the production of a photographic image on the X-ray film, superimposed over the autoradiographic image. This procedure, indirect autoradiography, is far more sensitive for ^{32}P and ^{125}I detection than direct autoradiography. As with fluorography, exposure is again carried out at $-70°$C.

Table 3 gives the sensitivities of detection of direct autoradiography, indirect autoradiography and fluorography with the relevant isotopes. Note, however, that although fluorography or the use of an intensifying screen for autoradiography gives maximum sensitivity for

Table 2. Half-life activity corrections for selected radioisotopes

^{32}P (half-life 14.3 days)		^{35}S (half-life 87.1 days)		^{131}I (half-life 8.1 days)		^{125}I (half-life 60 days)		^{45}Ca (half-life 165 days)		^{51}Cr (half-life 28 days)	
Time (days)	% activity remaining	Time (days)	% activity remaining	Time (days)	% activity remaining	Time (days)	% activity remaining	Time (days)	% activity remaining	Time (days)	% activity remaining
1	95.3	2	98.4	0.2	98.3	4	95.5	10	95.9	2	95.2
2	90.8	5	96.1	0.4	96.6	8	91.2	20	91.9	4	90.6
3	86.5	10	92.3	0.6	95.0	12	87.1	30	88.2	6	86.2
4	82.4	15	88.7	1.0	91.8	16	83.1	40	84.5	8	82.0
5	78.5	20	85.3	1.6	87.2	20	79.4	50	81.1	10	78.1
6	74.8	25	82.0	2.3	81.2	24	75.8	60	77.7	12	74.3
7	71.2	31	78.1	3.1	76.7	28	72.4	70	74.5	14	70.7
8	67.8	37	74.5	4.0	71.0	32	69.1	80	71.5	16	67.3
9	64.7	43	71.0	5.0	65.2	36	66.0	90	68.5	18	64.0
10	61.5	50	67.0	6.1	59.3	40	63.0	100	65.7	20	61.0
11	58.7	57	63.6	7.3	53.4	44	60.2	110	63.0	22	58.0
12	55.9	65	59.6	8.1	50.0	48	57.4	120	60.4	24	55.2
13	53.2	73	56.0			52	54.8	130	57.9	26	52.5
14	50.7	81	52.5			56	52.4	140	55.5	28	50.0
14.3	50.0	87.1	50.0			60	50.0	150	53.3		
								160	51.1		

Table 3. Sensitivities of film detection methods for commonly used radioisotopes

Isotope	Method	Detection limit d.p.m. cm^{-2} for 24 h	Relative performance compared to direct autoradiography
^{3}H	Direct autoradiography	$> 8 \times 10^6$	1
	Fluorography using PPO	8000	> 1000
^{14}C or ^{35}S	Direct autoradiography	6000	1
	Fluorography using PPO	400	15
^{32}P	Direct autoradiography	525	1
	Intensifying screen	50	10.5
^{125}I	Direct autoradiography	1600	1
	Intensifying screen	100	16

the detection of ^{14}C/^{35}S and ^{32}P/^{125}I respectively, these methods decrease resolution by secondary scattering. Hence for maximum resolution, direct autoradiography should be chosen.

5. SCINTILLATION COUNTING

5.1. Cerenkov counting

High-energy (> 0.5 MeV) β-particles traveling through water cause the polarization of molecules along their trajectory, which then emit photons of light (350–600 nm) as their energy returns to the ground state (the Cerenkov effect). Data relevant to counting in aqueous buffer in a scintillation counter without added scintillation fluid (Cerenkov counting) are given in *Table 4*.

A low-energy window must be used when counting under these conditions. The ^{3}H channel is frequently used for Cerenkov counting, but it is better to calibrate the counter specifically for each isotope.

Table 4. Counting efficiency using Cerenkov counting

Radioisotope	E_{max} (MeV)	% of β-spectrum above 0.5 MeV	Counting efficiency (%)
^{24}Na	1.39	60	40
^{32}P	1.71	80	50
^{36}Cl	0.71	30	10
^{42}K	3.58	90	80

5.2. Counting of finely dispersed or solvent-soluble substances

Radioactive samples that are soluble in hydrocarbon solvents or finely dispersed (e.g. on filters) can be directly introduced into commercial scintillants or mixtures such as:

(i) scintillation-grade toluene or xylene containing 5 g l^{-1} 2, 5-diphenyloxazole (PPO) and 0.1 g l^{-1} 1, 4-di-(2-(5-phenyloxazolyl))-benzene (POPOP) (*Figure 1*);

(ii) scintillation-grade toluene or xylene containing 10 g l^{-1} 2-phenyl-5-(4-biphenylyl)-1,3,4-oxadiazole (PBD) or 15 g l^{-1} 2-(4'-t-butylphenyl)-5-(4"-biphenylyl)-1,3,4,-oxadiazole (butyl PBD) (*Figure 1*).

Chemical quenching and color quenching can both occur. Particulate material will also tend to reduce the counting efficiency.

Figure 1. Structures of compounds used in scintillants.

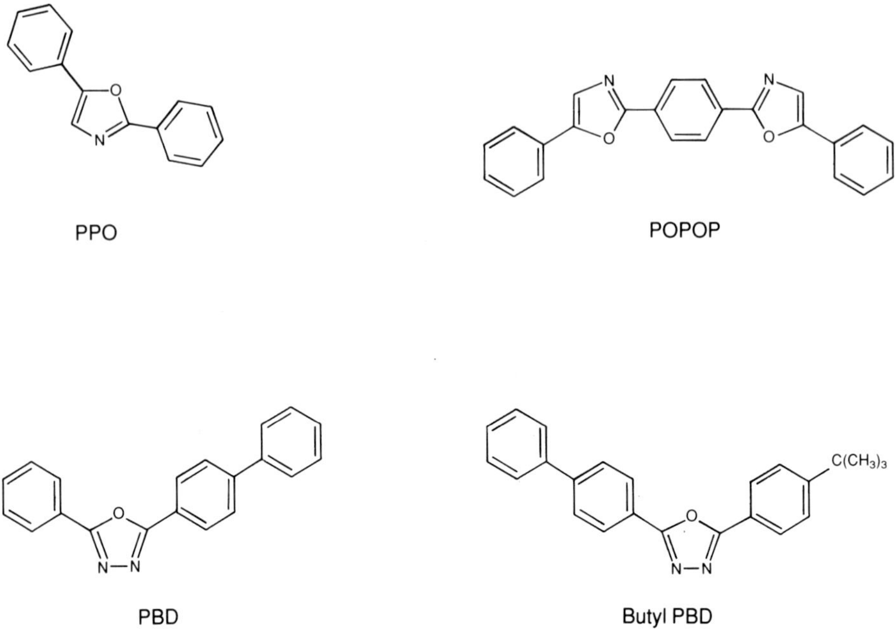

PPO

POPOP

PBD

Butyl PBD

5.3. Counting of aqueous samples

Aqueous samples can be counted in hydrocarbon-based scintillation fluids, using a polyethoxylate surfactant to produce an emulsion. For example:

340 ml surfactant (Triton X-100 or Triton X-114)
600 ml toluene or xylene (sulfur-free or scintillation grades)
5.0 g PPO, 0.1 g POPOP — can be replaced by 15 g butyl PBD or 10 g PBD, an advantage for scintillation counters containing bialkali photocathodes.

The disadvantage of such systems is that separation into two phases can occur, with a marked reduction in counting efficiency (1). Sometimes the phase separation cannot be observed visually and it cannot be detected by external standard measurements. Consequently, each scintillation fluid must be examined for the stability of counting rate with time at the counting temperature, with various percentages of aqueous sample incorporated. A recipe for a modified Tritosol cocktail has been published (2) that can accept up to 23% of water with no phase separation and that gives a linear relationship between counting efficiency and external standard ratio over wide quenching ranges. The recipe for this is:

Ethylene glycol	35 ml
Ethanol	140 ml
Triton X-100 (X-114)	250 ml
Xylene	575 ml
PPO	3.0 g
POPOP	0.2 g

Many commercial emulsifying cocktails are also available, but it is wise to examine the counting efficiency and emulsion stability under the precise conditions being used, for example the percentage of aqueous sample incorporated, the presence of solutes, and the pH; avoid highly alkaline or acidic solutions.

Solid tissues and other solid biological samples should be solubilized before counting, by incubating with quaternary bases such as hyamine hydroxide. However, this often produces strong phosphorescence with tissue proteins. Various commercial stabilizers have been developed based on quaternary bases that do not have this disadvantage. During digestion of the sample, frequent agitation or ultrasonic vibration is required: heating to 50°C helps, but higher temperatures cause excessive coloration and quenching. Alternatively, some scintillation fluids (e.g. Soluene) contain the digestion agent and can be added directly to samples.

5.4. Gamma counting

Isotopes which emit γ-rays or energetic β-particles can be measured in a gamma counter. In this case the scintillator is solid (usually NaI containing thallium) and surrounds the sample; as in the case of the liquid scintillation counting, the light (350–500 nm) is detected by photomultipliers. No preparation of the sample is required and the counting efficiency is not affected by the color of the sample nor by chemical quenching. Laboratories that only use γ-emitting isotopes often buy a gamma counter for its cheapness and simplicity, but otherwise the flexibility of liquid scintillation counters makes them much more attractive.

6. USE OF RADIOISOTOPES AS TRACERS

One of the most important applications of radioisotopes in biochemistry is their use in determining metabolic pathways and for measuring the flow of metabolites through pathways. This section summarizes some of the most important aspects of the use of tracers.

The term *isotopic abundance* describes the percentage of radioactive atoms present with respect to the total. The theoretical maximum specific activity of ^{14}C at 100% isotopic abundance is 2.31 GBq mg^{-1} atom while that of ^{3}H is 1.07×10^3 GBq mg^{-1} atom. Thus, radiochemicals containing a single ^{14}C or ^{3}H atom per molecule have maximum specific activities of 2.31 and 1.07×10^3 GBq mmol^{-1}, respectively. Multiple intramolecular labeling with ^{14}C is difficult to achieve and is rarely undertaken. However, if, for example, phenylalanine was prepared by successive radioactive syntheses and every one of the eight carbon atoms in the molecule was ^{14}C, its specific activity would be 18.5 GBq mmol^{-1}. In contrast, multiple intramolecular labeling with ^{3}H is much easier and very high specific activities can be achieved. ^{14}C-Labeled steroids are commercially produced with ^{14}C at C-4 only and are supplied therefore at specific activities of up to 2.2 GBq mmol^{-1}. ^{3}H-Labeled steroids can be prepared commercially with up to 10 ^{3}H atoms per molecule, and are supplied at specific activities up to 5×10^3 GBq mmol^{-1}. Such differences in the specific activity of commercially available ^{14}C- and ^{3}H-labeled compounds is of importance when working at nanomolar or picomolar concentrations. If a ^{14}C-labeled compound was diluted to such low concentrations, the amount of radioactivity may fall below detection limits. In such cases, ^{3}H-labeled compounds, which are available at 1000-fold higher specific activities, must be used.

A basic assumption when using radioisotopes as tracers is that the radiolabeled substance behaves chemically and physiologically exactly like the natural substance. In this respect, extra care should be taken when using isotopes of hydrogen in metabolic studies because of the so-called *isotope effect*. An isotope effect is defined as any difference in the chemical or physical behavior of two compounds which differ only in the isotopic composition of one (or more) of their chemical elements. In biological systems, by far the most important isotope effects are on rates of reactions in successive steps. Deuterium atoms (^{2}H) are twice as heavy as hydrogen atoms (^{1}H) and therefore the C—D bond has a lower frequency of vibration than the C—H bond. As a result, breakage of a C—D bond requires a higher activation energy than breakage of a C—H bond and several-fold differences in reaction rates have been observed. Thus, if a metabolic study involves monitoring the rupture of C—H bonds, the use of ^{2}H or ^{3}H to monitor the rate of this process can give misleading results. For isotopes of higher atomic mass, the isotope effect is far less serious because the mass difference is much less.

In addition, labeling of proteins with ^{125}I can sometimes cause the radiolabeled protein to behave differently from the natural substance. Such a consideration is important for antibody or receptor binding of the radiolabeled protein. Radioiodination of tyrosine residues by the Chloramine T method is more likely to cause tracer damage than radioiodination of primary amino groups by the Bolton and Hunter method (3).

Radioisotopic methods have been developed to trace and assay enzymic reactions. If the enzyme reaction is simple, either utilization of radiolabeled substrate or formation of radiolabeled product may be monitored. However, for a two-stage enzymic reaction in which intermediates accumulate, only substrate utilization will measure the true rate. This condition also applies when the reaction product(s) are further utilized by other enzymes in a metabolic pathway. Thus, a typical radioisotopic enzyme assay involves incubating the radiolabeled substrate with the enzyme under optimal conditions for a fixed period of time. The residual radiolabeled substrate is then separated from the radiolabeled product(s) prior to determination of radioactivity in either substrate or product(s). The success of any radioisotopic enzyme assay is critically dependent on the complete separation of the radiolabeled substrate and products. Several methods have been used and include precipitation, solvent extraction, ion-exchange and paper chromatography, electrophoresis and charcoal adsorption. It is imperative that the reaction rate remains constant throughout the incubation period and this should be validated when setting up a radioisotopic enzyme assay. A further useful validation is that the reaction rate should show a linear relationship to the enzyme concentration.

Radioisotopes are also used for tracing metabolic pathways. Such experiments usually involve pulse–chase techniques where the radioactive substance is presented to the biological system as a pulse and the radiolabel chased through various metabolites or cellular compartments. For metabolic studies, samples are taken at various times and the radiolabeled metabolites are separated by chromatographic techniques. These metabolites can often be identified by comparison of their chromatographic behavior to authentic reference compounds, although mass spectrometry is often necessary to elucidate their structure. Thus, by comparing the time course of appearance of radiolabeled metabolites, the sequence of metabolites in a metabolic pathway can often be elucidated. Tracers in metabolic studies are invariably radiolabeled in one or more specific positions in the molecule. Chemical degradation of the radiolabeled metabolites can be used to locate the specific position(s) of the radiolabel in the metabolites. Such an approach not only gives information concerning the path of the radiolabel through a metabolic pathway, but often gives information concerning reaction mechanisms in the pathway.

Radioactive tracers can be used to determine the kinetics of many processes *in vivo* (4). In such applications, complex compartmental analysis is often required as in, for example, glucose kinetics (5). A more simple tracer application is the measurement of blood flow kinetics by the *indicator dilution technique* (6). This relies on administering a radiolabeled tracer into the circulation and measuring its dilution at a sampling site downstream from the site of administration. The method assumes that the tracer is inert, is well mixed in the blood, does not leave the circulation between the administration and sampling sites and does not disturb the hemodynamics.

Considering blood flowing in a vessel at F ml min^{-1}. If a radiolabeled tracer is administered at a constant infusion rate of I ml min^{-1} and at an activity of D_{in} c.p.m. ml^{-1}, then,

$$\text{input rate} = I \cdot D_{in} \text{ c.p.m. min}^{-1}.$$

Blood samples are taken at a sampling site downstream from the infusion site until a constant, maximum steady-state concentration of radioactivity is achieved (D_{out} c.p.m. ml^{-1}). Now,

$$\text{output rate} = (F + I) \cdot D_{out} \text{ c.p.m. min}^{-1}.$$

At steady state, the input rate must equal the output rate, that is:

$$I \cdot D_{in} = (F + I) \cdot D_{out}.$$

If the infusion rate (I) is small compared to blood flow (F), then $(F + I) \approx F$, that is:

$$F = \frac{I \cdot D_{in}}{D_{out}} \text{ ml min}^{-1}. \qquad \text{Equation 1}$$

Thus, knowing the specific activity of the infused tracer (D_{in} c.p.m. ml^{-1}), its infusion rate (I) and the specific activity at the downstream sampling site (D_{out} c.p.m. ml^{-1}), the flow rate can be calculated.

Finally, another application of tracers is in *isotope dilution analysis* (7). Such a technique can measure accurately small quantities of compounds provided the pure compound is available in a radiolabeled form. The radiolabeled tracer is simply diluted with the unlabeled compound and the new specific activity of the tracer determined. The amount of the compound, W_X, can then be calculated from the equation:

$$W_X = W_0 \cdot (S_0/S_1 - 1),$$

where S_0 is the specific activity of the pure radiolabeled compound; W_0 is the weight of the added radiolabeled compound; and S_1 is the new specific activity of the compound after dilution into the unknown amount. Although the weight of the isolated compound is required for the determination of S_1, this need not involve weighing the sample. Microgram amounts of the isolated compound can be determined, for example, by spectrophotometry or other suitable methods.

7. RADIOLOGICAL PROTECTION

All radiation should be considered as hazardous and all practical measures should be taken to avoid exposure to ionizing radiations. This can be done by avoiding the ingestion of radioisotopes, and by minimizing exposure to isotopes by minimizing the time of exposure and maximizing the distance from the radioactive source. Shielding can also be used to reduce exposure to radiation (*Tables 5* and *6*). When working with strong β-particle emitters or sources of γ- or X-rays, it is usual to wear a personal dosimeter, a film badge or a thermoluminescent dosimeter.

7.1. Shielding for β-particles and γ-rays

The attenuation of β-particles depends on the density of the shielding material and is largely independent of its atomic weight.

Table 5. Shielding for β-particle emitters

Energy of β-particles (MeV)	Weight of material cm^{-2} to reduce intensity by one-half (mg)	Thickness of material (mm) required to reduce intensity by one-half			
		Water	Glass (d=2.5)	Lead	Plexiglass (Perspex)
0.1	1.3	0.013	0.005	0.0011	0.0125
1.0	48	0.48	0.192	0.042	0.38
2.0	130	1.3	0.52	0.115	1.1
5.0	400	4.0	1.6	0.35	4.2

Table 6. Shielding for γ-ray emitters

Energy of γ-rays (MeV)	Thickness of material (mm) required to reduce the intensity of a broad beam of γ-rays by a factor of 10		
	Aluminum	Iron	Lead
0.5	203	61	18
1.0	254	82	38
2.0	320	110	59
3.0	370	120	64
4.0	400	127	63
5.0	440	130	61

8. METHODS FOR DECONTAMINATING LABORATORIES

A survey should indicate the areas requiring decontamination and such areas should be clearly delineated; be sure to select the correct type of monitor to be used. The areas should be decontaminated promptly to prevent the spread of contamination. The decontamination procedures should be carefully planned and correct materials selected. Personnel should wear appropriate protective clothing, including plastic gloves and overshoes.

In general, wet methods should always be used to prevent the dispersal of dust. If, because of special circumstances, dry methods cannot be avoided, special precautions will be necessary (e.g. special vacuum cleaners with filters).

8.1. Methods

Spilt liquid should be absorbed on paper tissue or Vermiculite. Where dry material has been spilt and there is loose particulate contamination, the best method of decontamination may well be the application of a strippable coating by brush or spray. This coating will hold the contaminant and prevent the dispersal (provided that the method chosen does not disturb the loose contaminant). When dry, the coating is stripped off, taking the adhering contaminant with it. Self-adhesive tapes can also be applied to non-porous surfaces for the removal of loosely held dust. *Table 7* summarizes the decontamination procedures that can be used.

Table 7. Decontamination methods

Type of surface or equipment	Decontamination agent	Treatment
Walls, floors, etc., contaminated clothing	Suitable detergents, for example Decon 90 or RBS 25, or wetting agents (it is preferable to add a little EDTA)	A first method for particularly greasy or dirty surfaces; use a 5% solution with swabbing action, or in a washing machine for clothing
Textiles, plastics, paints, rubber, metals	Detergents or combinations of citric acid, EDTA, etc.	Use in the form of a cream or immerse articles in tanks for periods up to several hours (0.8% solution at elevated temperatures)
Linoleum	Organic solvents, detergents	To remove the normal waxed coating
Machine tools	Solvents (proprietary grease or emulsifying solvents)	Apply directly to heavily oiled and greasy surfaces with a cloth or brush, emulsifying solvents may be rinsed off with water
Glassware	Detergents, concentrated acids, chromic acid	Use in the normal way
Stainless steel, mild steel and light alloys, ferrous metals	Sulfuric acid, sulfuric acid with inhibitors, nitric acid/sodium fluoride, proprietary rust removers	These dissolve the contaminated surface, taking it into solution
Painted surfaces	Paint removers, solvent strippers, alkaline strippers	Use in difficult cases where the paint itself has to be removed

Repeat any of the above if necessary, but if this does not remove the contamination satisfactorily, apply mild abrasive cleaners (proprietary brands) with a cloth to the affected surface. Frequently the inclusion of unlabeled compounds of the same type as the contaminant in the decontamination solution will facilitate decontamination.

For the second stage of decontamination, the use of damp swabs is preferable to uncontrolled sluicing, to prevent the spread of contamination. The actual method to be used and the appropriate agent will depend to some extent on the ease of removal of the remaining contamination and the methods set out below are arranged in order, to deal with increasingly difficult circumstances:

(i) Treat with a suitable detergent, which may be in the form of a cream to prevent the spread of the contamination by splashing. Swabbing or light scrubbing may also be necessary.
(ii) Scrub lightly with a complexing solution. If an application of a thickened complexing agent is used, it should be left on the surface for a few hours before rinsing off. The addition of pigment will help to identify the areas to which the decontamination procedure has been applied.
(iii) Swab or scrub with mild abrasive pastes containing complexing agents.
(iv) Where none of the above methods is successful and the contamination still remains, it will be necessary to treat the surface by more vigorous scrubbing and abrasion or more severe treatment, for example by planing off wood surfaces and chipping away concrete and brick surfaces. In such cases it will be necessary to restore the original surface before further work is recommenced.

9. REFERENCES

1. Pande, S.V. (1976) *Anal. Biochem.*, **74**, 25.
2. Fricke, U. (1975) *Anal. Biochem.*, **63**, 555.
3. Bailey, G.S. (1990) in *Radioisotopes in Biology: A Practical Approach* (R.J. Slater, ed.). IRL Press, Oxford, p. 191.
4. Shipley, R.A. and Clark, R.E. (1972) *Tracer methods for in vivo Kinetics: Theory and Applications*. Academic Press, London.
5. Katz, J., Rostami, H. and Dunn, A. (1974) *Biochem. J.* **142**, 161.
6. Griffiths, R. (1990) in *Radioisotopes in Biology: A Practical Approach* (R.J. Slater, ed.). IRL Press, Oxford, p. 109.
7. Tolgyessy, J., Braun, T. and Krys, M. (1972) *Isotope Dilution Analysis*. Pergammon Press, Oxford.

10. FURTHER READING

1. Billington, D., Jayson, G.G. and Maltby, P.J. (1992) *Radioisotopes*. BIOS Scientific Publishers, Oxford.
2. Chapman, J.M. and Ayrey, G. (1981) *The Use of Radioactive Isotopes in the Life Sciences*. George Allen and Unwin, London.
3. Chard, T. (1987) *An Introduction to Radioimmunoassay and Related Techniques*. Elsevier, Amsterdam.
4. Evans, E.A. and Oldham, K.G. (1988) *Radiochemicals in Biomedical Research: Critical Reports on Applied Chemistry*, Vol. 24. Wiley, Chichester.
5. Geary, W. (1986) *Radiochemical Methods*. Wiley, Chichester.
6. Kricka, L.J. (1985) *Ligand-Binder Assays: Labels and Analytical Strategies: Clinical and Biochemical Analysis*, Vol. 17. Marcel Decker, New York.
7. Parker, R., Smith, P. and Taylor, D. (1978) *Basic Science of Nuclear Medicine*. Churchill Livingstone, Edinburgh.
8. Sampson, C.B. (1990) *Textbook of Radiopharmacy*. Gordon and Breach, London.
9. Sharp, P., Gemmell, H. and Smith, F. (1989) *Practical Nuclear Medicine*. IRL Press, Oxford.
10. Slater, R.J. (1990) *Radioisotopes in Biology: A Practical Approach*. IRL Press, Oxford.
11. Webb, S. ed. (1988) *The Physics of Medical Imaging*. Adam Hilger, Bristol.

CHAPTER 3
CHROMATOGRAPHIC FRACTIONATION MEDIA
D. Patel

1. INTRODUCTION

Chromatography has been one of the most important methods used by biochemists for the separation of both large and small molecules. Many manuals are available that describe the methodology of chromatographic separations, and the reader should consult them for detailed protocols. There are four main types of chromatographic separations:

(i) separations dependent on molecular size;
(ii) separations based on the overall charge or charge distribution on molecules;
(iii) reversed-phase separations; and
(iv) affinity separations in which the chromatographic medium has a particular affinity for a molecule.

In the first case, separations are based on the use of media with pores of specific sizes. The pore size is chosen such that small molecules pass in and out of the pores essentially without hindrance, while large molecules, which do not enter the pores, elute from the column first. Molecules of intermediate size comparable to the size of the pores in the matrix pass through the column with difficulty, resulting in differing retention times. Separations may be made under denaturing conditions (e.g. 6 M guanidine hydrochloride, sodium dodecyl sulfate, urea, or organic solvents such as acetonitrile) or non-denaturing conditions. In the former the separation is based on the molecular weight of the molecule; in the latter the conformation of the molecule (e.g. protein folding) will affect the separation obtained.

A range of media are available for the separation of molecules on the basis of their charge, with media for both low-pressure and high-pressure liquid chromatography (HPLC) separations. The separation proceeds because ions of opposite charge are retained to different extents. The resolution is influenced by the pH of the eluant, which affects the selectivity, and by the ionic strength of the buffer, which mainly affects the retention.

The third type of separation, reversed-phase HPLC is a rapid and powerful separation technique of unusual specificity. The typical conditions used, low pH and high concentrations of organic solvents, favour denaturation. This technique is used in many circumstances in which denaturation may be unimportant, such as in structural studies, or where structure and activity are retained despite the high concentrations of organic solvents used. Solute retention is mainly due to hydrophobic interactions between the solutes and the hydrocarbon stationary-phase surface. Solutes are eluted in order of decreasing polarity (increasing hydrophobicity). Increasing the polar (aqueous) component of the eluant increases the retention of the solutes.

In the case of affinity separations, the molecule to be purified is specifically and reversibly adsorbed by a complementary binding substance (ligand) immobilized on an insoluble support (matrix). The choice of medium is usually very dependent on the type of molecule being separated. The affinity ligands can vary from simple chemical groups (e.g. sulfydryl groups) to specific proteins. Many of the affinity media required are now available commercially; in the case of proteins, activated matrices can be mixed with the desired protein under fairly mild conditions to form the affinity medium. The availability of matrices with bound streptavidin is also extremely useful since it is possible to attach biotin to a wide range of molecules, hence enabling many different types of molecule to be immobilized without difficulty.

This chapter lists the different types of media used for the major types of chromatography. The choice of media depends on the type of separation that is required and whether the separation is to be carried out using a HPLC system.

The following tables are not intended to be exhaustive and there may well be column support materials, manufacturers and suppliers other than those listed. Where media have been given as a series the reader should consult the manufacturers for the individual types of media and their properties in that series.

The manufacturers and suppliers are given in abbreviated form, consult Section 6 for the complete names and addresses.

2. MEDIA FOR SIZE EXCLUSION (also termed gel filtration, gel permeation or molecular sieving)

The original medium for this type of separation was Sephadex and for the separation of larger molecules, Sepharose. The low flow-rate of the latter medium has led to it being replaced by Sephacryl. The choice of media depends on the size of molecules to be separated (*Figure 1*). The media based on carbohydrates can be affected by highly denaturing agents and they are not able to withstand the pressure in modern HPLC systems. Hence other families of separation media have been devised which can be used in all kinds of eluant solutions and are able to withstand high pressures. *Tables 1–4* list the media that are available.

3. MEDIA FOR ION-EXCHANGE SEPARATIONS

As in the case of the media used for size exclusion chromatography, most of the original media were based on polysaccharides (e.g. cellulose, *Table 5*). Due to their hydrophilic nature, they had little tendency to denature proteins; however, due to their irregular shape, many cellulose ion exchangers had low capacities and poor flow properties and were replaced by media with improved flow properties and high capacities for macromolecules. Modern high-performance media are usually derived from silica, hydroxylated polyethers, polystyrene and cross-linked, non-porous polyacrylates. These are chemically bonded with acidic groups (such as sulfonic acid or carboxylic acid) for the separation of cations, or basic groups (such as amine or quaternary amine) for the separation of anions. *Tables 6–8* list the media available.

4. MEDIA FOR REVERSED-PHASE SEPARATIONS

The media for this type of separation are essentially based on silica chemically bonded with an alkylsilyl compound to give a non-polar, hydrophobic surface. These are presented in *Table 9*. A smaller range of non-silica and polystyrene-based media is available, as shown in *Table 10*. The media are stable in the pH range 2–7 and at elevated temperatures.

5. MEDIA FOR AFFINITY CHROMATOGRAPHY

The usual solid supports for this type of separation are based on polysaccharides, in particular beaded agarose. Media based on silica and synthetic hydrophilic gels are available. The main drawback of silica is its high solubility at pH 8 and above, in contrast to the other media which can be used between pH 1 and 14. Separation media are listed in *Table 11*.

Figure 1. Fractionation ranges of commercially available gel filtration matrices.

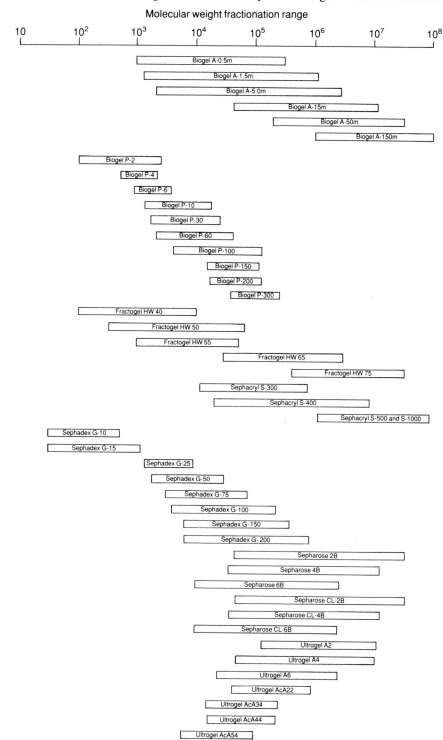

Table 1. Carbohydrate-based column support materials for separation by size exclusion

Names and types of matrix	Fractionation range ($\times 10^3$)		Dry bead diameter (μm)	Approx. bed volume (ml g^{-1})	Comments
	Peptides, globular proteins	Dextrans			
Sephadex					
Beaded gel of dextran cross-linked with epichlorohydrin					pH range, 2; temperature stability, 120°C; stability to solvents, subject to microbial degradation
G-10	<0.7	<0.7	40–120	2–3	
G-15	<1.5	<1.5	40–120	2.5–3.5	
G-25 coarse	1–5	0.1–5	100–300	4–6	
G-25 medium	1–5	0.1–5	50–150	4–6	
G-25 fine	1–5	0.1–5	20–80	4–6	
G-25 superfine	1–5	0.1–5	20–50	4–6	
G-50 coarse	1.5–30	0.5–10	100–300	9–11	
G-50 medium	1.5–30	0.5–10	50–150	9–11	
G-50 fine	1.5–30	0.5–10	20–80	9–11	
G-50 superfine	1.5–30	0.5–10	20–50	9–11	
G-75	3–80	1–50	40–120	12–15	
G-75 superfine	3–70	1–50	20–50	12–15	
G-100	4–150	1–100	40–120	15–20	
G-100 superfine	4–100	1–100	20–50	15–20	
G-150	5–300	1–150	40–120	20–30	
G-150 superfine	5–150	1–150	20–50	18–22	
G-200	5–600	1–200	40–120	30–40	
G-200 superfine	5–250	1–200	20–50	20–25	

Table 1. Continued

Names and types of matrix	Fractionation range ($\times 10^3$)		Dry bead diameter (μm)	Approx. bed volume (ml g^{-1})	Comments
	Peptides, globular proteins	Dextrans			
PDX					
Cross-linked dextran beads					
GF-25	1–5	—	50–150	4–6	
GF-25	1–5	—	100–300	4–6	
GF-50	1.5–30	—	50–150	9–11	
GF-50	1.5–30	—	100–300	9–11	
Sephacryl					
Cross-linked copolymer of allyl dextran and N,N'-methylenebis(acrylamide)					pH range, 2–11; temperature stability, 120°C; stable in solvents
S-100 high resolution	1–100	—	25–75	Supplied preswollen	
S-200 superfine	5–250	1–80	40–105	Supplied preswollen	
S-200 high resolution	5–250	1–80	25–75	Supplied preswollen	
S-300 superfine	10–1500	1–400	40–105	Supplied preswollen	
S-300 high resolution	10–1500	1–400	25–75	Supplied preswollen	
S-400 superfine	20–8000	10–2000	40–105	Supplied preswollen	
S-400 high resolution	20–8000	10–2000	25–75	Supplied preswollen	
S-500 superfine	40–20 $\times 10^3$	—	40–105	Supplied preswollen	
S-500 high resolution	40–20 $\times 10^3$	—	25–75	Supplied preswollen	
S-1000 superfine	5000–10^4	—	40–105	Supplied preswollen	

Table 1. Continued

Names and types of matrix	Fractionation range ($\times 10^3$)		Dry bead diameter (μm)	Approx. bed volume (ml g^{-1})	Comments
	Peptides, globular proteins	Dextrans			
Sepharose and Sepharose CL					Sepharose: pH range, 4–9; stable at 40°C; avoid urea, organic solvents and chaotropic salts
Beaded agarose					Sepharose CL: pH range, 3–14; stable at 120°C; avoid strong oxidizing agents
6B	10–4000	10–1000	45–165	Supplied preswollen	
6B cross-linked	10–4000	10–1000	45–165	Supplied preswollen	
4B	60–20 × 10^3	30–5000	60–140	Supplied preswollen	
4B cross-linked	60–20 × 10^3	30–5000	60–140	Supplied preswollen	
2B	70–40 × 10^3	100–20 × 10^3	60–200	Supplied preswollen	
2B cross-linked	70–40 × 10^3	100–20 × 10^3	60–200	Supplied preswollen	
Superose					
Highly cross-linked agarose matrix					
6 Prep. grade	5–5000	—	20–40	Supplied preswollen	
12 Prep. grade	1–300	—	20–40	Supplied preswollen	
Triacryl Plus					
Beaded poly(*N*-tris[hydroxymethyl]methyl methacrylamide)					
GF2-M	1–15	—	40–80	Supplied preswollen	
GF4-M	5–25	—	40–80	Supplied preswollen	
GF10-M	5–65	—	40–80	Supplied preswollen	
GF20-M	10–130	—	40–80	Supplied preswollen	

Table 2. Silica-based column support materials available for size exclusion HPLC of proteins and peptides

Names of matrices	Pore size (nm)	Mol. wt × 10³ exclusion limit	Particle size (μm)	Column sizes Length (cm)	Column sizes i.d. (mm)	Availability bulk material	Manufacturer
Aquapore OH-100	10	90	10	25	4.6	No	BL
Aquapore OH-300	30	800	10	25	4.6	No	BL
Aquapore OH-500	50	5000	10	25	4.6	No	BL
Aquapore OH-1000	100	20 × 10³	10	25	4.6	No	BL
LiChrospher DIOL	10	70	5, 7	25	4.0	Yes	MB
LiChrospher DIOL	50	800	10	25	4.0	No	MB
LiChrospher DIOL	100	3000	10	25	4.0	No	MB
Polyol Si300	30	600	3, 5, 10	25, 50	4.6, 7.1, 9.5	Yes	SFG
Polyol Si500	50	1000	10	25, 50	4.6, 7.1, 9.5	Yes	SFG
SynChropak 60	6	30	5	25, 30	4.6, 7.8	Yes	SC
SynChropak 100	10	300	5	25, 30	4.6, 7.8	Yes	SC
SynChropak 300	30	1500	5	25, 30	4.6, 7.8	Yes	SC
SynChropak 500	50	10 × 10³	7	25, 30	4.6, 7.8	Yes	SC
SynChropak 1000	100	50 × 10³	7	25, 30	4.6, 7.8	Yes	SC
SynChropak 4000	400	—	10	25, 30	4.6, 7.8	Yes	SC
SOTA GF200	20	700	10	30	7.1	No	SOTA
TSK 2000 SW	12.5	70	10	30, 50	7.5, 21.5	No	TSM
TSK 3000 SW	25	300	10	30, 60	7.5, 21.5	No	TSM
TSK 4000 SW	50	1000	13	30	7.5	No	TSM
Zorbax Bio GF-250	25	400	4	25	9.4	No	DuP

Abbreviation: i.d., internal diameter.

Table 3. Polymer-based column support materials available for size exclusion HPLC of proteins and peptides

Name of matrices	Pore size (nm)	Mol. wt × 10^3 exclusion limit	Particle size (μm)	Column sizes Length (cm)	Column sizes i.d. (mm)	Availability bulk material	Manufacturer
MCI Gel CQS10	10	400	9–11	30	7.5	Yes	MCI
MCI Gel CQS30	30	1000	9–11	30	7.5	Yes	MCI
MCI Gel CQP06	6	1	9–11	30	7.5	Yes	MCI
MCI Gel CQP10	10	400	9–11	30	7.5	Yes	MCI
MCI Gel CQP30	20	1000	7–10	30	7.5	Yes	MCI
Micropak TSK Gel H Series	4–1000	1–400	10	—	—	—	VA
	$1 \times 10^4 – 1 \times 10^6$	$4000 – 40 \times 10^4$	17	—	—	—	VA
PLgel Series	$5 – 1 \times 10^5$	$0.5 – 40 \times 10^3$	5, 10	30	7.5	Yes	PL
PLgel MIXED-A		$1 – 40 \times 10^3$	20	30	7.5	—	PL
PLgel MIXED-B		$0.5 – 10 \times 10^3$	10	30	7.5	—	PL
PLgel MIXED-C		0.2–3000	10	30	7.5	—	PL
PLgel MIXED-D		0.2–400	5	30	7.5	—	PL
PLgel MIXED-E		up to 30	3	30	7.5	—	PL
RoGel SEC Series	3–17	0–200	5, 8, 10	25	7, 10	Yes	BR
TSK 2000 PW	5	4	10	30, 50	7.5, 21.5	No	TSM
TSK 3000 PW	20	50	13	30, 60	7.5, 21.5	No	TSM
TSK 4000 PW	50	300	13	30, 60	7.5, 21.5	No	TSM
TSK 5000 PW	100	1000	17	30, 50	7.5, 21.5	No	TSM
TSK 6000 PW	>100	8000	25	30, 50	7.5, 21.5	No	TSM

Table 4. Controlled pore glass for permeation chromatography

Nominal pore size (Å)	Unmodified. Mesh[a]	Glyceryl-controlled pore glass. Mesh
75	120–200	120–200
75	200–400	200–400
120	120–200	120–200
120	200–400	200–400
170	120–200	120–200
170	200–400	200–400
240	120–200	120–200
240	200–400	200–400
350	120–200	120–200
350	200–400	200–400
500	120–200	120–200
500	200–400	200–400
700	120–200	120–200
700	200–400	200–400
1000	120–200	120–200
1000	200–400	200–400
1400	120–200	120–200
1400	200–400	200–400
2000	120–200	120–200
2000	200–400	200–400
3000	80–120	120–200
3000	120–200	200–400

[a] Most beads are also available in 20–80 and 80–120 mesh.
Manufacturer: Electro-Nucleonics (EN).

Table 5. Ion-exchange cellulose media: physical and chemical properties of Whatman cellulose media

Physical form	Functional group	Normal pH range	Small ion capacity (meq dg^{-1})	Protein capacity (mg dg^{-1})	Bed volume (mg ml^{-1})	Amount of exchanger required per litre bed volume (kg)	Packing density (dg ml^{-1})
Anion exchange media							
Pre-swollen microgranular							
DE-51	Diethylaminoethyl	2–9	0.20–0.25	175	30	1.20	0.17
DE-52	Diethylaminoethyl	2–9.5	0.88–1.08	700	130	0.90	0.19
DE-53	Diethylaminoethyl	2–12	1.8–2.2	750	150	1.05	0.20
CA-52	Quarternary ammonium	2–12	1.0–1.2	750	150	1.20	0.20
Dry microgranular							
DE-32	Diethylaminoethyl	2–9.5	0.88–1.08	700	140	0.24	0.20
Dry fibrous							
DE-23	Diethylaminoethyl	2–9.5	0.88–1.08	425	60	0.19	0.15
Cation exchange media							
Pre-swollen microgranular							
CM-52	Carboxymethyl	3–10	0.90–1.15	1180	210	1.05	0.18
SE-52	Sulphoxyethyl	2–12	0.9–1.1	1300	195	1.05	0.15
SE-53	Sulphoxyethyl	2–12	2.1–2.6	1300	210	1.15	0.16
Dry microgranular							
CM-32	Carboxymethyl	3–10	0.90–1.15	1180	200	0.21	0.17
Dry fibrous							
CM-23	Carboxymethyl	3–10	0.55–0.70	675	85	0.16	0.13
P-11	Orthophosphate	3–10	3.2–5.3	—	—	0.22	0.17
Hydrogen bonding medium							
Pre-swollen microgranular							
HB-1	Hydroxyl	2–12	0	375	—	0.70	0.15
Cell debris remover							
Pre-swollen fibrous							
CDR	Diethylaminoethyl	2–9.5	0.25–0.35	—	—	1.15	0.19

Abbreviation: dg, dry gram.

Table 6. Ion exchange: anion exchangers on polystyrene

Cross-linkage (%)	Gel or macroret[a]	Mesh size	Ionic form	Moisture content (%)	Max. oper. temp. (°C)	Total exchange capacity (meq ml^{-1})	Total exchange capacity (meq g^{-1})	pH range
Amberlite weakly basic anion exchangers								
—	M	16–50	Free base	57	100	1.2	4.7	0–7
—	M	16–50	Free base	60	100	1.2	4.7	0–7
—	G	16–50	Free base	60	60	1.6	5.6	0–7
Amberlite strongly basic anion exchangers								
—	M	16–50	Cl$^-$	59	60(OH); 80(Cl)	1.0	4.2	0–14
—	M	20–50	Cl$^-$	57	60(OH); 80(Cl)	0.6	2.3	0–14
8	G	16–50	Cl$^-$	46	60(OH); 80(Cl)	1.4	3.8	0–14
8	G	16–50	OH$^-$	46	60(OH); 80(Cl)	1.15	4.0	0–14
8	G	16–45	Cl$^-$	46	60(OH); 80(Cl)	1.4	3.8	0–14
6	G	16–50	Cl$^-$	53	60(OH); 80(Cl)	1.25	4.1	0–14
—	G	20–50	Cl$^-$	42	40(OH); 80(Cl)	1.4	3.4	0–14
—	G	16–45	Cl$^-$	53	60(OH); 80(Cl)	1.25	4.1	0–14
—	G	16–50	Cl$^-$	58	60(OH); 80(Cl)	1.2	4.4	0–14
—	G	100–200	Cl$^-$	50–58	60(OH); 80(Cl)	1.3	1.8	0–14

Table 6. Continued

Cross-linkage (%)	Gel or macroret[a]	Mesh size	Ionic form	Moisture content (%)	Max. oper. temp. (°C)	Total exchange capacity (meq ml^{-1})	Total exchange capacity (meq g^{-1})	pH range
Dowex 1 strongly basic anion exchangers								
2	G	50–100	Cl$^-$	65–75	150	0.7	3.5	0–14
2	G	100–200	Cl$^-$	70–80	150	0.6	3.5	0–14
2	G	200–400	Cl$^-$	70–80	150	0.6	3.5	0–14
4	G	20–50	Cl$^-$	≥50	150	1.0	3.5	0–14
4	G	50–100	Cl$^-$	≥50	150	1.0	3.5	0–14
4	G	100–200	Cl$^-$	55–63	150	1.0	3.5	0–14
4	G	200–400	Cl$^-$	55–63	150	1.0	3.5	0–14
8	G	20–50	Cl$^-$	43–48	150	1.2	3.5	0–14
—	G	20–50	OH$^-$	≤60	50	1.1	3.9	0–14
—	G	16–40	OH$^-$	60–70	50	1.0	4.4	0–14
8	G	50–100	Cl$^-$	43–48	150	1.2	3.5	0–14
8	G	100–200	Cl$^-$	39–45	150	1.2	3.5	0–14
8	G	200–400	Cl$^-$	39–45	150	1.2	3.5	0–14
8	G	50–100	Cl$^-$	≥38	150	1.2	—	0–14
8	G	100–200	Cl$^-$	34–40	150	1.2	—	0–14
8	G	200–400	Cl$^-$	34–40	150	1.2	—	0–14

Abbreviation: Max. oper. temp., maximum operating temperature.
[a] Gel is in the form of beads where ion exchange occurs on the surface of the beads while macroret is a porous matrix where ion exchange occurs in the pores.

Table 7. Ion exchange: cation exchangers on polystyrene

Cross-linkage (%)	Gel or macroret[a]	Mesh size	Ionic form	Moisture content (%)	Max. oper. temp. (°C)	Total exchange capacity (meq ml^{-1})	Total exchange capacity (meq g^{-1})	pH range
Amberlite weakly acidic cation exchangers								
4	M	16–50	Na$^+$	67	120	2.5	8.1	5–14
4	M	16–50	H$^+$	48	120	3.5	10.0	5–14
4	M	100–200	H$^+$	5	120	3.5	10.0	5–14
4	M	100–500	H$^+$	10	120	3.5	10.0	5–14
4	M	100–500	K$^+$	10	120	2.5	10.0	5–14
Amberlite strongly acidic cation exchangers								
20	M	16–50	Na$^+$	48	150	1.7	4.2	0–14
4.5	G	16–50	H$^+$	65	120	1.3	4.9	0–14
8	G	16–50	Na$^+$	45	120	1.9	4.4	0–14
10	G	16–50	Na$^+$	40	120	2.1	—	0–14
—	G	16–50	Na$^+$	46	120	1.9	4.4	0–14
8	G	100–200	Na$^+$	8	120	1.9	4.4	0–14
8	G	100–500	Na$^+$	10	120	1.9	4.3	0–14
Dowex 50W strongly acidic cation exchangers								
1	G	50–100	H$^+$	51–54	150	1.8	4.8	0–14
2	G	50–100	H$^+$	74–82	150	0.6	4.8	0–14
2	G	100–200	H$^+$	74–82	150	0.6	4.8	0–14
2	G	200–400	H$^+$	74–82	150	0.6	4.8	0–14
4	G	50–100	H$^+$	64–72	150	1.1	4.8	0–14
4	G	100–200	H$^+$	64–72	150	1.1	4.8	0–14
4	G	200–400	H$^+$	64–72	150	1.1	4.8	0–14
8	G	20–50	H$^+$	51–54	150	1.8	4.8	0–14
8	G	50–100	H$^+$	50–56	150	1.7	4.8	0–14
8	G	100–200	H$^+$	50–58	150	1.7	4.8	0–14
8	G	200–400	H$^+$	50–58	150	1.7	4.8	0–14
12	G	50–100	H$^+$	51–54	150	1.8	4.8	0–14

Abbreviation: Max. oper. temp., maximum operating temperature.
[a] See footnote [a] in *Table 6*.

Table 8. Some commercially available column support materials for ion-exchange HPLC of proteins

Name	Surface	Functional group	Pore size (nm)	Particle size (μm)	Column dimensions length (cm) × i.d. (mm)	Manufacturer
Silica-based columns						
Aquapore AX-300	Weak anion	Diethylaminoethyl	30	10	10, 22 × 4.6, 25 × 7.0	BL
Aquapore AX-1000	Weak anion	Diethylaminoethyl	10	10	10, 22 × 4.6, 25 × 7.0	BL
Aquapore CX-300	Weak cation	Carboxymethyl	30	10	10, 22 × 4.6, 25 × 7.0	BL
Aquapore CX-1000	Weak cation	Carboxymethyl	10	10	10, 22 × 4.6, 25 × 7.0	BL
Bakerbond PEI	Weak anion	Polyethyleneimine, CH_2CH_2NH	30	5	5 × 4.6	JTB
Bakerbond CBX	Weak cation	Carboxyethyl	30	5	5 × 4.6	JTB
Daltosil			30,10	3,5,10		SFG
Dynamax ~300 Å AX			30	5,12		RI
LiChrosorb NH_2		Primary amino (bonded)	6	5		MB
Nucleogen 60		Tertiary amino (bonded)	6	7		MNG
Nucleogen 500		Tertiary amino (bonded)	50	10		MNG
Nucleogen 4000		Tertiary amino (bonded)	400	10		MNG
Nucleosil AN/CAT			10	5,10		MNG
Partisil SAX		Quaternary amino (bonded)	10	10		WLS
PEI widepore		Primary and secondary amino (pellicular)	33	5		JTB
RSil AN/CAT			9	10	25 × 4.6	BR
RoSil AN/CAT			9	5	15 × 4.6	BR
Spherisorb AN			8	5,10		PS
Supelcosil			30	5		Supel
SynChropak AX300	Weak anion	Polyamine	30	6.5	10, 25 × 4.6, 10, 22.5	SC
SynChropak AX1000	Weak anion	Poly	30	10	10, 25 × 4.6, 10, 22.5	SC
SynChropak Q300	Strong anion	Quaternary amine	30	6.5	10, 25 × 4.6, 10, 22.5	SC

Table 8. Continued

Name	Surface	Functional group	Pore size (nm)	Particle size (μm)	Column dimensions length (cm) × i.d. (mm)	Manufacturer
SynChropak Q1000	Strong anion	Quaternary amine	30	10	10, 25 × 4.6, 10, 22.5	SC
SynChropak S300	Strong cation	Sulfonic acid	30	6.5	10, 25 × 4.6, 10	SC
SynChropak CM300	Weak anion	Carboxymethyl	30	6.5	10, 25 × 4.6, 10	SC
TSK DEAE-2/SW	Weak anion	$-N^+HEt_2$	13	5	30 × 4.6	TSM
TSK DEAE-3/SW	Weak anion	$-N^+HEt_2$	25	10	15 × 6.0, 21.5	TSM
TSK CM-2/SW	Weak cation	$-COO-$	13	5	30 × 4.6	TSM
TSK CM-3/SW	Weak cation	$-COO-$	25	10	15 × 6.0, 21.5	TSM
TSK CM-5/SW	Weak cation	$-COO-$	50	10	15 × 4.6, 21.5	TSM
Waters Accell	Cation	Carboxymethyl	50	37–55	Bulk material only	Mp
Zorbax AN			5	7		DuP
Polymer-based columns						
Mono Q	Anion	Quaternary amine	70	10	50 × 5	P-LKB
Mono S	Cation	Sulfonate	—	10	50 × 5	P-LKB
PL SAX	Strong anion	Quaternary amine	100	8, 10	5, 15 × 4.6, 5, 15 × 7.5, 25	PL
TSK DEAE 5 PW	Weak anion	$-N^+HEt_2$	100	10	7.5 × 7.5, 15 × 21.5	TSM
TSK SP 5 PW	Strong cation	$-SO_3^-$	100	10	7.5 × 7.5, 15 × 21.5	TSM
TSK Gel DEAE-NPR	Weak anion	$-N^+HEt_2$	100	2.5	3.5 × 4.6	TSM
TSK Gel SP-NPR	Weak cation	$-(CH_2)_3SO_3^-$	100	2.5	3.5 × 4.6	TSM

Abbreviation: i.d., internal diameter.

Table 9. List of silica-based reversed-phase column support materials

Name	Particle diameter (μm)	Mean pore diameter (nm)	Suppliers
Bonded phase: C_1			
Astec 300 A Cl[a]	5	30	Tcol
Spherisorb S5(10)X Cl[a]	5(10)	30	PS
Ultropak TSK TMS-250	10	25	P-LKB
Bonded phase: C_3			
Bakerbond wide pore			
HI propyl[a]	5	30	JTB
Ultrapore RPSC	5	30	Ana, BI
Bonded phase: C_4			
Apex WP C4[a]	7	30	JC
Aquapore BU-300	10	30	Ana, P
Astec 300A C4[a]	5	30	Tcol
Bakerbond wide pore			
Butyl[a]	5	30	JTB
Hipore RP-304	5, 10	33	BR
Hypersil WP Butyl[a]	5	30	SS
Nucleosil 300 C4[3]	5, 7, 10	30	MNG
Supelco LC-304	5	30	Supel
Vydac 214 TP[a]	5, 10	30	TSG
Bonded phase: C_6			
Spherisorb S5(10)X C6[a]	5(10)	30	PS
Bonded phase: C_8			
Apex I	3, 5, 10	10	JC
Apex WP C8[a]	7	30	JC
Aquapore RP-300	10	30	P
Astec 300A C8[a]	5	30	Tcol
Bakerbond wide pore			
Octyl[a]	5	30	JTB
Hypersil	5, 10	12, 30	SS
LiChrosorb	10	—	MB
Lichrospher CH8/II	10	50, 100, 400	MB
Nucleosil	5, 10	10	MNG
Partisil	5, 10	8.5	WLS
Pro RPC	5	30	P-LKB
Protesil 300 Octyl	10	30	WLS
RoSil[a]	3, 5, 8	9	BR
Spherisorb[a]	3, 5, 10	8	PS
Supelco LC-308	5	30	Supel
Vydac 228 TP[a]	5, 10	3	TSG
Zorbax	5	7	DuP
Bonded phase: C_{18}			
Apex I	3, 5, 10	10	JC
Apex WP C18[a]	7	30	JC
Astec 300A C18	5	30	Tcol
Bakerbond wide pore			
Octadecyl[1]	5	30	JTB

Table 9. Continued

Name	Particle diameter (μm)	Mean pore diameter (nm)	Suppliers
Hypersil	5, 10	12, 30	SS
LiChrosorb	10	—	MB
Nucleosil	3, 5, 10	10	MNG
Partisil	10	8.5	WLS
RoSil[a]	3, 5, 8	9	BR
Serva Si100[a]	5, 10	10	SFG
Serva Si300[a]	5, 10	30	SFG
Serva Si500[a]	10	50	SFG
Spherisorb[a]	3, 5, 8, 10	8	PS
Supelco LC-318	5	30	Supel
Vydac 201 TP[a]	5, 10	30	TSG
Vydac 218 TP[a]	5, 10	30	TSG
Bonded phase: Phenyl			
Apex I	5	10	JC
Apex WP Phenyl[a]	7	30	JC
Aquapore PH-300	10	30	P
Astec 300A Diphenyl[a]	5	30	Tcol
Bakerbond wide pore Diphenyl[a]	5	30	JTB
Hypersil	5, 10	12, 30	SS
Nucleosil	5, 10	10	MNG
Protesil 300 Diphenyl	10	30	WLS
RoSil[a]	3, 5	9	BR
Spherisorb[a]	3, 5, 10	8	PS
Supelco LC-3DP	5	30	Supel
Vydac 219 TP[a]	5	30	TSG
Zorbax	5	7	DuP
Bonded phase: Cyano			
Apex I	5	10	JC
Apex WP CN	7	30	JC
Bakerbond wide pore cyanopropyl[a]	5	30	JTB
Hypersil	5, 10	12, 30	SS
Nucleosil	7	10	MNG
RoSil[a]	3, 5	9	BR
Spherisorb[a]	3, 5, 8, 10	8	PS
Zorbax	5	7	DuP
Bonded phase: Amino			
Apex I	5	10	JC
Hypersil	5, 10	12, 30	SS
Nucleosil	5, 10	10	MNG
RoSil[a]	3, 5	9	BR
Spherisorb[a]	3, 5, 10	8	PS

[a] Bulk materials available.

Table 10. List of non-silica based reversed-phase column support materials

Name	Function	Particle diameter (μm)	Mean pore diameter (nm)	Suppliers
Polyether based				
Biogel TSK RP+	'High-density' phenyl	10	100	BR
TSK Phenyl 5PW	Phenyl	10	100	Ana, BR
Polystyrene based				
PLRPS-300A	Phenyl	10	30	PL
RoGel RP Series[a]	—	5, 8, 10	3–17	BR

[a] Bulk materials available.

6. MANUFACTURERS AND SUPPLIERS OF CHROMATOGRAPHY COLUMN SUPPORT MATERIALS

Ana	Anachem Ltd, Charles St, Luton, Bedfordshire LU2 0EB, UK.
BI	Beckman Instruments Ltd, Sands Industrial Estate, High Wycombe, Buckinghamshire HP12 4JL, UK.
BL	Brownlee Laboratories Inc., 2045 Martin Av., Santa Clara, CA 95050, USA.
BR	BioRad Laboratories Ltd, BioRad House, Maylands Av., Hemel Hempstead, Hertfordshire HP2 7TD, UK.
DP	Dyno Particles A.S., PO Box 160, N-2001, Lillestrom, Norway.
DuP	Du Pont Ltd, Wedgewood Way, Stevenage, Hertfordshire SG1 4QN, UK.
EN	Electro-Nucleonics Inc., 368 Passaic Av., Fairfield, NY 07006, USA.
JC	Jones Chromatography Ltd, New Rd, Hengoed, UK.
JTB	J.T. Baker Chemicals, Hayes Gate House, Uxbridge Rd, Hayes, Middlesex UB4 JD, UK.
LIW	Laboratory Instrument Works, Prague, Czechoslovakia.
LSL	Life Science Laboratories, Sedgewick Rd, Luton LU4 9DT, UK.
MB	Merck Ltd (BDH Chemicals), PO Box 15, Freshwater Rd, Dagenham, Essex RM8 1RF, UK.
MCI	Mitsubishi Chemical Industries Ltd, Marunouchi 2-chome, Chiyoda-ku, Tokyo 100, Japan.
MNG	Machery-Nagel GmbH., Neumann-Neander-St., PO Box 307, D-5160 Dueren, Germany.
Mp	Millipore Ltd, The Boulevard, Ascot Rd, Croxley Green, Watford WD1 8YW, UK.
P	Pierce Ltd, Pierce & Warriner (UK) Ltd, 44 Upper Northgate St, Chester CH1 4EF, UK.
PL	Polymer Laboratories Ltd, Essex Rd, Church Stretton, Shropshire SY6 6AX, UK.
P-LKB	Pharmacia-LKB, Davy Av., Knowhill, Milton Keynes MK5 8PH, UK.
PS	Phase Separations Ltd, Deeside Industrial Park, Queensferry, Clwyd CH5 2LR, UK.
RI	Rainin Instrument Co. Inc., Mack Road, Woburn, MA 01801, USA.
RPG	Rohm Pharma GmbH., Westerstadt, PO Box 4347, D-6100, Darmstadt 1, Germany.
SC	SynChrom Inc., see Anachem (Ana).
SFG	Serva Feinbiochemica GmbH., D-6900, Heidelburg 1, Germany.
SOTA	SOTA Chromatography Inc., PO Box 693, Crompond, NY, USA.
SS	Shandon Southern Products Ltd, 93 Chadwick Rd, Runcorn, Cheshire WA7 1PR, UK.
Supel	Supelchem & Supelco, Shire Hill, Saffron Walden, Essex CB11 3AZ, UK.
Tcol	Technicol Ltd, Brook St., Higher Hillgate, Stockport, Cheshire SK1 3HS, UK.
TSG	The Separations Group, PO Box 867, 17434 Mojave St., Hesperia, CA 92345, USA.
TSM	Toyo Soda Manufacturing Co. Ltd, see Anachem (Ana).
VA	Varian Associates Ltd, 28 Manor Rd, Walton-on-Thames, Surrey KT12 2QF, UK.
WLS	Whatman Lab Sales Ltd, St. Leonards Rd, Maidstone, Kent ME16 0LS, UK.

Table 11. Some characteristics of commonly used matrices for affinity chromatography

Matrix type	Particle size (μm)	Measure of pore size (nm or exclusion limits)	Stability pH	Stability Max. pressure (bar)	Surface groups	Comments	Manufacturer
Natural							
Cross-linked agarose (12%)	3–10	450×10^3	2–14	60	Hydroxyl	Must not dry out	ref. 1
Cross-linked agarose (12%)	5–40	450×10^3	2–14	60	Hydroxyl		ref. 1
Dextran	40–125		2–12			Swells and shrinks depending on ionic strength	P-LKB
Superose 6	13	$5 \times 10^3 - 5 \times 10^6$	1–14	15	Hydroxyl		P-LKB
Superose 12	10	$< 3 \times 10^5$	1–14	30	Hydroxyl		P-LKB
Silica							
Hypersil WP 300	5 and 10	30 nm	2–8	1035	Silanol	Unstable above pH 8	SS
LiChrospher range	10	30–400 nm	2–8	—	Silanol		MB
SelectiSpher-10	10	50 nm	2–8	80	Tresyl-diol-bonded		P
Ultraffinity-EP	—	—	2–8	—	Epoxy-bonded		BI
Synthetic							
TSK-PW range	10 to 20	$10^3 - 3 \times 10^7$	1–13	130	Hydroxyl		TSM
Separon H 1000	10	$< 1 \times 10^6$	1–13	50	Hydroxyl		LIW
Dynospheres XP-2507	20	5–200 nm	1–13	140	Hydroxyl		DP
Eupergit C30 N	30	$10^3 - 25 \times 10^3$	1–12	100	Epoxy		RPG
Polyacrylamide			2–11			$N-N'$ bis is toxic	LSL, BR, P-LKB
Polystyrene	Up to 1100		Any pH			Non-ionic interaction	BR, SFG

7. REFERENCE

1. Hjerten, S. (1984) *Trends in Analyt. Chem.*, **3**, 87.

8. FURTHER READING

1. Dean, P.D.G., Johnson, W.S. and Middle, F.A. eds (1985) *Affinity Chromatography, A Practical Approach*. IRL Press at Oxford University Press, Oxford.

2. Harris, E.L.V. and Angal, S. eds (1989) *Protein Purification Methods, A Practical Approach*. IRL Press at Oxford University Press, Oxford.

3. Harris, E.L.V. and Angal, S. eds (1990) *Protein Purification Applications, A Practical Approach*. IRL Press at Oxford University Press, Oxford.

4. Knox, J.H. ed. (1978) *High-Performance Liquid Chromatography*. Edinburgh University Press, Edinburgh.

5. Lim, C.K. ed. (1986) *HPLC of Small Molecules, A Practical Approach*. IRL Press at Oxford University Press, Oxford.

6. Oliver, R.W.A. ed. (1989) *HPLC of Macromolecules, A Practical Approach*. IRL Press at Oxford University Press, Oxford.

CHAPTER 4
ELECTROPHORESIS OF PROTEINS AND NUCLEIC ACIDS

D. Rickwood

Many manuals offer detailed guidance to the use of polyacrylamide gel electrophoresis (PAGE) as a method for fractionating and analyzing both proteins and nucleic acids. This chapter does not seek to duplicate the protocols given in these manuals but rather seeks to provide a reference source of recipes for the gels, electrophoresis buffers and staining methods that are used.

1. PROTEINS

PAGE has been a routine method for analyzing complex mixtures of proteins and determining the purity of protein fractions for a whole variety of experimental procedures. The methods can be classified broadly into three types:

(i)   SDS-denaturing gels;
(ii) non-denaturing gels; and
(iii) isoelectric focusing gels.

1.1 Separations on denaturing gels

A common type of separation uses sodium dodecyl sulfate (SDS) denaturing gels (SDS-PAGE) to determine the molecular weight of proteins. The basis of the method is that proteins bind SDS, giving them an overall negative charge, and the proteins then have a mobility which is inversely proportional to their size. Caution must be exercised, however, because some proteins bind less SDS and so migrate anomalously; as a rule the accuracy of size estimates by SDS-PAGE is less than 10%. In order to maximize the resolution of protein bands it is best to use discontinuous gels in which the proteins are focused in a stacker gel and then separated in a smaller-pore gel. *Table 1* gives some recipes for SDS-PAGE discontinuous systems that are likely to be useful for most types of separation. In the first instance try separating your proteins on a 12% polyacrylamide gel, if the protein(s) of interest migrate too slowly then use a lower percentage gel.

1.2. Separations on non-denaturing gels

In contrast to SDS-PAGE, the direction and speed of migration of proteins in non-denaturing gels is dependent on the amino acid composition of the protein and the pH of the electrophoresis buffer. In this case it is necessary to choose the correct buffer to ensure optimal separation of the protein of interest. As in the case of SDS-PAGE it is advantageous to use a system of discontinuous buffers to obtain stacking of the proteins and so enhance the resolution of the protein bands. *Table 2* lists some of the buffers that can be used for non-denaturing gels. Recipes for non-denaturing gels are given in *Table 3*.

1.3. Separations of proteins by isoelectric focusing

In isoelectric focusing (IEF) proteins are separated on the basis of their isoelectric points, which in turn are related to their amino acid composition. The separation is achieved by establishing a pH gradient in the gel, following which proteins move under the influence of an

Table 1. Recipe for gel preparation using the SDS-PAGE discontinuous buffer system

Stock solution	Stacking gel[a]	Final acrylamide concentration in resolving gel(%)[a]				
		20.0	17.5	15.0	12.5	10.0
Acrylamide – bisacrylamide (30:0.8)	2.5	20.0	17.5	15.0	12.5	10.0
Stacking gel buffer stock[b]	5.0	–	–	–	–	–
Resolving gel buffer stock[c]	–	3.75	3.75	3.75	3.75	3.75
10% SDS	0.2	0.3	0.3	0.3	0.3	0.3
1.5% ammonium persulfate	1.0	1.5	1.5	1.5	1.5	1.5
Water	11.3	4.45	6.95	9.45	11.95	14.45
TEMED	0.015	0.015	0.015	0.015	0.015	0.015

Final concentration of buffers: stacking gel, 0.125 M Tris–HCl, pH 6.8; resolving gel, 0.375 M Tris–HCl, pH 8.8; reservoir buffer, 0.025 M Tris, 0.192 M glycine, pH 8.3.[d]
[a] The columns represent volumes (ml) of the various reagents required to make 30 ml of gel mixture.
[b] Stacking gel buffer stock: 0.5 M Tris–HCl (pH 6.8); 6.0 g of Tris is dissolved in 40 ml of water, titrated to pH 6.8 with 1 M HCl (~48 ml), and brought to 100 ml final volume with water. The solution is filtered through Whatman No. 1 filter paper and stored at 4°C.
[c] Resolving gel buffer stock: 3.0 M Tris–HCl (pH 8.8); 36.3 g of Tris and 48.0 ml of 1 M HCl are mixed and brought to 100 ml final volume with water. The buffer is then filtered through Whatman No. 1 filter paper and stored at 4°C.
[d] Reservoir buffer × 10 stock: 0.25 M Tris, 1.92 M glycine, 1% SDS (pH 8.3); 30.3 g of Tris, 144.0 g of glycine, and 10.0 g of SDS are dissolved in and made up to 1 liter with water. The solution is stored at 4°C.

Table 2. Buffers for non-denaturing discontinuous systems

High pH discontinuous
Stacks at pH 8.3, separates at pH 9.5

Stacking gel buffer: Tris–HCl (pH 6.8)	Dissolve 6.0 g of Tris base in 40 ml of water and titrate it to pH 6.8 with 1 M HCl (~ 48 ml). Adjust to 100 ml final volume.
Resolving gel buffer: Tris–HCl (pH 8.8)	Mix 36.3 g of Tris and 48.0 ml of 1 M HCl and bring to 100 ml final volume with water. Titrate the solution to pH 8.8, with HCl, if necessary.
Reservoir buffer: Tris–glycine (pH 8.3)	Dissolve 3.0 g of Tris base and 14.4 g of glycine in water and bring to 1 liter final volume.

Neutral pH discontinuous
Stacks at pH 7.0, separates at pH 8.0

Stacking gel buffer: Tris–phosphate (pH 5.5)	Dissolve 4.95 g of Tris base in 40 ml of water and titrate to pH 5.5 using 1 M orthophosphoric acid. Add water to 100 ml final volume.
Resolving gel buffer: Tris–HCl (pH 7.5)	Dissolve 6.85 g of Tris base in 40 ml water and titrate to pH 7.5 with 1 M HCl. Add water to 100 ml final volume.
Reservoir buffer: Tris–diethylbarbiturate (pH 7.0)	Dissolve 5.52 g of diethylbarbituric acid and 10.0 g of Tris base in water and make to 1 liter final volume.

Low pH discontinuous
Stacks at pH 5.0, separates at pH 3.8

Stacking gel buffer: acetic acid–KOH (pH 6.8)	Mix 48.0 ml of 1 M KOH and 2.9 ml of glacial acetic acid and then add water to 100 ml final volume.
Resolving gel buffer: acetic acid–KOH (pH 4.3)	Mix 48.0 ml of 1 M KOH and 17.2 ml of glacial acetic acid and add water to 100 ml final volume.
Reservoir buffer: acetic acid–β-alanine (pH 4.5)	Dissolve 31.2 g of β-alanine and 8.0 ml of glacial acetic acid in water and make to 1 liter final volume.

electric field to a position where they have no net charge. The pH gradient is formed by including amphoteric compounds in the gels, usually one of the commercially available mixtures is used. *Figure 1* shows the ranges of ampholyte mixtures that are available from different sources. Recipes for IEF gels are given in *Table 4*. Alternatively the pH gradient can be fixed in the gel using Immobilines.

1.4. Two-dimensional gel electrophoresis

When separating very complex mixtures of proteins the resolution obtained from one type of separation may not be sufficient. In this case, it is worth considering the use of two-dimensional gel electrophoresis to improve the separation. To do this the proteins are separated by IEF in the first dimension and then SDS-PAGE in the second dimension. Using two-dimensional separations it is possible to resolve several thousand proteins on a single gel.

Table 3. Recipe for gel preparation using non-denaturing continuous buffer systems

Stock solution	Final acrylamide concentration (%)[a]				
	20.0	17.5	15.0	12.5	10.0
Acrylamide–bisacrylamide (30:0.8)	20.0	17.5	15.0	12.5	10.0
Continuous buffer (5 × conc.)[b]	6.0	6.0	6.0	6.0	6.0
1.5% ammonium persulfate	1.5	1.5	1.5	1.5	1.5
Water	2.5	5.0	7.5	10.0	12.5
TEMED[c]	0.015	0.015	0.015	0.015	0.015

[a] The columns represent volumes (ml) of the various reagents required to make 30 ml of gel mixture.
[b] See *Table 2*, resolving gel buffers.
[c] The concentration of TEMED may need to be increased for low pH buffers.

Figure 1. Commercially available carrier ampholytes.

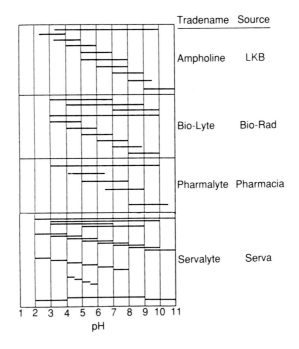

Table 4. Recipes for isoelectric focusing gels (4–7% T and 4–8 M urea)

Gel vol. (ml)	30% T monomer solution (ml)			2% Carrier ampholytes[a] (ml)		Urea (g)			TEMED (μl)	40% Ammonium persulfate[b] (μl)
	4% T[c]	5% T[c]	6% T[c]	7% T[c]	A	B	4 M	6 M	8 M		
30	4.00	5.00	6.00	7.00	1.50	1.88	7.20	10.40	14.40	9.0	30
25	2.34	4.17	5.00	5.83	1.25	1.56	6.00	9.00	12.00	7.5	25
20	2.66	3.34	4.00	4.67	1.00	1.25	4.80	7.20	9.60	6.0	20
15	2.00	2.50	3.00	3.50	0.75	0.94	3.60	5.40	7.20	4.5	15
10	1.33	1.66	2.00	2.33	0.50	0.63	2.40	3.60	4.80	3.0	10
8	1.06	1.33	1.60	1.86	0.40	0.50	1.92	2.88	3.84	2.4	8
7	0.93	1.17	1.40	1.63	0.35	0.44	1.68	2.52	3.36	2.1	7
6	0.80	1.00	1.20	1.40	0.30	0.38	1.44	2.16	2.88	1.8	6
5	0.66	0.83	1.00	1.16	0.25	0.31	1.20	1.80	2.40	1.5	5
4	0.53	0.66	0.80	0.93	0.20	0.25	0.96	1.44	1.92	1.2	4

[a] A, for 40% solution (Ampholine, Servalyte, Resolite); B, for Pharmalyte.
[b] To be added after degassing the solution and just before pouring it into the gel mould.
[c] Total acrylamide concentration (bisacrylamide and acrylamide); to prepare, dissolve 29.1 g acrylamide and 0.9 g bisacrylamide in 100 ml.

1.5. Marker proteins

In order to calibrate gels, whether they are SDS-PAGE or IEF gels, it is necessary to use standard proteins as markers. *Table 5* is a list of standard marker proteins that can be used. Usually it is best to select three or four proteins that migrate similarly to the protein of interest. Marker proteins can be obtained from a number of commercial sources.

1.6. Staining protein gels

Proteins separated on gels can be visualized using a number of different procedures (*Table 6*). Generally, staining using Coomassie blue remains as the most popular method, but if a more sensitive stain is required, silver staining is recommended. If the proteins are radioactive, they can be located by using autoradiography or fluorography (see *Chapter 2, Section 4*). Specific proteins can be detected by blotting the gel and then probing the blot with specific, labeled antibodies.

2. NUCLEIC ACIDS

2.1. Gels for separating nucleic acids and nucleoproteins

Nucleic acids can be separated on gels on the basis of their size and conformation. The gels used are either polyacrylamide or agarose gels; the former are usually used for separating smaller molecules. In some cases, especially for nucleoproteins, gels of a low concentration (approx 2%) of polyacrylamide are strengthened by the addition of a low concentration (0.5%) of agarose; these are known as composite gels. Recipes for non-denaturing polyacrylamide and composite gels are given in *Tables 7* and *8*. Agarose gels are very simple to prepare, the agarose is heated in the buffer of choice to obtain a molten gel (a microwave oven is useful for this) and then allowed to cool. Recipes for a selection of running buffers used for the electrophoresis of nucleic acids are given in *Table 9*. The conformation of the nucleic acids will affect separations on non-denaturing gels, and nicks in nucleic acids may be hidden by hydrogen bonding. These effects can be minimized by denaturing the sample prior to electrophoresis by treatment with formaldehyde, glyoxal, dimethylsulfoxide or formamide, or by heating. However, in order to be sure of avoiding these problems it is necessary to separate the molecules under denaturing conditions; to do this the denaturant is added to the gel and often to the running buffer. *Table 10* lists some of the denaturants that have been used and comments on their usage.

2.2. Markers for nucleic acids

A number of naturally occurring species of RNA can be used as markers, and a selection of these is listed in *Table 11*. DNA markers can be prepared from restriction nuclease digests of plasmids or viral DNA; *Tables 12* and *13* list two digests that can be used. Both RNA and DNA markers are available commercially from a number of sources.

2.3. Staining nucleic acid gels

Nucleic acids can be visualized in gels by staining them, usually with a solution of ethidium bromide (*Table 14*); this is applicable for a wide variety of nucleic acids. However, other methods have also been used. If the nucleic acids are radioactive then they can be detected by autoradiography or fluorography (see *Chapter 2, Section 4*).

Table 5. Standard marker proteins

Protein	Species	Tissue	Number of subunits	Subunit mol. wt ($\times 10^3$)	Isoelectric point (pI)
Adenine phosphoribo-syltransferase	Human	Erythrocyte	3	11	4.8
Nerve growth factor	Mouse	Salivary gland	2	13.259	9.3
Ribonuclease	Bovine	Pancreas	1	13.7	7.8
Hemoglobin	Rabbit	Erythrocyte	4	16	7.0
Micrococcal nuclease	S. aureus	—	1	16.8	9.6
β-Lactoglobulin	Bovine	Serum	2	17.5	5.2
Ceramide trihexosidase	Human	Plasma	4	22	3.0
Adenylate kinase	Rat	Liver (cytosol)	3	23	7.5
Trypsinogen	Cow	Pancreas	1	24.5	9.3
Chymotrypsinogen A	Bovine	Pancreas	1	25.7	9.2
Triosephosphate isomerase	Rabbit	Muscle	2	26.5	6.8
Galactokinase	Human	Erythrocyte	2	27	5.7
Arginase	Human	Liver	4	30	9.2
Deoxyribonuclease I	Cow	Pancreas	1	31	4.8
Uricase	Pig	Liver	4	32	6.3
Glycerol-3-phosphate dehydrogenase	Rabbit	Kidney	2	34	6.4
Malate dehydrogenase	Pig	Heart	2	35	5.1
Alcohol dehydrogenase	Yeast	—	4	35	5.4
Deoxyribonuclease II	Pig	Spleen	1	38	10.2
Aldolase	Yeast	—	2	40	5.2
Pepsinogen	Pig	Stomach	1	41	3.7
Hexokinase	Yeast	—	2	51	5.3
Lipoxidase	Soybean	—	2	54	5.7
Catalase	Cow	Liver	4	57.5	5.4

Table 5. Continued

Protein	Species	Tissue	Number of subunits	Subunit mol. wt ($\times 10^3$)	Isoelectric point (pI)
Alkaline phosphatase	Calf	Intestine	2	69	4.4
Acetylcholinesterase	*Electrophorus*	—	4	70	4.5
Glyceraldehyde-3-phosphate dehydrogenase	Rabbit	Muscle	2	72	8.5
β-Glucuronidase	Rat	Liver	4	75	6.0
Lysine decarboxylase	*E. coli*	—	10	80	4.6
Glycogen synthetase	Pig	Kidney	4	92	4.8
Phosphoenolpyruvate carboxylase	*E. coli*	—	4	99.6	5.0
Phosphoenolpyruvate carboxylase	Spinach	Leaf	2	130	4.9
Urease	Jack bean	—	2	240	4.9

Table 6. Staining procedures for proteins separated on polyacrylamide gels

Staining procedure	Destaining procedure	Additional information
1. 3–4 h in 0.1% Coomassie blue in methanol:water:acetic acid (5:5:1)†	Overnight by diffusion against methanol:water:acetic acid (5:5:1)	—
2. 20 min in 0.1% Coomassie blue in 50% TCA	Several changes of 7% acetic acid	Removal of commercial ampholytes
3. 15 min in 0.55% amido black in 50% acetic acid	40 h in 1% acetic acid	—
4. 3 h in 0.25% Coomassie blue in methanol:water:acetic acid (5:5:1)	Several changes of 5% methanol, 10% acetic acid	—
5. Overnight in 25% isopropyl alcohol, 10% acetic acid, 0.025–0.05% Coomassie blue, followed by 6–9 h in 10% isopropyl alcohol, 10% acetic acid, 0.0025–0.005% Coomassie blue	Several changes of 10% acetic acid	An additional optional staining step overnight in 10% acetic acid containing 0.0025% Coomassie blue helps intensify the gel pattern
6. 1–4 h in 0.1% Coomassie blue R-250 in 7.5% acetic acid, 50% methanol in water	Overnight in 7.5% acetic acid, 50% methanol in water	—
7. 3 h at 80°C, or overnight at room temperature in 0.1% amido black in 0.7% acetic acid, 30% ethanol in water	Several changes of 7% acetic acid, 20% ethanol in water	—
8. 60 min in 50% methanol, stain with fresh alkaline 0.8% AgNO$_3$ solution. Wash with water for 5 min. To develop soak in fresh 0.02% HCHO, 0.005% citric acid for 10 min	Wash gel in water, then transfer to 50% methanol	Much more sensitive than Coomassie blue methods

Table 7. Recipes for the preparation of polyacrylamide gels for the electrophoresis of nucleic acids

	Final polyacrylamide concentration (%)								
	2.2	2.4	2.5	2.6	3.0	4.0	5.0	7.5	10.0
Stock acrylamide solution (ml)[a]	5.0	5.0	5.0	5.0	5.0	5.0	10.0	10.0	10.0
Gel buffer (ml)[b]	6.8	6.25	6.0	5.8	5.0	3.75	6.0	4.0	3.0
TEMED (ml)[c]	0.025	0.025	0.025	0.025	0.025	0.025	0.025	0.025	0.025
10% ammonium persulfate (ml)[c]	0.25	0.25	0.25	0.25	0.25	0.25	0.25	0.25	0.25
Water (ml)	22.0	19.7	18.7	17.8	14.7	9.7	13.4	5.4	2.0

[a] Stock acrylamide solution: acrylamide, 15 g; bisacrylamide, 0.75 g; distilled water to 100 ml.
[b] Gel buffer: five times concentrated running buffer given in *Table 9*.
[c] Catalysts: N, N, N', N'-tetramethylethylenediamine (TEMED); freshly dissolved 10% ammonium persulfate.
Reservoir buffer: this is running buffer, prepared as described in *Table 9*.

Table 8. Recipes of gels used for the electrophoresis of polysomes and ribosomes

Gel composition (polyacrylamide/agarose)	Gel buffer	Agarose (g)	Water (ml)	20% Acrylamide-bisacrylamide solution[a] (ml)	Buffer stocks					
					6.4% DMAPN[b] (ml)	10 × Tris–borate–EDTA[c] (ml)	1 M Tris–HCl (ml)	3 M KCl (ml)	1 M MgCl$_2$ (ml)	1.6% Ammonium persulfate (ml)
2.25%/0.5%	25 mM Tris–HCl 60 mM KCl 10 mM MgCl$_2$	0.8	118	18	10	—	4	3.2	1.6	5
2.25%/0.5%	25 mM Tris–HCl 6 mM KCl 2 mM MgCl$_2$ (pH 7.6)	0.8	122	18	10	—	4	0.32	0.32	5
2.75%/0.5%	25 mM Tris–HCl 0.2 mM MgCl$_2$ (pH 7.6)	0.8	119	22	10	—	4	—	0.032	5
3%/0.5%	Tris–borate–EDTA (pH 8.3)	0.8	105	24	10	16	—	—	—	5

[a] 20% acrylamide solution: 19% acrylamide, 1% bisacrylamide.
[b] DMAPN; 3, dimethylaminopropionitrile.
[c] Composition of buffer, see *Table 9*.

Table 9. Running buffers for the electrophoresis of nucleic acids[a]

1. Tris base   3.63 g (30 mM)
   EDTA.Na$_2$·2H$_2$O   0.04 g (0.1 mM)
   SDS   1.00 g (0.1%)
   1.0 M HCl   16 ml
   Distilled water to 1 liter
   The buffer should be pH 8.0 at room temperature
   Changes in conformation/mobility may be observed if 2 mM magnesium acetate is added to this buffer
   The gels are eletrophoresed at 25°C to prevent the SDS precipitating

2. Tris base   0.44 g (3.6 mM)
   NaH$_2$PO$_4$·2H$_2$O   0.48 g (3.0 mM)
   EDTA.Na$_2$·2H$_2$O   0.04 g (0.1 mM)
   SDS   1.00 g (0.1%)
   Distilled water to 1 liter
   The buffer should be pH 7.7 at room temperature

3. Tris base   10.90 g (90 mM)
   Boric acid   5.56 g (90 mM)
   EDTA.Na$_2$·2H$_2$O   0.10 g (2.5 mM)
   SDS   1.00 g (0.1%)
   Distilled water to 1 liter
   The buffer should be pH 8.3 at room temperature

4. Tris base   21.7 g (180 mM)
   NaH$_2$PO$_4$·2H$_2$O   23.8 g (150 mM)
   EDTA.Na$_2$·2H$_2$O   1.85 g (5 mM)
   Distilled water to 1 liter
   The buffer should be pH 7.7 at room temperature

[a] The concentrations of reagents given in brackets are the final concentrations in the buffer. The corresponding buffer used to make the gels (*Table 7*) should be five times the above concentrations and without SDS. When using buffer 2, it is best to circulate the solution between the two buffer reservoirs during electrophoresis, to minimize pH changes.

Table 10. Denaturants used in denaturing gels for separating nucleic acids

Denaturant	Effective concentration	Comments
Alkali (>pH 12)	0.1 M	For DNA only, RNA degraded; deaminates polyacrylamide
Formamide	98%	Deionize before use, stops agarose gels setting
Urea	8 M at 60°C	
Methyl mercuric hydroxide	3–5 mM	Very toxic, use a fume hood, run gels in borate sulfate buffer

Table 11. Molecular weight markers for gel electrophoresis of RNA

RNA species	Mol. wt[a]	Number of nucleotides
Fibroin mRNA (silkworm)	57×10^6	19×10^3
Myosin heavy chain mRNA (chicken)	2.02×10^6	6500
28S rRNA (HeLa)	1.9×10^6	6333
25S rRNA (*Aspergillus*)	1.24×10^6	4000
23S rRNA (*E. coli*)	1.07×10^6	3566
18S rRNA (HeLa)	0.71×10^6	2366
17S rRNA (*Aspergillus*)	0.62×10^6	2000
16S rRNA (*E. coli*)	0.53×10^6	1776
A2 crystallin RNA (calf lens)	0.45×10^6	1460
Immunoglobulin light chain RNA (mouse)	0.39×10^6	1250
β-globin mRNA (mouse)	0.24×10^6	783
β-globin mRNA (rabbit)	0.22×10^6	710
α-globin mRNA (mouse)	0.22×10^6	696
α-globin mRNA (rabbit)	0.20×10^6	630
Histone H4 RNA (sea-urchin)	0.13×10^6	410
5.8S RNA (*Aspergillus*)	4.89×10^4	158
5S RNA (*E. coli*)	3.72×10^4	120
4S RNA (*Aspergillus*)	2.63×10^4	85

[a] Molecular weights are approximate only and based upon an average 'molecular weight' of 310 for each nucleotide.

Table 12. Sizes of the restriction fragments of pBR322[a]

AccI	RsaI	HaeIII	HpaII	AluI	HinfI	TaqI	ThaI (Fnu DII)	HhaI	HaeII	MboI	MnlI
2767	2117	587	622	910	1631	1444	581	393	1876	1374	611
1595	1565	540	527	659	517	1307	493	347	622	665	400
	680	504	404	655	506	475	452	337	439	358	314
		458	309	521	396	368	372	332	430	341	267
		434	242	403	344	315	355	270	370	317	262
		267	238	281	298	312	341	259	227	272	218
		234	217	257	221	141	332	206	181	258	206
		213	201	226	220		330	190	83	207	204
		192	190	136	154		145	174	60	105	199
		184	180	100	75		129	153	53	91	186
		124	160	63			129	152	21	78	186
		123	160	57			122	151		75	136
		104	147	49			115	141		46	130
		89	147	19			104	132		36	116
		80	122	15			97	131		31	111
		64	110	11			68	109		27	81
		57	90				66	104		18	68
		51	76				61	100		17	61
		21	67				27	93		15	60
		18	34				26	83		12	58
		11	34				10	75		11	57
		7	26				5	67		8	56
			26				2	62			38
			15					60			30
			9					53			27
			9					40			
								36			
								33			
								30			
								28			
								21			

[a] These sizes (in base pairs) do not include any single-stranded extensions which may be left by the restriction nuclease.

Table 13. Sizes of the restriction fragments of phage λ cI ts 857[a]

EcoRI	HindIII	SalI	BamHI	KpnI	HpaI	PvuI	PvuII	XhoI	SmaI	AvaI	AvrII	BglII	XbaI	SstI	SstII
21240	23150	32760	16840	29960	8680	14310	21110	33520	19420	14670	24340	22010	24530	24790	20340
7420	9420	15270	7230	17070	6910	12730	4775	15020	12220	8630	24120	13290	24010	22550	18780
5810	6560	500	6785	1500	5410	11960	4420		8630	6890	70	9700		1190	8130
5650	4380		6530		4540	9540	4270		8270	4740		2390			1080
4880	2320		5620		4490		3980			4720		650			210
3540	2020		5530		4350		3910			3730		435			
	560				3400		2300			1880		60			
	125				3380		1710			1670					
					3040		640			1600					
					2240		530								
					750		470								
					440		230								
					410		140								
					250		60								
					230										

[a] These sizes (in base pairs) do not include any single-stranded extensions which may be left by the restriction nuclease.

Table 14. Visualization of nucleic acids in gels. Gels containing high molecular weight RNA or DNA can be stained directly, gels with lower molecular weight nucleic acids should be fixed by soaking them in 1 M acetic acid or 10% trichloroacetic acid

Staining procedure	Destaining	Comments
Soak gel in 1–5 μg ml^{-1} ethidium bromide in 0.5 M ammonium acetate for 30–60 min	0.5 M Ammonium acetate or distilled water for 2 h in the dark view under UV light	RNA and DNA fluoresce pink
Soak gel in 0.1% pyronin Y in 0.5% acetic acid, 1 mM citric acid	0.5% Acetic acid	
Soak gel in 0.2% methylene blue in 0.2 M sodium acetate (pH 4.7)	Distilled water	Can also use toluidine blue, thionin or azure A

Abbreviation: UV, ultraviolet.

3. METHODS MANUALS ON ELECTROPHORESIS

1. Hames, B.D. and Rickwood, D. (1991) *Gel Electrophoresis of Proteins: A Practical Approach*. IRL Press at OUP, Oxford.

2. Rickwood, D. and Hames, B.D. (1991) *Gel Electrophoresis of Nucleic Acids: A Practical Approach*. IRL Press at OUP, Oxford.

CHAPTER 5
GENERAL CENTRIFUGATION DATA
D. Rickwood

1. CALCULATION OF CENTRIFUGAL FORCE

The centrifugal force (RCF) experienced by a particle depends on its radial distance from the center of rotation (r) in centimeters and the speed of rotation in r.p.m., this is true for all types of rotors.

$$\text{RCF} = 11.18 \times r \times \left(\frac{\text{r.p.m.}}{1000}\right)^2.$$

2. APPLICATIONS OF CENTRIFUGE ROTORS

Table 1. Applications of centrifuge rotors

Type of centrifuge rotor	Tube angle	Type of separation		
		Pelleting	Rate-zonal (size)	Isopycnic (density)
Swinging bucket	90	Inefficient	Excellent	Acceptable
Fixed-angle	20–45	Excellent	Poor	Good
Near vertical (NVT)	8–10	Poor	Poor	Good
Vertical tube	0	Do not use	Good	Good
Zonal	NA	Do not use	Excellent	Acceptable

Abbreviations: NA, not applicable.

3. CALCULATION OF k-FACTORS OF ROTORS AND PELLETING TIMES

The k-factor of a rotor is a measure of its pelleting efficiency. The k-factor is the time in hours required to pellet a particle with a sedimentation coefficient of one in a medium with the viscosity and density of water. It can be calculated from the minimum radius (r_{min}) and maximum radius (r_{max}) of the rotor, together with the speed of rotation (Q).

$$k = \frac{(\ln r_{max} - \ln r_{min})}{Q^2} \times 2.533 \times 10^{11}.$$

Note that as the k-factor becomes smaller the pelleting efficiency of a rotor increases, as a general rule the rotor with the smallest k-factor will give you the fastest (but not always the best) separation.

The time (T) in hours taken for particles with a sedimentation coefficient, S, to pellet in a medium with the same viscosity and density as water can be calculated from the k-factor of the rotor, from the equation:

$$T = \frac{k}{S}.$$

4. DERATING ROTORS FOR USE WITH DENSE SOLUTIONS

Rotors can only be used at their maximum speed if the sample liquid does not exceed the design density of the rotor, usually 1.2 g cm^{-3}. However, some rotors are designed for higher density samples (1.7 g cm^{-3} or even 4 g cm^{-3}) and other manufacturers define the maximum load in each pocket. Derating is particularly important when using swinging-bucket rotors.

The formula for calculating the amount of derating for a solution with a density of E g cm^{-3} is:

$$Q_d = Q_n \times \sqrt{\left(\frac{D}{E}\right)}$$

where Q_n is the normal maximum speed, Q_d is the maximum speed of the rotor with the dense solution and D is the design density of the rotor.

5. PROPERTIES OF CENTRIFUGE TUBES AND BOTTLES

The physical properties of materials used for centrifuge tubes and bottles are given in *Table 2*. *Table 3* describes the chemical characteristics and the methods used for sterilizing tubes and bottles. Details of the chemical resistance of rotor and tube materials are given in *Table 4*.

Table 2. Centrifuge tube and bottle materials

Tube/bottle material	Appearance	Max. temperature (°C)	Puncturable or sliceable	General characteristics
Polyallomer	Clear–translucent Colorless	80	Yes	Good chemical resistance, resists stress cracks, somewhat flexible; rather hydrophobic unless treated with chromic acid
Polycarbonate	Transparent Colorless	120	No	Strong, glass-like, reusable plastic but brittle; excellent temperature and speed range; very sensitive to alkaline (>pH 8) solutions
Polypropylene	Translucent Colorless	120	Yes	Very broad chemical resistance and good stress crack resistance; will deform under centrifugation (>50 000g) when not supported
Polysulfone	Transparent Yellow	120	Yes	Strong, temperature-resistant plastic; excellent chemical resistance including alkalis
Polyclear®/Ultraclear®	Transparent Colorless	60	Yes	Sensitive to many organic solvents including alcohols and DMSO as well as acids and strong bases
Cellulose acetate butyrate	Transparent Colorless	60	Yes	Strong thermoplastic; good chemical resistance; only moderate heat resistance; sensitive to chaotropic solutions
Stainless steel	Opaque	>180	No	Excellent for specific resistance, high temperature, and high speed but rotors usually need derating
Pyrex® glass	Clear	120	No	Excellent chemical resistance; life affected by use, handling, high centrifugal force; sterilization by autoclaving
Corex® glass	Clear	300	No	Four to six times the strength of conventional glass; excellent chemical resistance and durability; avoid using scratched tubes

Abbreviation: DMSO, dimethylsulfoxide.

Table 3. Centrifuge tube and bottle care and use

Tube/bottle material	Chemical resistance						Max. rotor speed	Sterilization and disinfection[a]			
	Weak acids	Strong acids	Weak bases	Strong bases	Organic solvents	Chloride & hypochlorite		Autoclave	Gas	Chemical	UV
Polyallomer	Yes	<60°C	Yes	<60°C	<60°C	Yes	80×10^3	Yes	Yes	Yes	No
Polycarbonate	Yes	No	No	No	No	Yes	65×10^3	Yes	Yes	Yes	No
Polypropylene	Yes	<60°C	Yes	<60°C	<60°C	Yes	60×10^3	Yes	Yes	Yes	No
Polysulfone	Yes	Yes	Yes	Yes	No	Yes	60×10^3	Yes	Yes	Yes	Yes
Polyclear®/Ultraclear®	Yes	No	Yes	No	No	Yes	80×10^3	No	Yes	Yes	—
Cellulose acetate butyrate	Yes	No	Yes	No	No	Yes	40×10^3	No	—	Yes	Yes
Stainless steel	No	No	Yes	Yes	Yes	No	Derate rotor	Yes	Yes	Yes	Yes
Pyrex® glass	Yes	No	Yes	No	Yes	Yes	5000	Yes	Yes	Yes	Yes
Corex® glass	Yes	Yes (except HF and H_3PO_4)	Yes	Yes	Yes	Yes	10×10^3	Yes	Yes	Yes	Yes

Abbreviation: UV, ultraviolet.
[a] For details see *Table 5*.

Table 4. Chemical resistance chart

Chemical	Aluminum	Anodic coating for aluminum	Carbon fiber	Cellulose acetate butyrate	Delrin®	Glass	Neoprene	Noryl®	Nylon	Polyallomer	Polycarbonate	Polyclear®/Ultraclear®	Polypropylene	Polysulfone	Polyvinyl chloride	Rulon A®, Teflon®	Silicone rubber	Stainless steel	Titanium	Tygon®	Viton®
Acetaldehyde	S	—	S	U	—	—	U	—	—	M	U	M	M	—	M	S	M	—	S	—	U
Acetic acid (5%)	S	S	S	S	—	S	S	S	S	M	U	S	M	S	M	S	S	S	S	S	M
Acetic acid (60%)	S	S	S	U	M	S	M	S	U	M	U	U	M	S	M	S	M	M	S	M	U
Acetic acid (glacial)	S	S	U	U	U	S	U	S	U	S	U	S	S	M	U	S	M	M	S	—	U
Acetone	M	S	U	—	M	S	S	U	S	—	U	S	S	U	U	S	S	S	S	U	U
Acetonitrile	S	S	U	U	S	S	S	U	S	—	U	S	S	—	M	S	S	—	S	—	—
Allyl alcohol	—	—	—	U	S	—	—	—	S	—	S	S	S	S	S	S	S	U	S	—	S
Aluminum chloride	U	U	M	—	U	S	M	S	M	S	S	S	S	S	M	S	M	U	S	S	S
Ammonium acetate	S	S	S	—	S	S	S	S	S	S	S	S	S	S	S	S	S	S	S	S	S
Ammonium carbonate	S	S	S	S	M	S	S	S	S	S	S	S	S	S	S	S	S	S	S	S	S
Ammonium hydroxide (10%)	U	U	U	U	M	—	U	U	U	—	—	S	S	U	U	S	M	U	U	M	U
Ammonium hydroxide (28%)	U	U	S	—	M	—	U	U	U	S	U	S	M	U	U	S	M	—	S	M	U
Ammonium hydroxide (conc.)	—	—	S	—	M	—	—	S	M	—	—	—	—	—	M	S	S	—	S	—	—
Ammonium phosphate	U	U	S	—	U	S	S	S	S	M	U	S	S	S	M	S	S	M	S	S	S
Ammonium sulfate	S	S	S	U	S	S	M	S	U	M	S	S	S	—	S	S	S	M	S	S	U
Amyl alcohol	S	S	S	U	S	S	—	—	—	—	U	S	S	S	S	S	U	—	S	—	S
Aniline	S	S	S	U	S	S	S	U	U	—	S	S	S	S	—	S	S	S	S	U	M
Aqua Regia	U	—	U	U	U	—	—	—	—	U	U	U	U	—	—	—	—	—	S	—	M

CENTRIFUGATION

Table 4. Continued

Chemical	Aluminum	Anodic coating for aluminum	Carbon fiber	Cellulose acetate butyrate	Delrin®	Glass	Neoprene	Noryl®	Nylon	Polyallomer	Polycarbonate	Polyclear®/Ultraclear®	Polypropylene	Polysulfone	Polyvinyl chloride	Rulon A®, Teflon®	Silicone rubber	Stainless steel	Titanium	Tygon®	Viton®
Barium salts	M	U	S	—	S	S	S	S	S	S	S	S	S	S	S	S	S	M	S	S	S
Benzene	S	S	U	U	S	S	U	U	S	M	U	M	M	U	U	S	U	S	S	U	S
Benzyl alcohol	S	—	U	U	M	—	M	—	S	U	—	U	U	—	M	S	M	—	S	—	S
Boric acid	U	S	U	M	U	S	S	S	S	S	S	S	S	S	S	S	M	S	S	S	S
N-Butyl alcohol	S	—	S	U	S	—	S	M	—	S	S	S	S	S	S	S	S	M	S	—	S
N-Butyl phthalate	S	S	—	—	S	S	S	U	—	M	U	U	U	S	U	S	S	U	S	U	S
Calcium chloride	M	—	S	—	S	S	S	—	S	S	S	S	S	S	S	S	S	M	S	S	S
Calcium hypochlorite	M	—	M	—	M	—	M	—	S	M	M	S	S	—	M	M	S	U	S	—	S
Carbon tetrachloride	U	U	U	—	S	S	U	U	S	S	S	S	M	S	M	S	S	M	S	S	S
Cesium acetate	M	S	S	—	S	S	S	S	S	S	S	S	S	S	S	S	S	S	S	S	S
Cesium bromide	M	S	S	—	S	S	S	S	S	M	M	S	S	S	S	S	S	M	S	S	S
Cesium chloride	M	S	S	—	S	S	S	S	S	S	S	S	S	S	M	S	S	M	S	S	S
Cesium formate	M	S	S	—	S	S	S	S	S	S	S	S	S	S	S	S	S	M	S	S	S
Cesium iodide	M	S	S	—	S	S	S	S	S	M	S	S	M	S	S	S	S	M	S	S	S
Cesium sulfate	M	S	S	—	S	S	S	S	S	M	S	M	M	U	M	S	U	M	U	M	S
Chloroform	U	U	U	U	M	S	U	U	M	S	U	U	M	S	U	U	M	U	U	S	S
Chromic acid (10%)	M	—	U	U	U	—	—	S	—	M	U	U	M	U	M	S	M	U	M	S	S
Chromic acid (50%)	U	—	U	U	U	—	—	S	—	S	U	U	M	U	M	S	—	U	S	—	S

90   BIOCHEMISTRY LABFAX

Chemical																					
Citric acid (10%)	S	S	S	S	S	S	—	S	S	S	S	S	S	S	S	S	S	S	S	S	S
Cresol mixture	S	S	S	S	S	S	—	S	U	U	S	S	S	U	S	S	—	S	S	S	S
Cyclohexane	S	U	S	S	S	S	U	S	M	S	M	S	M	M	M	S	S	S	M	S	S
Deoxycholate	S	S	S	S	S	S	S	S	S	S	S	S	S	S	S	S	S	S	S	S	S
Dextran	M	S	S	M	S	S	M	S	S	S	S	S	S	S	S	S	S	S	M	S	S
Diethyl ether	S	S	—	—	S	S	—	—	S	U	S	S	S	S	—	M	M	—	S	U	S
Diethyl ketone	S	S	S	S	—	S	S	M	—	S	S	M	M	S	S	M	S	S	S	S	S
Diethylpyrocarbonate	S	S	S	M	M	S	S	—	S	S	S	S	S	S	S	M	S	S	S	S	S
N,N-Dimethylformamide	S	S	S	M	M	S	—	S	S	S	S	S	M	S	S	M	S	S	S	U	S
Dimethylsulfoxide	S	S	S	M	M	S	—	—	U	S	U	S	S	S	S	S	S	M	S	U	U
Dioxane	M	S	S	S	M	S	—	S	S	S	S	S	M	S	S	U	S	S	S	U	S
Ethyl acetate	M	S	S	M	S	S	M	M	S	S	S	S	S	S	—	S	S	M	S	S	S
Ethyl alcohol (50%)	S	S	S	S	S	S	S	S	S	S	S	M	S	S	S	S	S	S	S	S	S
Ethyl alcohol (95%)	S	S	M	M	M	U	U	S	U	S	S	S	—	S	S	M	S	S	S	S	S
Ethylene dichloride	—	S	—	M	S	M	S	M	S	S	S	—	M	S	—	U	S	M	S	M	S
Ethylene glycol	S	S	S	S	S	S	—	M	S	S	S	S	S	S	S	S	S	S	S	S	S
Ethylene oxide vapor	S	S	S	S	S	S	M	U	—	S	S	S	—	S	S	S	S	S	S	S	S
Ferric chloride	—	—	M	—	S	—	S	S	—	S	S	—	—	—	—	S	S	S	S	—	S
Ficoll-Hypaque®	M	M	M	—	M	M	M	M	—	S	S	M	—	S	—	S	S	S	S	—	S
Formaldehyde (40%)	M	U	S	—	M	U	S	S	S	S	S	S	—	S	—	S	S	M	S	—	—
Formic acid (100%)	—	—	S	—	M	S	—	S	—	S	S	S	S	—	—	S	S	M	S	S	S
Glutaraldehyde	S	S	S	—	S	S	S	M	S	S	S	S	—	S	—	S	S	M	S	S	S
Glycerol	M	S	S	—	S	S	S	S	S	S	S	S	S	S	S	S	S	S	S	S	S
Guanidine hydrochloride	U	U	S	—	S	M	S	S	S	S	S	S	S	S	S	M	S	M	S	S	S
Guanidine thiocyanate	U	U	U	S	S	U	U	S	S	S	S	S	S	S	S	S	S	S	S	U	U
Hexane	S	S	S	M	S	—	S	S	S	S	S	S	S	S	S	S	S	S	S	S	S
Hydrochloric acid (10%)	U	M	S	—	U	M	M	S	U	—	U	S	S	S	S	M	S	S	S	M	S
Hydrochloric acid (50%)	U	U	U	S	U	U	U	M	U	M	U	S	U	M	M	S	M	M	S	—	S
Hydrochloric acid (conc.)	—	M	U	U	U	U	U	S	—	M	U	—	U	U	M	—	M	S	M	—	—
Hydrofluoric acid (10%)	—	U	U	S	S	U	U	U	—	M	U	S	U	S	—	S	M	S	S	—	S
Hydrofluoric acid (50%)	—	U	U	S	U	U	U	U	—	M	U	U	U	S	—	S	S	S	S	M	U
Hydrogen peroxide (3%)	S	—	S	S	U	M	S	S	—	S	S	S	S	S	S	S	S	M	S	—	S
Hydrogen peroxide (10%)	S	U	M	U	S	S	M	S	M	S	S	S	S	S	S	S	M	S	S	M	S

CENTRIFUGATION

Table 4. Continued

Chemical	Aluminum	Anodic coating for aluminum	Carbon fiber	Cellulose acetate butyrate	Delrin®	Glass	Neoprene	Noryl®	Nylon	Polyallomer	Polycarbonate	Polyclear®/Ultraclear®	Polypropylene	Polysulfone	Polyvinyl chloride	Rulon A®, Teflon®	Silicone rubber	Stainless steel	Titanium	Tygon®	Viton®
Iodoacetic acid	S	–	–	–	S	S	M	S	S	S	S	–	S	S	S	S	M	S	S	M	M
Isobutyl alcohol	–	–	–	U	S	–	U	–	S	–	S	U	S	–	S	S	S	–	S	–	S
Isopropyl alcohol	M	M	M	U	S	S	U	S	S	S	S	U	S	S	S	S	S	M	M	M	S
Kerosene	S	S	S	–	S	S	M	U	S	M	M	U	M	M	M	S	U	S	S	U	S
Lactic acid (20%)	–	–	S	S	M	–	M	S	U	S	S	U	S	–	M	S	M	S	S	–	S
Lactic acid (100%)	–	–	S	–	M	–	M	S	U	S	S	S	S	S	M	S	M	S	S	–	S
Magnesium chloride	M	S	S	–	S	S	S	S	S	S	S	U	S	S	S	S	U	M	S	S	S
Mercaptoacetic acid	U	S	U	–	–	S	M	S	U	S	M	U	U	S	M	S	U	M	S	S	S
2-Mercaptoethanol	S	S	U	–	S	S	S	S	S	S	S	U	S	S	U	S	U	S	S	S	S
Metrizamide	M	S	S	–	S	S	S	S	S	S	S	U	S	S	U	S	S	S	S	S	U
Methyl alcohol	S	S	S	–	S	S	S	S	S	S	M	U	S	U	U	S	S	S	S	M	U
Methyl ethyl ketone	S	S	–	U	M	S	U	U	U	S	U	U	U	U	U	S	U	S	S	U	U
Methylene chloride	U	U	S	U	M	S	U	–	U	U	U	U	U	U	U	U	S	S	S	S	S
Nickel salts	M	S	S	S	S	–	S	S	S	S	S	U	S	S	S	S	S	S	S	S	S
Nitric acid (10%)	M	S	U	S	U	S	U	U	–	S	M	U	M	U	S	S	M	S	S	M	S
Nitric acid (50%)	M	S	U	M	M	–	U	U	U	M	U	U	M	U	U	S	U	S	S	S	S
Nitric acid (95%)	M	–	U	U	U	–	U	U	U	U	U	U	M	U	U	S	U	S	S	–	S

Compound																						
Nycodenz®	S	S	S	M	U	—	—	S	S	S	S	M	S	S	S	S	S	S	S	S	S	S
Oils (petroleum)	S	S	S	U	S	S	S	S	S	S	S	U	S	M	S	S	S	S	S	S	S	S
Oleic acid	S	—	U	M	—	—	—	S	U	S	U	U	S	S	S	M	S	M	M	M	M	S
Oxalic acid	M	U	M	U	U	S	U	U	U	S	U	U	M	U	M	M	M	S	M	M	M	S
Perchloric acid (10%)	U	—	U	U	U	M	U	U	M	U	U	U	U	M	U	U	U	U	—	S	—	S
Perchloric acid (70%)	U	S	U	U	U	M	M	U	U	U	U	U	U	U	U	U	U	U	—	U	—	S
Phenol (5%)	M	S	U	M	S	M	M	M	U	S	M	U	M	U	M	U	M	S	S	U	M	S
Phenol (50%)	M	U	U	U	S	U	U	U	U	U	U	M	M	U	M	U	M	M	M	M	M	S
Phosphoric acid (10%)	U	S	U	S	S	S	S	S	M	U	M	S	S	S	S	S	S	U	S	S	S	S
Phosphoric acid (conc.)	U	U	M	M	U	S	M	M	U	U	U	U	M	U	U	U	S	U	—	U	—	S
Picric acid	—	—	M	M	M	S	S	—	U	M	U	U	M	S	S	S	M	—	—	—	S	S
Potassium bromide	S	S	S	S	S	S	S	S	S	S	S	S	S	S	S	S	S	S	S	M	S	S
Potassium carbonate	M	S	S	S	S	S	S	S	S	S	M	S	S	S	S	S	S	—	M	S	S	S
Potassium chloride	U	S	S	S	S	S	S	S	S	S	S	S	S	S	S	S	S	S	S	S	S	S
Potassium hydroxide (5%)	U	—	S	S	M	S	S	S	S	S	S	S	S	S	S	M	M	—	M	M	M	S
Potassium hydroxide (conc.)	U	—	U	M	M	S	M	S	M	S	S	S	U	S	S	U	S	—	M	U	M	M
Potassium permanganate	S	S	S	S	S	M	S	S	S	U	S	S	S	M	S	S	S	—	S	S	S	S
Pyridine (50%)	U	S	M	U	S	S	S	S	S	S	U	S	S	S	S	S	S	—	S	S	S	S
Rubidium bromide	M	S	S	S	S	S	S	S	S	S	S	S	S	S	S	M	S	—	—	S	S	S
Rubidium chloride	M	S	S	S	S	S	S	S	S	S	S	S	S	S	S	S	S	—	—	S	S	S
Sodium borate	S	S	S	S	S	S	S	S	S	S	S	S	S	S	M	S	S	—	S	S	S	S
Sodium bromide	U	—	S	S	S	S	S	S	S	S	S	S	S	S	S	S	S	—	S	S	S	S
Sodium carbonate (2%)	M	—	M	S	S	S	S	S	S	S	S	S	S	S	S	S	S	—	M	S	S	S
Sodium chloride (10%)	S	S	S	S	S	S	S	S	S	S	S	S	S	S	M	S	S	—	—	S	S	S
Sodium chloride (sat'd)	S	S	S	S	S	S	S	S	S	S	S	S	S	S	M	S	S	—	—	M	S	S
Sodium dodecyl sulfate	S	—	S	S	S	S	S	S	M	S	S	S	S	S	S	S	S	—	—	—	—	S
Sodium hydroxide (<1%)	—	S	U	S	M	S	S	S	U	U	M	U	S	—	S	S	S	—	—	S	—	S
Sodium hydroxide (10%)	U	M	M	U	M	M	U	U	U	U	U	U	U	U	U	M	S	S	U	M	U	S
Sodium hypochlorite (5%)	U	S	M	U	U	M	U	U	U	U	M	U	U	U	U	U	U	—	M	M	M	S
Sodium iodide	M	S	S	S	S	S	S	S	S	S	S	S	S	S	M	S	M	—	—	S	S	S
Sodium nitrate	S	S	S	S	S	S	S	S	S	S	S	S	S	S	S	S	S	—	S	S	S	S
Sodium sulfate	U	—	S	S	S	S	S	S	S	S	M	S	S	S	M	U	S	—	—	M	S	S
Sodium sulfide	S	S	S	S	S	S	S	S	S	S	S	S	S	S	S	S	M	—	—	—	—	S

CENTRIFUGATION

Table 4. Continued

Chemical	Aluminum	Anodic coating for aluminum	Carbon fiber	Cellulose acetate butyrate	Delrin®	Glass	Neoprene	Noryl®	Nylon	Polyallomer	Polycarbonate	Polyclear®/Ultraclear®	Polypropylene	Polysulfone	Polyvinyl chloride	Rulon A®, Teflon®	Silicone rubber	Stainless steel	Titanium	Tygon®	Viton®
Sodium sulfite	S	S	S	—	S	M	S	S	S	S	S	S	S	S	S	S	S	S	S	S	S
Stearic acid	S	—	S	—	S	S	S	S	S	S	S	S	S	S	S	S	M	M	S	S	S
Sucrose	M	S	S	—	S	S	S	S	S	S	S	S	S	S	S	S	S	S	S	S	S
Sucrose (alkaline)	M	S	S	S	S	S	S	S	S	S	S	S	S	—	S	S	S	M	S	S	S
Sulfosalicylic acid	U	U	U	U	U	S	S	M	U	S	U	S	M	S	S	S	U	U	M	M	S
Sulfuric acid (10%)	U	U	U	U	U	S	—	M	U	S	M	U	M	S	M	S	U	U	M	—	U
Sulfuric acid (50%)	U	U	U	U	U	S	U	M	S	M	S	U	S	U	U	S	U	S	S	U	S
Sulfuric acid (conc.)	S	—	U	U	U	S	U	U	S	M	M	U	M	U	U	S	U	S	U	U	M
Tetrahydrofuran	U	S	U	U	S	S	U	U	U	S	U	U	U	U	U	S	U	U	S	U	U
Tris buffer (neutral pH)	S	S	U	U	M	S	U	S	S	U	S	U	U	U	U	S	U	S	S	U	S
Toluene	S	S	S	U	U	S	U	S	U	U	U	U	U	U	U	S	U	U	S	M	M
Trichloroacetic acid	U	—	M	—	S	—	U	—	U	S	S	U	U	U	U	S	U	U	S	—	U
Trichloroethane	U	—	U	—	M	—	U	—	S	U	U	U	U	U	U	S	U	U	U	U	S
Trichloroethylene	—	—	U	U	—	—	U	—	S	U	U	U	U	U	U	S	U	—	S	U	S
Trisodium phosphate	—	—	S	S	M	—	U	—	—	S	—	—	U	—	—	S	—	—	—	—	U
Triton X-100®	S	S	S	—	S	S	S	S	S	S	S	S	S	S	S	S	S	S	S	S	S

Urea	S	–	S	S	S	–		–	S	S	–	S	S	–	S	–	S	–	S
Xylene	S	S	U	S	S	M	S		U	U	U	S	U	U	M	S	S	U	S
Zinc chloride	M	U	S	S	U		S	S	U	S	S	S	S	M	M	S	S	U	S
Zinc sulfate	U	S	S	–	S		S	S	S	S	S	S	S	S	S	S	S	S	M

NOTE: Chemical resistance data is included only as a guide to product use. Because no organized chemical resistance data exists for materials under the stress of centrifugation, when in doubt pretesting of sample lots is recommended.

Abbreviations: M, moderate attack, may be satisfactory for use in a centrifuge depending on length of exposure, speed involved, etc., suggest testing under actual conditions of use; S, satisfactory; U, unsatisfactory, not recommended; —, performance unknown, suggest testing, using sample to avoid loss of valuable material.

6. STERILIZATION AND DISINFECTION PROCEDURES

Techniques commonly used for the sterilization and disinfection of laboratory products are outlined below (see also *Chapter 14*).

6.1. Sterilization techniques

Autoclaving
A hot steam sterilization technique (generally 121°C, 15 psi, 15 min) is commonly used to destroy micro-organisms. (Separate individual pieces before autoclaving.)

Ethylene oxide gas
A gas sterilization technique compatible with most rotors, tubes, bottles and adaptors. Ethylene oxide is highly diffusive and permeates areas not reached by liquids or steam.

Formaldehyde gas
A gas sterilization technique compatible with most rotors, tubes, bottles and adaptors. Sterilization occurs after an 8 h exposure to 0.3 g ft^{-3} (10 g m^{-3}) at temperatures greater than 21°C and a relative humidity greater than 70%.

Glutaraldehyde solutions
These activated liquid sterilization solutions are compatible with almost all material that can be submerged. Sterilization is achieved after 10 h of immersion. (These solutions are available commercially as Glutarex®, Cidex® and Sonacide®.)

Ultraviolet (UV) radiation
UV sterilization techniques are compatible with most metals and some plastics. Both radiation intensity and exposure time must be controlled to ensure sterilization. Note that many plastics absorb UV light and so external illumination may not sterilize the interior containers made of these materials.

6.2. Biological disinfection

Alcohol
Alcohols (70% ethanol or 70% isopropanol) are effective for inactivating vegetative bacteria, lipid viruses and some nonlipid viruses. The minimum contact time is 10 min.

Hypochlorite
Sodium hypochlorite (a 1:8 dilution of household liquid chlorine bleach) inactivates vegetative bacteria, lipid and nonlipid viruses, and bacterial spores. The minimum contact time is 10 min. This reagent is a strong oxidizing agent and corrosive to metals.

Formaldehyde solutions
Formaldehyde gas in water is generally marketed in concentrations of approximately 37% and is referred to as Formalin. This solution is a broad-spectrum disinfectant and inactivates the same types of organisms as chlorine. The minimum contact time is 10 min. Formaldehyde is a toxic chemical with a strong, pungent odor.

Glutaraldehyde solutions
These activated liquid sterilization solutions are compatible with almost all materials that can be submerged. These solutions (available commercially as Glutarex®, Cidex® and Sonacide®) are broad-spectrum disinfectants similar to chlorine and formaldehyde. The minimum contact time is 10 min.

Phenolic compounds
Phenolic compounds (such as the phenolics in commercial disinfectants such as Lysol®) are effective against vegetative bacteria, lipid viruses and some nonlipid viruses. The minimum contact time is 10 min.

Table 5. Sterilization and disinfection procedures

Method	Aluminum	Anodic coating for aluminum	Carbon fiber	Cellulose acetate butyrate	Delrin®	Glass	Neoprene	Noryl®	Nylon	Polyallomer	Polycarbonate	Polyclear®/Ultraclear®	Polypropylene	Polysulfone	Polyvinyl chloride	Rulon A®, Teflon®	Silicone rubber	Stainless steel	Titanium	Tygon®	Viton®
Sterilization																					
Autoclaving (121°C)	U	–	S	U	M	S	U	S	U	S	M	U	S	M	M	S	S	S	S	M	S
UV irradiation	–	–	S	S	S	S	–	S	M	U	U	U	U	U	M	S	S	S	S	M	–
Ethylene oxide	S	S	U	–	S	S	–	S	S	S	S	S	S	S	S	S	S	S	S	S	S
Formaldehyde (gas)	S	S	S	–	S	S	S	S	S	S	S	S	S	S	S	S	S	S	S	U	S
Glutaraldehyde (2%)	S	S	S	S	S	S	S	S	S	S	S	S	S	S	S	S	S	S	S	S	S
Biological disinfection																					
Alcohol (70%)	S	S	S	M	M	S	S	S	S	S	M	U	S	M	M	S	S	M	S	M	M
Hypochlorite (5%)	U	U	M	S	U	S	M	S	M	S	S	S	S	S	M	S	M	M	M	M	S
Glutaraldehyde (2%)	S	S	S	S	S	S	U	S	S	S	S	S	S	S	S	S	S	S	S	S	S
Phenolic derivatives	U	–	U	–	U	S	U	S	U	M	U	S	M	U	U	S	U	U	U	U	S
Formaldehyde (40%)	M	M	S	S	S	S	S	S	S	S	S	–	S	M	M	S	S	S	S	M	S

Abbreviations: M, moderate attack, may be satisfactory, suggest testing; S, satisfactory; U, unsatisfactory, not recommended; –, performance unknown, suggest testing. UV, ultraviolet.

7. EQUATIONS RELATING THE REFRACTIVE INDEX TO THE DENSITY OF SOLUTIONS

When solutes are dissolved in water the refractive index of the resulting solution differs from that of water. The increase in the refractive index is proportional to the concentration of solute. In the case of gradient solutes, the density (ρ) of the solution is directly related to the solute concentration as well as the refractive index (η). This relationship can be expressed in the terms of the following equation

$$\rho = a\eta - b.$$

Tables 6 and 7 list the coefficients a and b for a number of ionic and non-ionic gradient media. However, before applying the equations it is essential that allowance is made for the presence of other solutes (e.g. EDTA, buffers, etc.). The following equation should be used:

$$\eta_{corrected} = \eta_{observed} - (\eta_{buffer} - \eta_{water}).$$

Table 6. Ionic gradient media

Gradient solute	Temperature (°C)		Coefficients		Valid density range (g cm^{-3})
	η	ρ	a	b	
CsCl	20	20	10.9276	13.593	1.2–1.9
	25	25	10.8601	13.497	1.3–1.9
Cs$_2$SO$_4$	25	25	12.1200	15.166	1.1–1.4
	25	25	13.6986	17.323	1.4–1.8
Cs(HCOO)	25	25	13.7363	17.429	1.7–1.8
	25	20	12.8760	16.209	1.8–2.3
NaBr	25	25	5.8880	6.852	1.0–1.5
NaI	20	20	5.3330	6.118	1.1–1.8
KBr	25	25	6.4786	7.643	1.0–1.4
KI	20	20	5.7317	6.645	1.0–1.4
	25	25	5.8356	6.786	1.1–1.7
RbCl	25	25	9.3282	11.456	1.0–1.4
RbBr	25	25	9.1750	11.241	1.1–1.7

Table 7. Non-ionic gradient media

Gradient solute	Temperature (°C)		Coefficients	
	η	ρ	a	b
Sucrose	20	0	2.7329	2.6425
Ficoll®	20	20	2.381	2.175
Metrizamide	20	20	3.350	3.462
Nycodenz®	20	20	3.242	3.323
Metrizoate	25	5	3.839	4.117
Renografin®	24	4	3.5419	3.7198
Iothalamate®	25	25	3.904	4.201
Chloral hydrate	4	4	3.6765	3.9066
Bovine serum albumin	24	5	1.4129	0.8814

8. PROPERTIES OF SUCROSE SOLUTIONS

Table 8. Dilution of stock solutions of sucrose

Desired final concentration % (w/w)	Volume (ml) of stock solution to be diluted to 1 liter[a]					Density (g cm^{-3})	Refractive index of the final solution at 20°C[b]
	60% (w/w)	66% (w/w)	70% (w/v)	80% (w/v)	2.0 M		
5	66	58	73	64	75	1.0179	1.3403
10	134	119	149	130	152	1.0381	1.3479
15	205	182	228	199	233	1.0592	1.3557
20	279	247	310	271	317	1.0810	1.3639
25	357	315	396	346	405	1.1036	1.3723
30	437	387	485	424	496	1.1270	1.3811
35	521	461	578	506	591	1.1513	1.3902
40	609	539	676	591	691	1.1764	1.3997
45	701	620	777	680	795	1.2025	1.4096
50	796	704	883	773	903	1.2296	1.4200
55	896	792	994	870	—	1.2575	1.4307
60	1000	884	—	971	—	1.2865	1.4418

[a] All values quoted are at 4°C.
[b] Refractive index of water at 20°C is 1.3330.

CHAPTER 6
ENZYMOLOGY
J. Qiu

1. INTRODUCTION

Enzymology has been at the heart of biochemistry since the early days of biochemical investigations. There are many books covering both the theory of enzymatic reactions and the methodology of the techniques used in measuring enzyme activities; some of the more important references are given in *Section 9*. There is not enough space within this chapter to cover all of the theory and details of the practical aspects of measuring enzyme kinetics and, indeed, it would not be appropriate to do so.

This chapter seeks to summarize the most important theoretical aspects of enzyme kinetics and provides detailed information on a selection of enzymes, coenzymes and enzyme inhibitors. It also offers advice on the best methods for the interpretation of experimental data.

2. ENZYME KINETICS

2.1. Enzyme kinetics of single substrate reactions

Michaelis–Menten equation
The relationship between initial velocity and substrate concentration is described by the Michaelis–Menten equation. The overall enzyme-catalyzed reaction is composed of two elementary reactions in which the substrate forms a complex with the enzyme that subsequently decomposes to products and enzyme:

$$E + S \underset{k_{-1}}{\overset{k_1}{\rightleftharpoons}} ES \underset{k_{-2}}{\overset{k_2}{\rightleftharpoons}} E + P. \qquad \text{Equation 1}$$

Here E, S, ES and P symbolize the enzyme, substrate, enzyme–substrate complex and products, respectively.

Experimentally, substrate concentrations are almost always very large compared to enzyme concentrations (e.g. 1 mM as compared with 1 nM). The substrate concentration is then high enough to convert all of the enzyme to the ES form, and in this case the second step of the reaction becomes rate limiting and the overall reaction rate becomes insensitive to further increases in substrate concentration, i.e.

$$E + S \underset{k_{-1}}{\overset{k_1}{\rightleftharpoons}} ES \overset{k_2}{\rightarrow} E + P. \qquad \text{Equation 2}$$

Thus the general expression for the velocity of this reaction is,

$$v_0 = \frac{d[P]}{dt} = k_2[ES].$$ Equation 3

The net rate of production of [ES] is the difference between the rate of formation of [ES] and the rate of its breakdown.

$$\frac{d[ES]}{dt} = k_1[E][S] - k_{-1}[ES] - k_2[ES].$$ Equation 4

To integrate this equation, the following simplifying assumption about the steady state of the reaction is required.

Figure 1 illustrates the progress curves of the various components of the preceding reaction model when the substrate is in great excess over the enzyme. With the exception of the initial transient stage of the reaction, which is usually over within milliseconds of mixing the enzyme and substrate, [ES] remains approximately constant until the substrate is nearly exhausted. Hence, the rate of synthesis of ES must equal its rate of consumption over most of the course of the reaction; that is, [ES] maintains a steady state. One can therefore assume with a reasonable degree of accuracy that [ES] is constant; that is,

$$\frac{d[ES]}{dt} = 0.$$ Equation 5

According to *Equation 4*, therefore,

$$k_1[E][S] - k_{-1}[ES] - k_2[ES] = 0$$ Equation 6

Figure 1. The progress curves for the components of a simple Michaelis–Menten reaction. Note that with the exception of the transient phase of the reaction, the slopes of the progress curves for [E] and [ES] are essentially zero so long as $[S] \gg [E_t]$.

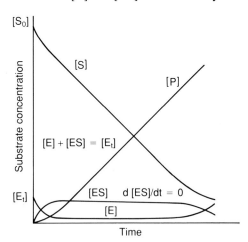

After a series of rearrangements, [ES] can be expressed as [Et], [S] and the rate constants k_1, k_{-1} and k_2.

$$[ES] = \frac{[Et][S]}{(k_{-1}+k_2)/k_1 + [S]} . \qquad \text{Equation 7}$$

The initial velocity of the reaction (v_0) can then be expressed in terms of the experimentally measurable quantities [Et] and [S].

$$v_0 = k_0[ES] = \frac{k_0[Et][S]}{K_m + [S]} \qquad \text{Equation 8}$$

where k_2 is written as k_0 (see below) and where K_m, which is known as the Michaelis constant, is defined as:

$$K_m = \frac{k_{-1}+k_2}{k_1} \qquad \text{Equation 9}$$

Equation 8, which is derived with the steady-state assumption, is the Michaelis–Menten equation, the fundamental equation of enzyme kinetics. A plot of v_0 against [S] is shown in *Figure 2*.

Although this is a useful model, it is not necessarily the true mechanism simply because of experimental adherence to the Michaelis–Menten equation. It is therefore best to represent its parameters as in *Equation 8*, with symbols k_0 and K_m that do not imply that they refer to particular steps of a mechanism, even though in the two-step case k_0 is in fact identical to k_2. There are several interesting features of the Michaelis–Menten equation that are worthwhile pointing out:

(i) Although k_0 may not refer to a single step of mechanism, it does have the properties of a first-order rate constant, defining the capacity of the enzyme–substrate complex, once

Figure 2. A plot of the initial velocity, v_0, of a simple Michaelis–Menten reaction versus the substrate concentration [S].

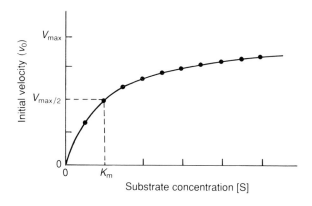

formed, to form the product, P. It is commonly called the catalytic constant of the enzyme (and often symbolized as k_{cat}); the alternative name, turnover number, is also sometimes used. Values of about 10^3 s^{-1} are typical for k_0, and some researchers always measure it, especially in the early stages of characterizing an enzyme, because the enzyme concentration is often unknown or difficult to measure. For this reason k_0 [Et] is often replaced by V_{max}, a quantity known as the limiting rate. Therefore,

$$V_{max} = k_0[Et] \qquad \text{Equation 10}$$

$$k_{cat} = k_0 = \frac{V_{max}}{[Et]}. \qquad \text{Equation 11}$$

(ii) The Michaelis constant, K_m, has a simple operational definition. It is the substrate concentration at which the reaction velocity is half-maximal.

(iii) If the substrate concentration, [S], is much smaller than K_m, it can be ignored by comparison with K_m in *Equation 8*, which thus simplifies to

$$v_0 = \frac{k_0}{K_m}[Et][S]; \qquad \text{Equation 12}$$

that is, with first-order dependence on both the enzyme and substrate, or second-order kinetics overall. The quantity k_0/K_m is therefore a measure of an enzyme's catalytic efficiency and is also known as the specificity constant and symbolized as k_A, where the subscript specifies which substrate is being considered.

(iv) At very high [S] ([S] $\gg K_m$), *Equation 8* simplifies to:

$$v_0 = k_0[Et] = V_{max}. \qquad \text{Equation 13}$$

It should be noted that V_{max} is not reached at any value of [S] and is in fact approached asymptotically.

Plots of the Michaelis–Menten equation
The most natural way of plotting steady-state kinetic data is to plot the initial velocity v_0 against the substrate concentration [S], as shown in *Figure 1*.

However, such a plot is not a good way to determine the kinetic parameters, because the line is curved and it does not approach the limit fast enough for one to be able to judge accurately where the limit is. Therefore, it is better to transform the Michaelis–Menten equation into a form corresponding to an equation for a straight line. There are a number of ways of doing this.

(i) The Lineweaver–Burke plot (*Figure 3*), which is also known as the double-reciprocal plot, is obtained by taking reciprocals of both sides of *Equation 8*:

$$\frac{1}{v_0} = \frac{1}{V_{max}} + \frac{K_m}{V_{max}} \cdot \frac{1}{[S]}. \qquad \text{Equation 14}$$

Therefore a plot of $1/v_0$ against $1/[S]$ is a straight line, with a slope of $1/k_A[Et]$, which is equal to K_m/V_{max}, and an intercept on the ordinate $(1/v_0)$ axis of $1/k_0[Et]$, which is equal to $1/V_{max}$, and an intercept on the abscissa $(1/[S])$ axis of $-k_A/k_0$, which is equal to $-1/K_m$.

This plot is used widely but is not recommended for determining kinetic parameters. The objection is that it distorts the appearance of any experimental error in the primary observations of v_0, so that one cannot judge which points are most accurate when drawing a straight line through the set of points. Neither of the other two straight-line plots is entirely free from distortion, but both are less severely affected than the double-reciprocal plot.

(ii) The Hane–Woolf plot (*Figure 4*) can be obtained by multiplying both sides of *Equation 14* by $[S]$:

$$\frac{[S]}{v_0} = \frac{K_m}{V_{max}} + \frac{1}{V_{max}}[S]. \qquad \text{Equation 15}$$

A plot of $[S]/v_0$ against $[S]$ is also a straight line. Now the slope, $1/k_0[Et]$, is equal to $1/V_{max}$; the ordinate intercept, $1/k_A[Et]$, is equal to K_m/V_{max}; and the abscissa intercept, $-k_0/k_A$, is

Figure 3. The Lineweaver–Burke plot (double-reciprocal plot) of reciprocal rate $1/v_0$ against reciprocal substrate concentration $1/[S]$.

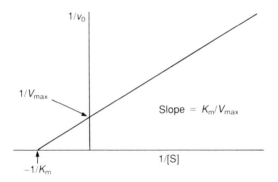

Figure 4. The Hane–Woolf plot of $[S]/v_0$ against $[S]$.

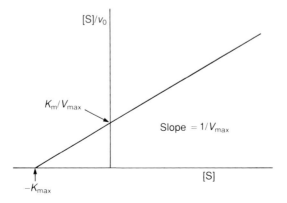

equal to $-K_m$. Note that the slope and ordinate intercept are interchanged from those of the double-reciprocal plot.

(iii) The Eadie–Hofstee plot (*Figure 5*) is generated by multiplying both sides of *Equation 15* by $V_{max} v_0/[S]$ and rearranging:

$$v_0 = V_{max} - K_m \frac{v_0}{[S]}.$$

Equation 16

A plot of v_0 against $v_0/[S]$ is also a straight line. The slope is $-K_m$ and the intercepts on the v_0 and $v_0/[S]$ axis are V_{max} and V_{max}/K_m, respectively.

(iv) The direct linear plot is conceptually rather different from the others described, as it requires each observation to be plotted as a straight line and the parameter values appear as a point rather than as the slope and intercept of a line. This plot is obtained by a rearrangement of *Equation 16*:

$$V_{max} = v_0 + K_m \frac{v_0}{[S]}.$$

Equation 17

If V_{max} and K_m are treated as variables and v_0 and [S] as constants, this equation defines a straight line with intercepts v_0 on the V_{max} axis and $-[S]$ on the K_m axis. This line relates all possible pairs of (K_m, V_{max}) values that exactly satisfy an observation of rate v_0 at substrate concentration [S]. A second line drawn in the same way for a second observation relates all parameter values that satisfy this observation. Only one point is on both lines, their point of intersection, and its coordinates define the unique pair of parameter values that satisfy both observations exactly.

In theory, n such lines for a set of n observations intersect at a unique point, the coordinates of which would give the values of K_m and V_{max}. Real experimental data, however, are subject to error, and give a family of intersection points, as shown in *Figure 6*. It is still easy to estimate the parameter values because each intersection point provides one estimate of K_m

Figure 5. The Eadie–Hofstee plot of v_0 against $v_0/[S]$.

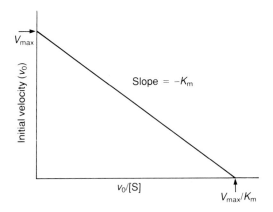

and one estimate of V_{max}. These can be marked on the axis and the best estimate of each parameter can be taken as the median value of the set. One takes the median rather than the mean because it is easier, requiring only counting rather than calculation, and also because it is much safer: some pairs of lines may be nearly parallel and so some of the intersection points may be very far from the correct values; such inaccurate estimates have a deleterious effect on the calculation of a mean, but hardly any effect on the median.

There are three related versions of the direct linear plot. The form shown in *Figure 6* is the most familiar, and perhaps the easiest to use, but the alternative shown in *Figure 7* has some advantages, of which the simplest is that it provides better-defined intersection points, which all occur in the first quadrant or very close to it.

2.2. Sigmoidal kinetics and allosteric enzymes

Many enzymes are oligomeric proteins and have more than one ligand-binding site. In these cases, the possibility of interaction between binding sites, which is called cooperativity, can occur. Such interactions could give rise to cooperative binding and would be reflected in departures from the expected linearity of normally linear plots if the ligand is a substrate. Four types of cooperativity can be identified:

(i) Positive cooperativity: the binding of one molecule of a substrate or ligand increases the affinity of the protein for other molecules of the same or a different substrate or ligand.
(ii) Negative cooperativity: the binding of one molecule of a substrate of ligand decreases the affinity of the protein for other molecules of the same or a different substrate or ligand.
(iii) Homotropic cooperativity: the binding of one molecule of a substrate or ligand affects the binding to the protein of subsequent molecules of the same substrate or ligand (i.e. the binding of one molecule of A affects the binding of further molecules of A).
(iv) Heterotropic cooperativity: the binding of one molecule of a substrate or ligand affects the binding of a different type of substrate or ligand molecule (e.g. the binding of one molecule of A affects the binding of B).

Figure 6. The direct linear plot. Each observation is represented not by a point as in most kinds of plot, but by a straight line, with intercepts $-[S]$ and v_0 on the abscissa and ordinate, respectively. Each point of intersection provides an estimate of K_m and an estimate of V_{max}. The best estimates are marked on the axes but complete lines are not shown. The best estimate of K_m is taken as the median (middle) of the ordered set; this is marked on the plot as K_m^*, and the best estimate V_{max}^* of V_{max} is obtained similarly.

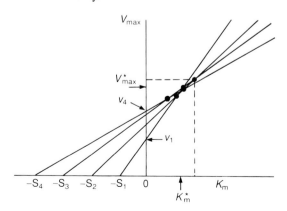

Figure 7. Alternative form of the direct linear plot. Here each observation is represented by a line making intercepts of $[S]/v_0$ and $1/v_0$ on the abscissa and ordinate axes, respectively.

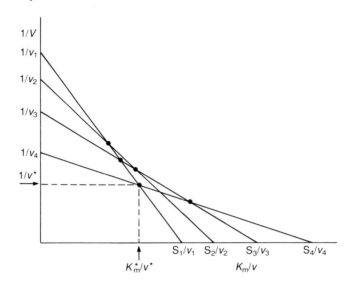

Enzymes showing cooperative ligand-binding are called allosteric enzymes.

Positive homotropic cooperativity and the Hill equation
In the simplest case of positive homotropic cooperativity in a dimeric protein: there are two identical ligand-binding sites, and when the ligand binds to one, it increases the affinity of the protein for the ligand at the other site, so the reaction sequence is:

$$M_2 + S \underset{}{\overset{slow}{\rightleftharpoons}} M_2S \qquad \text{Equation 18}$$

$$K_1 = \frac{[M_2][S]}{[M_2S]} \qquad \text{Equation 19}$$

$$M_2S + S \underset{}{\overset{fast}{\rightleftharpoons}} M_2S_2 \qquad \text{Equation 20}$$

$$K_2 = \frac{[M_2S][S]}{[M_2S_2]} \qquad \text{Equation 21}$$

where K_1 and K_2 are the dissociation constants of the first and second step of the reactions; M is the monomeric subunit, termed a protomer, and M_2 is the dimeric protein. If the increase in affinity is large enough, M_2S will react with S almost immediately it is formed. The overall reaction can be written as:

$$M_2 + 2S \rightleftharpoons M_2S_2 \qquad \text{Equation 22}$$

and the dissociation constant for the overall reaction is the product of K_1 and K_2, that is:

$$K = K_1 K_2 = \frac{[M_2][S]^2}{[M_2S_2]} \qquad \text{Equation 23}$$

Figure 8. The Hill plot of $\log(\bar{Y}/1-\bar{Y})$ against $\log[S]$, at fixed protein concentration, where the binding shows positive homotropic cooperativity.

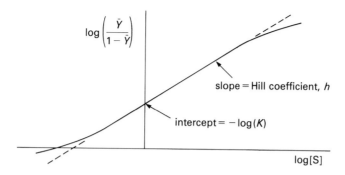

Similarly, for complete positive homotropic cooperativity of a protein with n identical binding sites, then,

$$M_n + nS \rightleftharpoons M_nS_n \qquad \text{Equation 24}$$

$$K = \frac{[M_n][S]^n}{[M_nS_n]}. \qquad \text{Equation 25}$$

The dissociation of M_n may be characterized by its fractional saturation, $\bar{Y}$, defined as the fraction of the substrate binding sites occupied by the substrate,

$$\bar{Y} = \frac{[M_nS_n]}{[M_n] + [M_nS_n]}. \qquad \text{Equation 26}$$

Combining *Equations 25* and *26* yields,

$$\bar{Y} = \frac{[M_n][S]^n/K}{[M_n](1 + [S]^n/K)}, \qquad \text{Equation 27}$$

which, upon algebraic rearrangement and cancellation of terms, becomes the Hill Equation,

$$\bar{Y} = \frac{[S]^n}{K + [S]^n} \qquad \text{Equation 28}$$

which describes the degree of saturation of a multisubunit protein as a function of ligand concentration.

The Hill constant, n, and the dissociation constant, K, that best describe a saturation curve can be graphically determined by rearranging *Equation 28* to the following *Equation 29*:

$$\frac{\bar{Y}}{1-\bar{Y}} = \frac{[S]^n}{K} \qquad \text{Equation 29}$$

Taking the log of both sides to yield a linear equation gives:

$$\log\left(\frac{\bar{Y}}{1-\bar{Y}}\right) = n \log[S] - \log K.$$ Equation 30

The linear plot of $\log[\bar{Y}/(1-\bar{Y})]$ versus $\log[S]$, the Hill plot, has a slope of n and an intercept on the $\log[\bar{Y}/(1-\bar{Y})]$ axis of $-\log K$ (*Figure 8*).

At values of $\bar{Y}$ below 0.1 and above 0.9, the slopes of Hill plots tend to 1, indicating an absence of cooperativity. This is because at very low ligand concentrations there is not enough ligand present to fill more than one site on most protein molecules, regardless of affinity; similarly at high ligand concentrations there are extremely few protein molecules present with more than one binding site remaining to be filled.

The Hill coefficient is therefore taken to be the slope of the linear, central portion of the graph, where the cooperative effect is most marked (*Figure 8*). For systems where cooperativity is complete, the Hill coefficient (h) is equal to the number of binding sites (n). In the cases where cooperativity is less complete, the linear central section is likely to shorten and h will be less than n.

In the case where S is a substrate and the reaction proceeds to yield products in such a way that the Michaelis–Menten equilibrium assumption is valid, a Hill plot of $\log(v_0/V_{max} - v_0)$ against $\log[S_0]$ may be substituted for the one shown in *Figure 8*. The slopes of the two graphs will have the same value and meaning.

One of the main problems in constructing a Hill plot from kinetic data is obtaining an accurate estimate of V_{max}: this is particularly true for cooperative systems, since the primary plots are not linear. Nevertheless, an estimate of V_{max} can be obtained from an Eadie–Hofstee or other plot, enabling a Hill plot to be constructed and a Hill coefficient (h) determined. The primary plot can then be redrawn substituting $[S]^h$ for $[S]$, which should give more linear results and a more accurate estimate of V_{max}. If this differs markedly from the initial estimate of V_{max}, the Hill plot should then be redrawn, incorporating the new estimate of V_{max}.

Negative homotropic cooperativity
Negative cooperativity results in the second step of the binding process being slower than it would be if there were no interaction between the binding sites. For negative homotropic cooperativity, a plot of $\bar{Y}$ against $[S]$ is neither sigmoidal nor a true rectangular hyperbola (*Figure 9*).

2.3. Enzyme kinetics of bisubstrate reactions

The Michaelis–Menten equation strictly applies to enzymes that require a single substrate (e.g. isomerases) only. In cases where the reaction involves two substrates and two products, such as transfer or redox reactions:

$$A + B \underset{}{\overset{E}{\rightleftharpoons}} P + Q.$$ Equation 31

Such reactions are also called Bi Bi reactions, since they involve two substrates and yield two products. Almost all of these bisubstrate reactions are either transferase reactions in which the enzyme catalyzes the transfer of a specific functional group, X, from one of the substrates to the other:

$$P - X + B \underset{}{\overset{E}{\rightleftharpoons}} P + B - X$$ Equation 32

or oxidation–reduction reactions in which reducing equivalents are transferred between the two substrates.

Figure 9. The effects of positive cooperativity (curves A) and negative cooperativity (curves C) on the common kinetic plots. Curves B show the plots for no cooperativity.

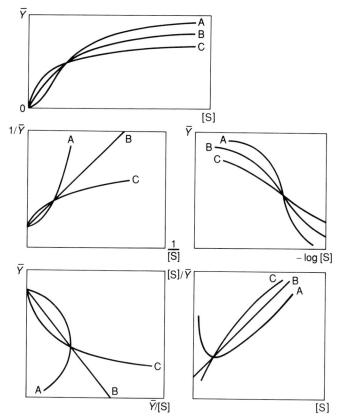

Types of Bi Bi reactions
Table 1 summarizes the different types of Bi Bi reactions, their mathematical expressions and the diagnostic plots.

The ordered Bi Bi mechanism may be experimentally distinguished from the random Bi Bi mechanism through product inhibition studies. If only one product of the reaction, P or Q, is added to the reaction mixture, the reverse reaction still cannot occur. Nevertheless, by binding to the enzyme, this product will inhibit the forward reaction. For an ordered Bi Bi reaction, Q (= B − X, the second product to be released) directly competes with A (= P − X, the leading substrate) for binding to E and hence is a competitive inhibitor of A when [B] is fixed (the presence of X in Q = B − X interferes with the binding of A = P − X). However, since B combines with EA, not E, Q is a mixed inhibitor of B when [A] is fixed (Q interferes with both the binding of B to enzyme and with the catalysis of the reaction). Similarly, P, which combines only with EQ, is a mixed inhibitor of A when [B] is held constant and of B when [A] is held constant. In contrast, in a rapid equilibrium Bi Bi reaction, since both products, as well as both substrates, can combine directly with E, both P and Q are competitive inhibitors of A when [B] is constant and of B when [A] is constant. These product inhibition patterns are summarized in *Table 2*. It can be seen that for an ordered reaction, a series of incubations where substrate A is held constant and substrate B is varied, a group of parallel lines is observed in a reciprocal plot, but for the random model these lines intersect on the y-axis.

Table 1. A summary of different types of Bi Bi reaction kinetics

Types of Bi Bi	Mathematical expression and diagnostic plot
Sequential reactions[a] Ordered mechanism: there is a compulsory order of substrate addition to the enzyme	**1. Mathematical expression** $$\frac{1}{v_0} = \frac{K_{ma}}{k_0[Et]} \left(1 + \frac{K_{sa}K_{mb}}{K_{ma}[b]}\right) \frac{1}{[a]} + \frac{1}{k_0[Et]} \left(1 + \frac{K_{mb}}{[b]}\right) \quad \text{Equation 33}^b$$ or $$\frac{1}{v_0} = \frac{K_{mb}}{k_0[Et]} \left(1 + \frac{K_{sa}}{[a]}\right) \frac{1}{[b]} + \frac{1}{k_0[Et]} \left(1 + \frac{K_{ma}}{[a]}\right) \quad \text{Equation 34}$$ **2. Diagnostic plots** a) Plot of $1/v_0$ vs $1/[a]$: Increasing constant [b], lines intersect at $-1/K_{ma}$; Slope $= K_{ma} + \frac{K_{sa}K_{mb}}{[b]} / V_{max}$; Intercept $= \frac{1 + K_{mb}/[b]}{V_{max}}$ b) Plot of $1/v_0$ vs $1/[b]$: Increasing constant [a]; Slope $= \frac{K_{mb} + \frac{K_{sa}K_{mb}}{[a]}}{V_{max}}$; $-\frac{K_{ma}}{K_{mb}K_{sa}}$; Intercept $= \frac{1 + K_{ma}/[a]}{V_{max}}$

Sequential reactions[a]
Random mechanism: with no preference for the order of substrate addition

1. Mathematical expression

$$\frac{1}{v_0} = \frac{K_{mb}}{k_0[Et]}\left(1 + \frac{K_{sa}}{[a]}\right)\frac{1}{[b]} + \frac{1}{k_0[Et]}\left(1 + \frac{K_{sa}K_{mb}}{K_{sb}[a]}\right) \qquad \text{Equation 35}$$

or

$$\frac{1}{v_0} = \frac{K_{sa}K_{mb}}{k_0[Et]}\left(\frac{1}{K_{sb}} + \frac{1}{[b]}\right)\frac{1}{[a]} + \frac{1}{k_0[Et]}\left(1 + \frac{K_{mb}}{[b]}\right) \qquad \text{Equation 36}$$

2. Diagnostic plots

a)

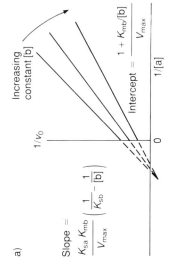

Slope =
$\dfrac{K_{sa}K_{mb}}{V_{max}}\left(\dfrac{1}{K_{sb}} - \dfrac{1}{[b]}\right)$

Intercept = $\dfrac{1 + K_{mb}/[b]}{V_{max}}$

b)

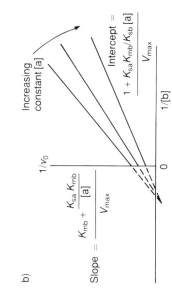

Slope = $K_{mb} + \dfrac{K_{sa}K_{mb}}{[a]}$ / V_{max}

Intercept = $\dfrac{1 + K_{sa}K_{mb}/K_{sb}[a]}{V_{max}}$

Table 1. Continued

Types of Bi Bi	Mathematical expression and diagnostic plot
Ping Pong reactions Mechanisms in which one or more products are released before all substrates have been added	1. Mathematical expression $$\frac{1}{v_0} = \frac{1}{k_0[\text{Et}]}\left(\frac{K_{ma}}{[a]} + \frac{K_{mb}}{[b]} + 1\right)$$ Equation 37 2. Diagnostic plots: a linear plot of $1/v_0$ versus $1/[b]$ (or $1/[a]$) for different [a] (or [b]) yields a family of parallel lines, which are diagnostic for a Ping Pong mechanism. a) b) 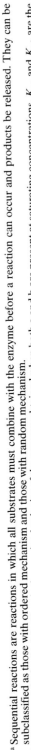

[a] Sequential reactions are reactions in which all substrates must combine with the enzyme before a reaction can occur and products be released. They can be subclassified as those with ordered mechanism and those with random mechanism.
[b] The term $k_0[\text{Et}]$, which is V_{max}, is the maximal velocity of the enzyme obtained when both a and b are present at saturating concentrations. K_{ma} and K_{mb} are the respective concentrations of a and b necessary to achieve $1/2\ V_{max}$ in the presence of a saturating concentration of the other, and K_{sa} and K_{sb} are the respective dissociation constants of a and b from the enzyme, E.

Table 2. Patterns of product inhibition for sequential bisubstrate mechanisms

Mechanism	Product inhibitor	A variable	B variable
Ordered Bi Bi	P	Mixed	Mixed
	Q	Competitive	Mixed
Rapid equilibrium random Bi Bi	P	Competitive	Competitive
	Q	Competitive	Competitive

3. ENZYME INHIBITORS

3.1. Reversible inhibition

Table 3 summarizes the characteristics, mathematical expressions and diagnostic plots for various types of reversible inhibition.

Allosteric inhibition
An allosteric inhibitor, by definition, binds to the enzyme at a site distinct from the substrate-binding site. Therefore, some of the types of inhibition considered in *Table 3* may be regarded as forms of allosteric inhibition. However, the term allosteric inhibition is usually reserved for the situation where the inhibitor, rather than forming a dead-end complex with the enzyme, influences conformational changes which may alter the binding characteristics of the enzyme for the substrate or the subsequent reaction characteristics (or both). The Michaelis–Menten plot becomes less hyperbolic and more sigmoidal (S-shaped), which means that the rate of reaction is reduced at low substrate concentrations but not necessarily at others. If the binding characteristics alone are affected, V_{max} will usually remain unchanged, so the inhibition pattern could be regarded as competitive. Similarly, other forms of allosteric inhibition, where V_{max} is altered, could be regarded as giving non-competitive or mixed inhibition, depending on whether K_m is changed or not. However, in most cases the Michaelis–Menten equation is not obeyed in the presence of allosteric inhibitors, nor are linear Lineweaver–Burke plots obtained, and so the terms competitive, non-competitive and mixed inhibition are not strictly applicable.

3.2. Irreversible inhibition

An irreversible inhibitor binds to the active site of the enzyme (often covalently) by an irreversible reaction:

$$E + I \longrightarrow EI \qquad \text{Equation 65}$$

and hence cannot subsequently dissociate from it. A covalent bond is usually formed between inhibitor and enzyme. The inhibitor may act by preventing substrate binding or it may destroy some component of the catalytic site. Compounds which irreversibly denature the enzyme protein or cause non-specific inactivation of the active site are not usually regarded as irreversible inhibitors. *Table 4* gives a selection of some important enzyme inhibitors.

Irreversible inhibitors effectively reduce the concentration of enzyme present. An inhibitor of initial concentration $[I_0]$ will reduce the concentration of active enzyme from an initial value of $[Et]$ to $[Et] - [I_0]$, assuming that the inhibitor is not in excess. If a substrate is introduced after the reaction between inhibitor and enzyme has gone to completion, a system which obeys the Michaelis–Menten equation in the absence of inhibitor will still do so. The value of K_m will be the same as for the uninhibited reaction, but V_{max} will be reduced because there is less available active enzyme. *Figure 10* shows the diagnostic plot for irreversible inhibition, which is similar to that of non-competitive inhibition (*Table 3*).

Modifying *Equation 10*, then in the presence of inhibitor,

$$V'_{max} = k_0 ([Et] - [I_0]).$$ Equation 66

Therefore,

$$\frac{V'_{max}}{V_{max}} = \frac{[Et] - [I_0]}{[Et]}$$ Equation 67

$$V'_{max} = V_{max}[Et]\left(1 - \frac{[I_0]}{[Et]}\right).$$ Equation 68

Figure 10. Diagnostic plot for irreversible inhibition.

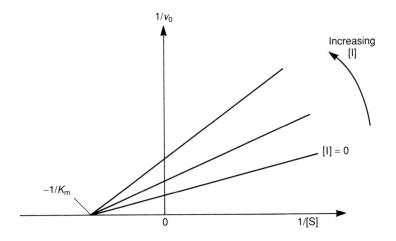

Similar results would be obtained even if the reaction between enzyme and inhibitor had not gone to completion, provided the degree of inhibition was relatively constant over the period when initial velocity studies were being carried out.

Therefore, patterns resembling those for reversible noncompetitive inhibition, with unchanged K_m and reduced V_{max}, may be obtained with irreversible inhibitors, even when the inhibitor binds to the same site as the substrate. However, the very real differences between the two forms of inhibition should be clear from the above discussion, and any attempt to calculate K_i from initial velocity measurements would be a totally meaningless exercise, since the relationship between V_{max} and V'_{max} does not involve K_i in this case. Hence, if a pattern of noncompetitive inhibitor is obtained during the investigation of a system, it is important to establish whether the inhibition is reversible or irreversible before the results can be interpreted.

Table 3. Types of reversible inhibition

Types of reversible inhibition	Model of inhibition reaction, mathematical expression, apparent parameters and diagnostic plots	
Competitive inhibition Inhibitor competes directly with a normal substrate for an enzymatic binding site. Such an inhibitor usually resembles the substrate, to the extent that it specifically binds to the active site, but differs enough from it so as to be unreactive	1. Model of inhibition reactions $E + S \underset{k_{-1}}{\overset{k_1}{\rightleftharpoons}} ES \xrightarrow{k_2} P + E$ $+$ I $\Updownarrow K_i$ $EI + S \longrightarrow$ no reaction 2. Mathematical expression $\dfrac{1}{v_0} = \dfrac{1}{k_0[Et]} + \dfrac{\alpha K_m}{k_0[Et]} \cdot \dfrac{1}{[S]}$ 3. Apparent parameters $k_0^{app} = k_0$ $K_m^{app} = K_m \alpha$ $k_s^{app} = k_0^{app}/K_m^{app} = k_s/\alpha$	Equation 38[a] Equation 39[b] Equation 40 Equation 41 Equation 42[c]

Table 3. Continued

Types of reversible inhibition	Model of inhibition reaction, mathematical expression, apparent parameters and diagnostic plots

4. Diagnostic plot

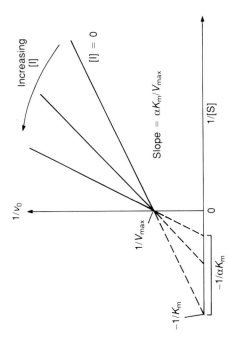

1. Model of inhibition reactions

$$E + S \underset{k_{-1}}{\overset{k_1}{\rightleftharpoons}} ES \overset{k_2}{\rightarrow} P + E$$
$$+$$
$$I$$
$$\Updownarrow K_I$$
$$ESI \rightarrow \text{no reaction}$$

Equation 43[d]

Uncompetitive inhibition
Inhibitor binds directly to the enzyme–substrate complex but not to the free enzyme

2. Mathematical expression

$$\frac{1}{v_0} = \frac{\alpha'}{k_0[\text{Et}]} + \frac{K_m}{k_0[\text{Et}]} \cdot \frac{1}{[S]}$$ Equation 44e

3. Apparent parameters

$k_0^{\text{app}} = k_0/\alpha'$ Equation 45

$K_m^{\text{app}} = K_m/\alpha'$ Equation 46

$K_s^{\text{app}} = K_0^{\text{app}}/K_m^{\text{app}} = k_s$ Equation 47

4. Diagnostic plot

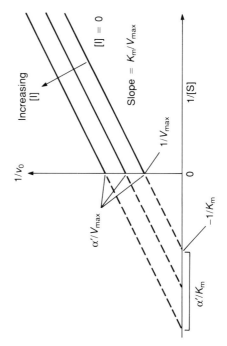

Table 3. Continued

Types of reversible inhibition	Model of inhibition reaction, mathematical expression, apparent parameters and diagnostic plots	
Non-competitive inhibition In the non-competitive type of inhibition, the combination of the enzyme with either S or I, by definition, does not affect the affinity for the other, that is, $K_i = K_I$, and therefore $\alpha = \alpha'$. As a result, the inhibitor does not affect the combination of the substrate with the enzyme, but affects only V_{max}	1. Model of inhibition reactions $$E + S \underset{k_{-1}}{\overset{k_1}{\rightleftharpoons}} ES \xrightarrow{k_2} P + E$$ $+ \quad\quad +$ $I \quad\quad I$ $\Updownarrow K_i \quad \Updownarrow K_I$ $EI \quad\quad EIS \rightarrow \text{no reaction}$	
	2. Mathematical expression $$\frac{1}{v_0} = \frac{\alpha'}{k_0[Et]} + \frac{\alpha K_m}{k_0[Et]} \cdot \frac{1}{[S]}$$	Equation 48 Equation 49
	3. Apparent parameters $k_0^{app} = k_0/\alpha'$ $K_m^{app} = K_m \alpha/\alpha' = K_m$ $k_s^{app} = k_0^{app}/K_m^{app} = k_s/\alpha$	Equation 50 Equation 51 Equation 52

Equation 53

Equation 54

4. Diagnostic plot

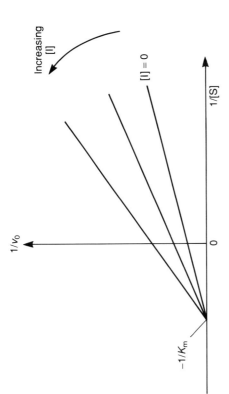

Mixed inhibition
Both the enzyme and the enzyme–substrate complex bind inhibitor but, unlike non-competitive inhibition, the combination of the enzyme with either S or I affects its affinity for the other. It is seen from *Equation 54* that the name 'mixed inhibition' arises from the fact that the denominator has the

1. Model of inhibition reactions

$$E + S \underset{k_{-1}}{\overset{k_1}{\rightleftharpoons}} ES \overset{k_2}{\rightarrow} P + E$$
$$+ \quad\quad +$$
$$I \quad\quad I$$
$$\updownarrow K_i \quad \updownarrow K_I$$
$$EI \quad\quad ESI \rightarrow \text{no reaction}$$

2. Mathematical expression

$$\frac{1}{v_0} = \frac{\alpha'}{k_0[E_t]} + \frac{\alpha K_m}{k_0[E_t]} \cdot \frac{1}{[S]}$$

Table 3. Continued

Types of reversible inhibition	Model of inhibition reaction, mathematical expression, apparent parameters and diagnostic plots	
factor α multiplying K_m as in competitive inhibition (*Equation 55*) and the factor α' multiplying [S] as in uncompetitive inhibition (*Equation 56*). Mixed inhibitors are therefore effective at both high and low substrate concentration	3. Apparent parameters	
	$k_0^{app} = k_0/\alpha'$	Equation 55
	$K_m^{app} = K_m \alpha/\alpha'$	Equation 56
	$k_s^{app} = k_0^{app}/K_m^{app} = k_s/\alpha$	Equation 57
	4. Diagnostic plot	

Diagnostic plot showing $1/v_0$ vs $1/[S]$ with lines of increasing [I], intersecting at a point in the second quadrant. Key labels: α'/V_{max}, $\dfrac{1-\alpha'}{(\alpha-1)K_m}$, $\dfrac{\alpha-\alpha'}{(\alpha-1)V_{max}}$, $-1/K_m$, $-\alpha'/\alpha K_m$, [I] = 0, Slope = $\alpha K_m/V_{max}$.

Partial inhibition
  The inhibition is only partial and the ESI complex could break down to yield product

1. Model of inhibition reactions

$$\begin{array}{c} E + S \underset{k_{-1}}{\overset{k_1}{\rightleftharpoons}} ES \overset{k_2}{\longrightarrow} E + P \\ + \qquad\quad + \\ I \qquad\quad I \\ \Big\Updownarrow K_i \quad\;\; \Big\Updownarrow K_I \\ EI \qquad ESI \underset{k'_2}{\longrightarrow} E + P + I \end{array}$$

Equation 58

2. Mathematical expression

$$\frac{1}{v_0} = \frac{1}{k_0[Et]} \cdot \frac{\alpha'}{\beta} + \frac{K_m}{k_0[Et]} \cdot \frac{1}{[S]} \cdot \frac{\alpha}{\beta}$$

Equation 59[f]

3. Apparent parameters

$k_0^{app} = k_0 \beta / \alpha'$

Equation 60

$K_m^{app} = K_m \alpha / \alpha'$

Equation 61

$k_s^{app} = k_0^{app} / K_m^{app} = k_s \beta / \alpha$

Equation 62

4. Diagnostic plot

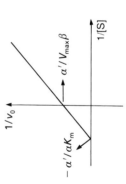

Table 3. Continued

Types of reversible inhibition	Model of inhibition reaction, mathematical expression, apparent parameters and diagnostic plots	
Substrate inhibition Occurs when a molecule of substrate binds to one site on the enzyme and then another molecule of substrate binds to a separate site on the enzyme to form a dead-end complex. This can be regarded as a form of uncompetitive inhibition, the extra substrate molecule being the inhibitor. v_0 decreases as [S] increases	1. Model of inhibition reactions $$E + S \rightleftharpoons ES \rightarrow E + P$$ $$+$$ $$S'$$ $$\Updownarrow K_1$$ $$ESS' \rightarrow \text{no reaction}$$ 2. Mathematical expression $$v_0 = \frac{k_0[Et]}{1 + [S]/K_1}$$ 3. Diagnostic plot	Equation 63 Equation 64

(a) Plot of initial velocity v_0 vs. substrate concentration [S]

(b) Plot of $\frac{1}{v_0}$ vs. $\frac{1}{[S_0]}$, showing intercepts at $-\frac{1}{K_m}$ and $\frac{1}{V_{max}}$

[a] K_i is the dissociation constant of the enzyme–inhibitor complex, EI.
[b] $\alpha = 1 + [I]/K_i$.
[c] k_0^{app}, K_m^{app} and k_s^{app} are apparent values of catalytic constant, Michaelis constant and specificity constant, respectively.
[d] K_1 is the dissociation constant of the enzyme–substrate–inhibitor complex, ESI.
[e] $\alpha' = 1 + [I]/K_1$.
[f] $\beta = 1 + k_2'[I]/k_2 K_1$.

Table 4. Inhibitors of specific enzymes[a]

Name	Mol. wt	Solubility	Effective conc. (M)	Inhibition	Ref.
Actinomycin	1255	V. sol in acetone, $CHCl_3$; sol. in MeOH, EtOH, mineral acids; sl. sol. in water		Inhibits DNA-primed RNA polymerase by complexing with DNA, via deoxyguanosine residues; at higher concentrations, DNA polymerases are inhibited	1
Alloxan (mesoxalylurea) anhyd. hydrate	142 160	Sol. in water, EtOH	10^{-3}	Inhibits skeletal muscle hexokinase; inhibits succinic dehydrogenase	1
α-Amanitin	903	Sol. in water, MeOH, EtOH	10^{-9}–10^{-8}; 10^{-5}–10^{-4}	Inhibits eucaryote RNA polymerase II; inhibits eucaryote RNA polymerase III	2
Amethopterin (methotrexate; 4-amino-N^{10}-methyl-pteroyl-glutamic acid)	454	Sol. in dilute HCl		Inhibits dihydrofolate reductase; inhibits purine syn.	3
Amphetamine (benzedrine; 1-phenyl-2-aminopropane, (+)-α-methylphenyl-ethylamine; (+)-isomer: dexamphetamine sulfate	135 368	Sol. in EtOH, acids, $CHCl_3$; sl. sol. in water Sulfate, sol. in water; sl. sol. in EtOH		Strong inhibitor of amine oxidase	1
Atractyloside (potassium atractylate) atractyloside K_2 salt carboxy atractyloside	803 771	Sol in EtOH; insol. in water		Displaces nucleotides from translocase proteins in the membrane	4

Name		Solubility	Action	Ref
Aurovertin B	460	Sol. in EtOH; insol. in water	Binds to mitochondrial ATPase, preventing phosphoryl group transfer and thereby oxidative phosphorylation	5
Azaserine (o-diazoacetyl-1-serine)	173	V. sol. in water; aqueous solns are most stable at pH 8, can be heated at 100°C in neutral aqueous soln for 5 min	Inactivates phosphoribosylformyl-glycinamidine synthetase irreversibly by covalent attachment to —SH group	1
Bongkrekic acid	487	Sol. in dilute alkali	Prevents release of nucleotides from translocase	4, 6
Caffeine (1,3,7-trimethyl-xanthine; theine; methyltheobromine)	194	Sol. in hot water, EtOH, acetone and $CHCl_3$; sl. sol in cold water	Inhibits cAMP phosphodiesterase	1
hydrate	212			
HCl	267			
sulfate	292			
Cerulenin (2,3-epoxy-4-oxo-7,10-dodecadienamide)	223	Sol. in EtOH acetone; sl. sol. in water; stable in neutral and acidic solns	Inhibits fatty acid synthetase, HMG-CoA synthetase: inhibition of peptidase reported	7
Chloramphenicol (CAP; chloromycetin)	323	0.25 in water; v. sol. in MeOH, EtOH	Inhibits bacterial protein synthesis by blocking peptidyl transferase reaction	8
Diisopropylphosphofluoridate (diisopropylfluoro-phosphate: DFP)	184	Sol. in organic solvents; 1.54 water; aqueous solns unstable	Inhibits serine esterases by covalent attachment to active-site serine; pseudocholinesterase inhibitor 10^{-9}–10^{-7}	1

Table 4. Continued

Name	Mol. wt	Solubility	Effective conc. (M)	Inhibition	Ref.
Distamycin A (stallimycin) HCl	481 518.1	Sol. in water		Inhibits various DNA and RNA polymerases *in vitro* by binding to DNA	9
Elastinal [*N*-(S)-1-carboxy-isopentyl]-carbamoyl-α-[2-iminohexahydro-4(S)-pyrimidyl]-L-glycyl-L-glutaminyl-L-alaninal)	513	Sol. in water, DMSO; sl. sol. in EtOH, acetone; unstable in alkaline soln	10^{-5}	Strong inhibitor of choline dehydrogenase	10
Ephedrine (adrenalin; α-(1-methylaminoethyl)-benzyl alcohol; 1-1-phenyl-2-methylaminopropanol HCl	165 202	Sol. in water, EtOH, $CHCl_3$, oils; aqueous solution, stable to light, air and heat HCl and sulfate, sol. in water, EtOH		Competitive inhibitor of choline dehydrogenase	1
5-Fluorouracil (5-fluoro-2,4-dihydroxypyrimidine)	130	Sol. in water		Metabolized to fluorodeoxy-uridylic acid which inhibits thymidylic acid synthetase	1
Hydrogen cyanide (hydrocyanic acid)	27	Free acid; infinite in water, EtOH		Inhibits by complexing with metals in metalloenzymes, e.g. cytochrome oxidase	1
Hydrogen cyanide, K salt (potassium cyanide)	65	Salt, v. sol. in water; sol. in EtOH	10^{-5}–10^{-3}	Inhibits by complexing with metals in metalloenzymes, e.g. cytochrome oxidase	1

Oleic acid	282	Infinite in EtOH; insol. in water	Inhibits 2,4-dinitrophenol-stimulated ATPase	1
Oligomycin B	805	Sol. in EtOH; 0.002 in water	Inhibits mitochondrial ATPase (F_1); inhibits Na^+-K^+ ATPase	11
Pyrazole (1,2-diazole)	68	Sol. in water and in EtOH	Inhibitor of alcohol dehydrogenase	12
Rifampicin (rifampin)	823	V. sl. sol. in water	Binds to β-subunit of bacterial RNA polymerases and inhibits chain initiation	1
Rotenone	394	Sol. in EtOH and in acetone; almost insol. in water	Inhibits NAD-linked substrate oxidation by mitochondria at oxygen side of NADH dehydrogenase	13
Theophylline (3,7-dihydro-1,3-dimethyl-1-purine-2,6-dione)	180	Sol. in hot water, alkaline hydroxides, dilute acids; 0.9 water; 1.25 EtOH	Competitively inhibits cAMP phosphodiesterase	14

[a] This table excludes inhibitors of proteases and nucleases, which are covered in *Chapter 7*.
Abbreviations: Anhyd., anhydrous; ATPase, adenine triphosphatase; cAMP, cyclic adenosine 3',5'-monophosphate; $CHCl_3$, trichloromethane; concs, concentrations; DMSO, dimethylsulfoxide; DNA, deoxyribonucleic acid; EtOH, ethanol; insol., insoluble; HMG-CoA, hydroxymethylglutarate-CoA; MeOH, methanol; NAD^+, nicotinamide adenine dinucleotide; NADH, nicotinamide adenine dinucleotide (reduced form); RNA, ribonucleic acid; sl. sol., slightly soluble; sol. soluble; soln(s), solution(s); syn., synthesis; v. sol., very soluble; v. sl. sol., very slightly soluble.

Table 5. Amino acid side-chain pK_a values in solution and in enzyme active centres

Side-chain	Free amino acid	Active center	Enzyme
Glu (γ-carboxyl)	3.9	6.5	Lysozyme
His (imidazole)	6.0	5.2, 6.8	Ribonuclease
Cys (thiol)	8.3	4.0	Papain
Lys (ε-ammonio)	10.8	5.9	Acetoacetate decarboxylase

4. EFFECTS OF pH ON ENZYMES

Enzymes have properties that are quite pH sensitive. Most enzymes, in fact, are active only within a narrow pH range, typically 5–9. This is a result of the effects of pH on a combination of factors: (i) the binding of substrate to enzyme, (ii) the catalytic activity of the enzyme, (iii) the ionization of substrate, and (iv) the variation of protein structure.

The side-chains of the free amino acids that ionize within the physiological range are given in *Table 5*, with some values that have been determined for the same groups in the active centers of enzymes. Quite large perturbations are possible, that is, active-center groups can have pK_a values substantially different from those of the corresponding amino acids in free solution. These may result from hydrophobicity, hydrogen bonding or electrostatic interactions. The optimal pH of some enzymes, i.e. the values of pH at which the enzymes are most catalytically active, are shown in *Table 6*.

5. COENZYMES — STRUCTURE AND FUNCTIONS

Coenzymes are prosthetic compounds required by enzymes for catalytic activity; they are often vitamins or derivatives of vitamins. Sometimes they can act as catalysts in the absence of enzymes, but not so effectively as in conjunction with an enzyme. *Table 7* lists the structure and function of some of the most important coenzymes.

6. ENZYME ASSAYS

The purpose of an enzyme assay is to determine how much of a given enzyme, of known characteristics, is present in a tissue homogenate or partially purified preparation. It is important that the enzyme preparations are treated carefully prior to assay, if the results are to have any meaning.

By far the most convenient way to estimate the concentration of a particular enzyme in a preparation is to determine its catalytic activity; that is, to find out how much substrate it is capable of converting to product in a given time under specified conditions. It is known from the Michaelis–Menten equation that for any system where the steady-state assumptions are valid, the initial velocity of a reaction catalysed by an enzyme is directly proportional to the concentration of that enzyme at fixed substrate concentration. Hence there is a linear relationship between enzyme concentration and catalytic activity:

$$v_0 = \frac{k_0 [\text{Et}][\text{S}]}{K_m + [\text{S}]} \qquad \text{Equation 8}$$

where k_0 is the rate constant relating to product formation (i.e. k_{cat}).

▶ p. 139

Table 6. A summary of the optimal pH of some important enzymes

Enzyme	Optimal pH	Enzyme	Optimal pH
Acetylcholinesterase	7.0	β-Glucuronidase (E. coli)	4.5–7.0
Acid phosphatase (wheat germ)	5.0	Glutamic acid decarboxylase	4.5–6.0
Alcohol dehydrogenase (yeast)	8.0	Glutamic oxaloacetic transaminase	8.0–8.5
Aldolase	7.0	Glutaminase	3.5–5.6
Alkaline phosphatase (chicken intestine)	8.0–9.0	Glycerol dehydrogenase	9.0
		Histidase	9.5
Alkaline phosphatase (E. coli)	8.0	Hyaluronidase	4.5–6.0
		Inorganic pyrophosphatase	7.0–7.2
D-Amino acid oxidase	9.0	L-Lactate dehydrogenase	7.2
L-Amino acid oxidase	7.5	Lecithinase C	7.0–7.6
α-Amylase	6.9	Leucine aminopeptidase	9.1
β-Amylase	4.0–5.0	Lipase (pancreatic)	7.5–8.0
Arginase	9.2–10.2	Lipoxidase	6.5–7.0
Arginine decarboxylase	4.5–6.0	Luciferase	6.8
Asparaginase	7.5	Lysine decarboxylase	4.5–6.0
Butyrylcholinesterase	6.0–8.0	Lysozyme	9.2
Carboxypeptidase B	7.5	Malate dehydrogenase	7.4
Catalase	7.0	Mercuripapain	6.0–7.0
Cellulase	4.2–5.2	Neuraminidase	5.0–5.1
Cholinesterase, acetyl	7.0	Nuclease (Staph. aureus)	9.2
Chymopapain	7.0–7.5	Oxalate decarboxylase	3.0
Collagenase	7.0–9.0	Papain	6.0–7.0
Cytochrome b_2	7.2	Pectin methyl esterase	8.5
Decarboxylases	4.5–6.0	Peroxidase	7.0
Deoxyribonuclease	7.0	Phosphodiesterase II (bovine spleen)	7.0
Deoxyribonuclease II	4.8		
Dextranase	5.0–7.0	Phospholipase C	7.0–7.6
Diaphorase	8.5	Polynucleotide phosphorylase	9.0–10.0
Elastase	8.8		
Emulsin	4.4–5.0	Polyphenol oxidase	6.0–7.0
Fructose 1,6-bisphosphatase	9.5	Pyruvate kinase	7.5
Galactose oxidase	7.0	Ribonuclease A	7.0–7.5
Glucose oxidase	5.5	Ribonuclease T1	7.5
Glucose 6-phosphate dehydrogenase:		Transaminase	8.0–8.5
		Tyrosinase	6.0–7.0
NAD linked	7.8	Tyrosine decarboxylase	4.5–6.0
NADP linked	>9.0	Urease (B. pasteurii)	6.5–7.0
β-Glucosidase (almond)	4.8	Urease (jack bean meal)	6.0
β-Glucuronidase (beef liver)	4.8	Uricase	9.0
		Xanthine oxidase	4.6

Table 7. Structure and functions of some important coenzymes

Coenzymes	Properties and function
NAD$^+$ and NADP$^+$	Hydrogen carriers. Can be readily and reversibly reduced, either by chemical reducing agents such as dithionite or by those dehydrogenases for which they are specific. In the reduction two equivalents of hydrogen per molecule are required. The oxidized form shows only a band at 260 nm, but the reduced form shows in addition a band at 340 nm. As an example, they are coenzymes of alcohol dehydrogenase in the reaction: alcohol + NAD$^+$(P) = aldehyde + NAD(P)H$_2$
Lipoate	Hydrogen carrier. Oxidized lipoic acid is yellowish and has a band at 335 nm. It is reduced by pyruvate and by 2-oxoglutarate in the presence of their respective dehydrogenases and thiamine pyrophosphate (e.g. pyruvate + oxidized lipoate = 6-S-acetylhydrolipoate + CO$_2$); in these cases the reduction is accompanied by the transfer of an acyl group from the substrate through thiamine to lipoate, giving acetylhydrolipoate and succinylhydrolipoate, respectively. The reduced lipoate is oxidized by NAD$^+$ in the presence of lipoamide dehydrogenase

Glutathione (GSH)

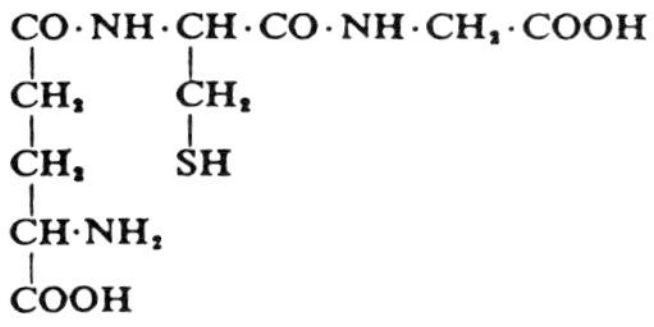

CO·NH·CH·CO·NH·CH₂·COOH
| |
CH₂ CH₂
| |
CH₂ SH
|
CH·NH₂
|
COOH

Hydrogen carrier. GSH acts as specific coenzyme for the glyoxalase system, for maleylacetoacetate and maleylpyruvate isomerases and for formaldehyde dehydrogenase. Functional group is the thiol group. Reduced form is oxidized to the disulfide by mild oxidizing agents, e.g. iodine or ferricyanide: $2GSH = GSSH + 2H$. It is also oxidized by molecular oxygen under suitable conditions in the presence of traces of catalytic metals, and by cytochrome c. It is oxidized enzymatically by dehydroascorbate in the presence of glutathione dehydrogenase. Powerful reducing agents are needed to reduce the oxidized form back to the thiol form; but enzymatically it can be reduced by NAD^+ or $NADP^+$ in the presence of glutathione reductase

Ascorbate

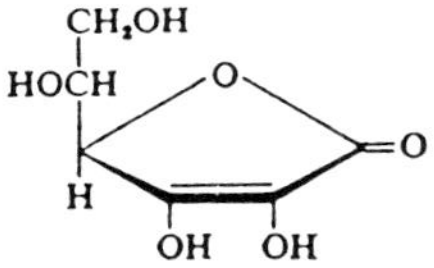

Hydrogen carrier. Ascorbate acts as the coenzyme of ascorbate oxidase in plants (e.g. 2 L-ascorbate + O_2 = 2 dehydroascorbate + $2H_2O$) and the cytochrome system in animals. It rapidly reduces many dyes and even neutral silver nitrate to metallic silver. Also oxidized by the usual oxidizing agents, e.g. iodine, and by oxygen in the presence of traces of catalytic metals. Its oxidation product can be reduced back readily by H_2S, or by glutathione in the presence of glutathione dehydrogenase

Ubiquinones

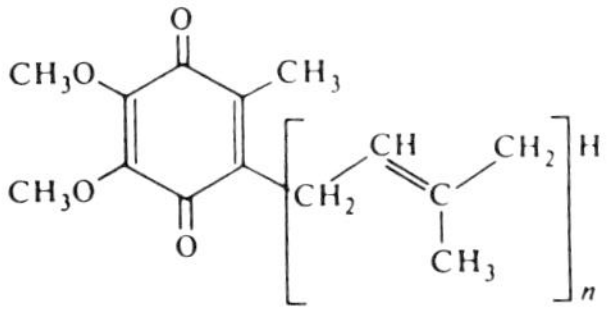

Hydrogen carrier. Forms an essential link between succinate and cytochrome c in the mitochondrial respiratory chain. Ubiquinone undergoes reduction to a hydroquinone form. The oxidation of the reduced form depends on cytochrome c and cytochrome oxidase, it is inhibited by antimycin A (unlike the reduction of ubiquinone) and coupled with phosphorylation. Antimycin contrasts with amytal, which inhibits the reduction of ubiquinone by $NADH_2$, though not by succinate. Ubiquinone has a band at 275 nm, while hydroquinone has a band at 288 nm

Table 7. Continued

Coenzymes	Properties and function
Cytochromes	Hydrogen carriers. The name includes all intracellular hemoproteins with the exception of hemoglobin, myoglobin, peroxidase and catalase. They have many different functions, all act by undergoing oxidation-reduction. The typical system of cytochromes (a, b and c) has a four-band visible absorption spectrum. For example, cytochrome c is a coenzyme of $NADH_2$ cytochrome c reductase: $NADH_2$ + 2 oxidized cytochrome c = NAD^+ + 2 reduced cytochrome c
2-Oxoglutarate $HOOC-CH_2-CH_2-CO-COOH$	Amino-group carrier. The aminated form of 2-oxoglutarate is L-glutamate. The great majority of the known transaminases react with this carrier pair, e.g. L-leucine + 2-oxoglutarate = 2-oxoisocaproate + L-glutamate, catalysed by leucine aminotransferase

Phosphate and glycosyl carriers. Some nucleoside phosphate 5′-diphosphates, particularly UDP, CDP and GDP, and glycosyls have specialized functions as carriers in an unusual type of glycosyl transfer, which involves phosphate transfer to complete the catalytic cycle, e.g. UDP acts as a carrier of both phosphate and glucose in the system for the biosynthesis of sucrose. Other examples include CDP in the synthesis of phosphatides and UDP in the interconversion of galactose and glucose

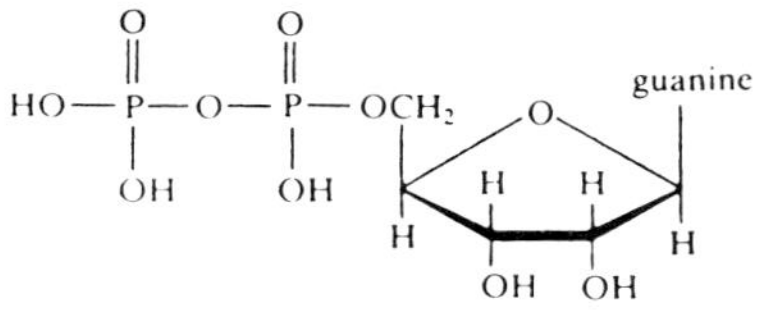

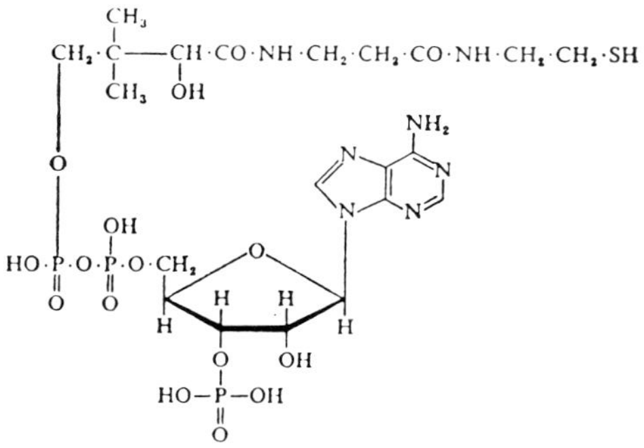

Acyl-group carrier. Coenzyme A is essential for the initiation of the citrate cycle, for the oxidative breakdown of fatty acids, and for various biosynthetic processes. For example, acetyl-CoA + L-aspartate = CoA + N-acetyl-L-aspartate, catalyzed by aspartate transferase. It is a colorless substance having an absorption band at 257 nm. Acyl groups can be added to the thiol group of coenzyme A either by transfer of the acyl group from another molecule or by a synthetase reaction. Acetyl-CoA is also involved in reactions of a different type involving lyases. Many of the enzymes involved in these reactions can use simple analogs of coenzyme A, e.g. N-acetyl-cysteamine; others, however, are specific for coenzyme A itself

Mixed phosphate glycosides

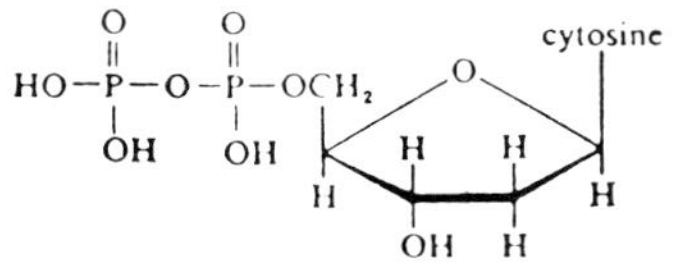

Coenzyme A

Table 7. Continued

Coenzymes	Properties and function
Tetrahydrofolate (THFA)	Carrier of one-carbon groups. THFA acts as a carrier of methyl, hydroxymethyl, formyl or formimino groups, e.g. L-serine + THFA = glycine + 5,10-methylene-THFA, in which the enzyme is serine hydroxymethyltransferase. It is a colorless compound, but it and its derivatives have characteristic absorption spectra in the ultraviolet. It is readily autoxidizable and is rapidly converted into dihydrofolate (DHFA) by O_2. It is also oxidized enzymatically by $NADP^+$. Folate is reduced to DHFA and THFA by $NADH_2$, and the liver enzyme THFA dehydrogenase (Dh); the bacterial enzyme DHFA Dh is specific for the reduction of DHFA to THFA
Adenosylhomocysteine	Carrier of one-carbon group. Acts as a methylhomocysteine carrier, linking pairs of methyltransferases in the transmethylation of methionine to other substances. Two enzymes are involved: methionine adenosyltransferase, catalyzing ATP + L-methionine + H_2O = orthophosphate + pyrophosphate + S-adenosylmethionine, and a specific methyltransferase, catalyzing the transfer of a methyl group from the sulfonium compound to other substances such as nicotinamide or guanidinoacetate, e.g. S-adenosylmethionine + a thiol = S-adenosylhomocysteine + a thioether, which is catalyzed by thiol methyltransferase

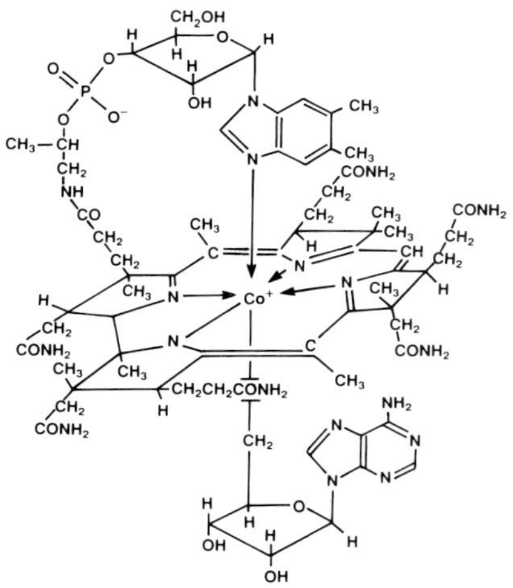

Carriers of one-carbon group. These are coenzymes related to vitamin B_{12}, the greater part of which is in the form of coenzyme B_{12} or closely related coenzymes. Coenzyme B_{12} acts as a coenzyme of isomerizations involving intramolecular transfer at C—C bonds. It acts as an essential cofactor for the enzymes methylaspartate mutase and methylmalonyl-CoA mutase in the respective reactions: L-*threo*-3-methylaspartate = L-glutamate and methylmalonyl-CoA = succinyl-CoA. Vitamin B_{12} can also act as a carrier of methyl groups: 5-methyl-THFA + B_{12} = THFA + methyl-B_{12}; methyl-B_{12} + homocysteine = B_{12} + methionine

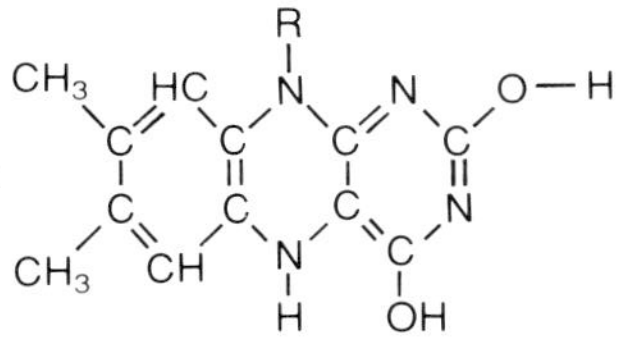

Hydrogen group carriers. These bind to many different enzymes to form flavoprotein enzymes. A few of these enzymes (e.g. lactate oxidase) have riboflavin 5′-phosphate (FMN) as the prosthetic group, but the majority (e.g. glucose oxidase, pyruvate dehydrogenase, etc.) contain flavin adenine dinucleotide (FAD). The oxidized forms are yellow and fluorescent, resulting in a band at 450 nm; they are readily reduced to the leuco-forms by chemical reducers, and on reduction lose their yellow color and fluorescence, owing to the disappearance of the absorption band at 450 nm

Cobamide

Flavin groups

Oxidized form

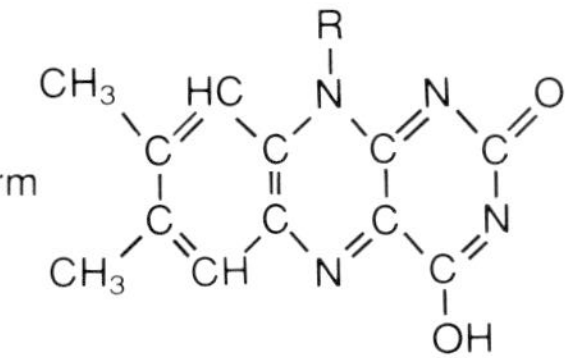

Reduced form

Table 7. Continued

Coenzymes	Properties and function
Pyridoxal phosphate	Prosthetic group. This forms the prosthetic phosphate group of a number of enzymes which catalyze a wide variety of reactions involving amino acids, e.g. lysine decarboxylase, threonine dehydratases, threonine synthase. In some of its transamination reactions it acts as an amino-carrier; in the other cases it forms a reactive compound with the amino acid. It also forms the prosthetic group of one enzyme of an entirely different type, i.e. α-glucan phosphorylase from muscle. Pyridoxal phosphate and pyridoxamine phosphate show absorption bands at 395 nm, and 330 nm, respectively
Thiamine pyrophosphate	Prosthetic group. It is involved in the decarboxylase system, pyruvate oxidase system, the 2-oxoglutarate oxidase system and in the formation of actoin. An intermediate α-hydroxyethyl-thiamine pyrophosphate is formed in these systems. The fate of this compound depends on which enzyme is involved. Transketolase has been shown to form an intermediate which is α,β-dihydroxyethothiamine pyrophosphate
Biotin	Prosthetic group. A cofactor of enzymes, e.g. methylmalonyl-CoA carboxytransferase, pyruvate carboxylase, acetyl-CoA carboxylase, etc. in enzymatic reactions involving the transfer and incorporation of CO_2

Although in theory this relationship is true provided the substrate concentration is fixed, regardless of the actual value, in practice it is more reliably valid when the substrate concentration is high enough to be essentially saturating. If a reaction involves more than one substrate, then the concentrations of each must be fixed, preferably at near-saturating levels. Another prerequisite for a successful assay system is that the reaction being catalyzed should be capable of being accurately monitored; that is, there should be a change in optical, electrical or other properties as substrate is converted to product (*Table 8*). If this is not the case, it may be possible to follow the course of the reaction indirectly by coupling it to another enzyme reaction where some such change does take place. Regardless of this, the reaction should be carried out under fixed and suitable conditions of pH, ionic strength and temperature.

Enzyme activity is measured as the amount of substrate lost (or product gained) per unit time, and it should also be related in some way to the amount of specimen used for assay. Two systems of enzyme unit are in current usage. The first is the International Unit (IU), which is defined as the amount of enzyme causing loss of 1 μmol substrate per minute under specified conditions. The second is the System International (SI) unit of enzyme activity, which is the katal (kat), defined as the amount of enzyme causing loss of 1 mmol substrate per second under specified conditions.

Obviously there must be a direct relationship between the number of units of activity found and the amount of sample assayed. Therefore, in order for the result of an assay to have any meaning, it must be expressed in terms of the amount of sample assayed, and, where appropriate, extrapolated back to the amount of the original tissue. However, even these units may not be ideal. For example, liver is likely to contain irregular deposits of glycogen and other nonprotein material, so any results expressed could vary according to which part of the same liver was cut off, weighed, homogenized and assayed. To minimize problems of this type, units of enzyme activity may be related to the total protein content of the sample being assayed, rather than to its weight or volume: this is termed specific activity, and may be expressed as, for example, IUs per mg protein, or as katals per kg protein. In the example being considered, a liver homogenate would be prepared and an aliquot of the same homogenate would be analyzed for its total protein content, and so the total amount of protein in the sample used for enzyme assay could be calculated and related to the observed activity.

Table 8. A summary of the main enzyme assay methods

Assays	Principles and comments	Examples
Spectrophotometry	Suitable when there is a difference between substrate and product in the absorbance light of a particular wavelength, according to the Beer–Lambert law. Often used with an absorbance less than 0.8. Can use either natural or synthetic chromogenic substrates	Lipoxidase (15), (deoxy)ribonucleases (16, 17), pepsin (18), trypsin (19), α-chymotrypsin(ogen) (20), lysozyme (21) and most dehydrogenases, kinases and phosphatases
Spectrofluorimetry	Used when there is a difference in fluorescence between substrate and product. More sensitive than spectrophotometry. Main problems: (i) fluorescence varies with temperature; (ii) quenching	Firefly luciferase (22)
Manometry	For reactions where there is gas uptake or output, and also those involving the production or consumption of acid or alkaline substances, if they are carried out in the presence of bicarbonate in equilibrium with a gas mixture containing a defined percentage of CO_2	Oxidase, decarboxylases
Electrode methods	For enzymes catalyzing redox reactions. Potential generated can either be measured by potentiometric techniques or be related to the concentration of a specific substance, using ion-selective electrodes. Advantages: contaminating substances can be present in suspension without affecting the analysis	Lipase (23), chymopapain (24), papain (25), carbonic anhydrase (26), butyrylcholinesterase (27)
Microcalorimetry	For reactions in which heat (enthalpy) is gained or lost. Advantages: sensitivity, freedom from interference and wide range of applications	Hexokinase

Table 8. Continued

Assays	Principles and comments	Examples
Radiochemical labeling	By using radioactively labeled substrate, the product concentration can be determined indirectly by measuring the radioactivity of the product fraction. Extremely sensitive. Disadvantages: (i) possible health hazard; (ii) multi-steps; (iii) difficult to monitor reactions continuously; (iv) quenching	Many types of enzyme, restricted only by the availability of radio-labeled compounds
Polarimetry	For enzymes involved in conversion between optically inactive and active isomers of a substance. Not very convenient	Lactate dehydrogenase
Chromatographic methods	Chromatography can be used to detect the formation of a product when other methods fail. Disadvantage: time consuming	Chloramphenicol acetyl transferase (CAT)

Enzyme assays based on the determination of catalytic activity can distinguish between active and inactive forms of enzymes, for they do not detect the presence of the latter. On the other hand, they cannot distinguish between isoenzymes, so the activity measured will be the sum of the contributions of all the active forms of the enzyme being assayed. Since these will not necessarily have the same molar activity, the relationship between activity and enzyme concentration could be less straightforward than suggested above. Even if it is known that only one isoenzyme is present, it cannot be assumed that its molar activity in, say, a homogenate is the same as that obtained in a pure preparation. Hence kinetic assays are usually used to give an indication of the relative concentration of an enzyme in a preparation, without any attempt being made to interpret the results in terms of actual molar concentrations: the results are usually expressed simply in terms of units of activity.

7. HANDLING AND STORAGE OF ENZYMES AND COENZYMES

Enzymes easily become denatured and lose catalytic activity, so careful storage and handling is essential if this is to be prevented. In general, temperatures higher than 40°C and extremes of pH (i.e. below pH 5 and above pH 9) should be avoided at all times. If the pH of an enzyme solution is being changed, continuous mixing helps to prevent denaturation taking place around the added drops of acid or alkali. However, too vigorous mixing causing the formation of froth is counterproductive, since many enzymes are denatured at surfaces.

At room temperature, most enzymes are denatured by organic solvents such as ethanol or acetone (see *Section 3.3* of *Chapter 1*), except where these are present at very low concentrations. As with changes of pH, changes in organic solvent concentration should be performed with extreme care to avoid localized denaturation of enzymes.

Most dry enzymes are quite stable at 0–4°C but care is necessary to avoid condensation on the container when opening it. This can be done by allowing it to warm up before opening it; some enzymes need to be stored at lower temperatures. Enzyme solutions should also be stored at low temperatures and should not be too dilute. Freezing and thawing may cause loss of activity, this depends on the nature of the preparation. Storage at $-20°C$ in the presence of 30–50% glycerol often helps to retain enzyme activity. Enzymes precipitated with ammonium sulfate solutions (see *Section 8* of *Chapter 1*) are best stored as a precipitate at 0–4°C, as are lyophilized enzyme preparations; however, the latter may be stored frozen if it is certain that they contain no moisture. In contrast, solutions of enzymes should always be stored frozen at temperatures as low as possible, but repeated freezing and thawing should be avoided to avoid loss of activity. For this reason, such solutions are best divided into small portions for storage. Enzyme preparations are sometimes stored in aqueous glycerol, since this enables them to be kept at temperatures below 0°C without freezing.

Most coenzymes must be stored as solids at 0–4°C, preferably in a desiccator, since many are subject to hydrolysis. Some coenzymes must also be protected from light or oxygen. NADH, for example, is destroyed by light to quite an appreciable extent, while also reacting with any atmospheric water to form inhibitors of dehydrogenases. In solution, NADH is particularly unstable in acid conditions, whereas NAD^+ is alkali-labile. The sulfhydryl group of coenzyme A can be oxidized by atmospheric oxygen, and its pyrophosphate bond is easily hydrolyzed if moisture is present. Some of the more stable coenzymes may be stored in frozen solution, but, as with enzyme solutions, repeated freezing and thawing should be avoided.

8. REFERENCES

1. Elliott, D.C., Elliot, W.H. and Jones, K.M. eds (1986) *Data for Biochemical Research (3rd edn)*. Oxford Univeristy Press, Oxford.

2. Wieland, T. and Faulstich, H. (1978) *Crit. Rev. Biochem.*, **5**, 185.

3. Nicolini, C. (1976) *Biochim. Biophys. Acta*, **458**, 264.

4. Vignais, P.V. (1976) *Biochim. Biophys. Acta*, **456**, 1.

5. Penefsky, H.S. (1979) *Adv. Enzymol.*, **49**, 223.

6. Lauquin, G.J.M., Duplaa, A., Klein, G., Rousseau, A. and Vignais, P.V. (1976) *Biochemistry*, **15**, 2323.

7. Omura, S. (1976) *Bacteriol. Rev.*, **40**, 681.

8. Monro, R.E. and Vasquez, D. (1967) *J. Mol. Biol.*, **28**, 161.

9. Zimmer, C. (1975) *Prog. Nucleic Acid Res. Mol. Biol.*, **15**, 285.

10. Okura, A., Marishima, H., Takita, T., Aoyagi, T., Takeuchi, T. and Umegzawa, H. (1973) *J. Antibiotics* **28**, 337.

11. Lardy, H., Reed, P. and Lin, C.C. (1975) *Fed. Proc.*, **34**, 1707.

12. Teschke, R., Matsuzaki, S., Ohnishi, K., Decarli, L.M. and Lieber, C.S. (1977) *Clin. Exper. Res.*, **1**, 7.

13. Singer, T.P. and Gutman, M. (1971) *Adv. Enzymol.*, **4**, 79.

14. Butcher, R.W., and Sutherland, E.W. (1962) *J. Biol. Chem.*, **327**, 1244.

15. Tappel, A.L. (1962) *Methods Enzymol.*, **V**, 539.

16. Kunitz, M. (1950) *J. Gen. Physiol.*, **33**, 349.

17. Zimmerma, S.B. and Sandeen, G. (1965) *Anal. Biochem.*, **10**, 444.

18. Jackson, W.T., Schlamowitz, M. and Shaw, A. (1965) *Biochemistry*, **4**, 1537.

19. Hummel, B.C.W. (1959) *Can. J. Biochem. Physiol.*, **37**, 1393.

20. Shuger, D. (1952) *Biochim. Biophys. Acta* **8**, 302.

21. Strehler, B.L. and Totter, J.K. (1954) *Methods of Biochemical Analysis* (D. Glick ed.). Interscience Publishers, New York, Vol. 1, p. 341.

22. Mattson, F.H. and Latton, E.S. (1958) *J. Biol Chem.*, **219**, 735.

23. Cayle, T., Saletan, L.T. and Lopez-Ramos, B. (1964) *Proc. Ann. Meeting, Am. Soc. Brewing Chem.*, p. 142.

24. Smith, E.L. and Kimmel, J.R. (1960) in *The Enzymes* (P.D. Boyer, H. Lardy and K. Myrback eds). Academic Press, New York, Vol. 4, p. 133.

25. Wilbur, K.M. and Anderson, N.G. (1948) *J. Biol. Chem.*, **176**, 147.

26. Augustinsson, K.B. (1960) in *The Enzymes* (P.D. Boyer, H. Lardy and K. Myrback eds). Academic Press, New York, Vol. 4, p. 521.

27. Kramer, D.N., Cannon, P.L. and Guilbault, G.G. (1962) *Anal. Biochem.*, **34**, 842.

9. FURTHER READING

1. Cornish-Bowden, A. (1988) *Enzyme Kinetics*. IRL Press, Oxford.

2. Cornish-Bowden, A. (1976) *Principles of Enzyme Kinetics*. Butterworth, London.

3. Dixon, M (1979) *Enzymes (3rd edn)*. Longman Group, London.

4. Fersht, A. (1985) *Enzyme Structure and Mechanism (2nd edn)*. W.H. Freeman, New York.

5. Palmer, T. (1985) *Understanding Enzymes (2nd edn)*. Ellis Horwood, New York.

6. Price, N.C. (1982) *Fundamentals of Enzymology*. Oxford University Press, Oxford.

7. Segel, I.H. (1975) *Enzyme Kinetics*. John, Wiley and Sons, USA.

CHAPTER 7
HYDROLYTIC ENZYMES
J. A. A. Chambers

1. INTRODUCTION

The following tables cover hydrolases of greater or lesser specificity for the hydrolysis of macromolecules for their removal as contaminants and the hydrolysis of conjugates (e.g. glycoproteins).

Three major groups of hydrolases are covered: proteases, nucleases and carbohydrases. For nucleases and proteases specific or effective inhibitors are also listed. Coverage of the proteases is extensive, but that of nucleases is limited to a few enzymes that are generally useful (*Table 1*). More data on nucleic acid metabolizing enzymes are found in *Molecular Biology Labfax*. The carbohydrases covered are limited to those useful in the removal of interfering carbohydrates, and some useful as analytical reagents.

2. INHIBITION OF NUCLEASES AND PROTEASES

2.1. Deoxyribonucleases
These are not generally known for their robust nature and can be inhibited by chelating agents if metal-dependent, by moving the pH or ionic strength outside the working range, by low concentrations of denaturants or by gentle heating. Many nucleic acid binding antibiotics (e.g. ethidium bromide, daunomycin) also inhibit DNases (*Table 2*).

2.2. Ribonucleases
Ribonucleases are renowned for their robustness and can be very difficult to inhibit. Two commercially available inhibitors, vanadyl ribonucleosides (transition state analogs) and rat placental ribonuclease inhibitor, are generally effective, but because they do not covalently modify the protein they are of little use in denaturing systems. If the reactive group at the active site is known, then it may be inhibited by modification with a group-specific reagent, e.g. iodoacetamide for thiol nucleases. Metal-dependent enzymes may be inhibited by chelating agents. The highly reactive alkylating agent diethyl pyrocarbonate is also widely used, but its great reactivity means that it is incompatible with some buffers and its reactivity towards nucleic acids is frequently underestimated (12–16).

A number of chemical inhibitors of ribonucleases are known (*Table 2*), many of these bind to the nucleic acids and prevent the enzyme from acting simply by steric effects, or they are active against a number of other nucleic acid metabolizing enzymes; the concentrations used must be carefully titrated.

The best approach to inhibiting ribonucleases during the isolation of RNA is the use of high concentrations of powerful denaturing agents, e.g. SDS, guanidinium salts, or 6 M urea, 3 M LiCl. Some ribonucleases will remain active in the presence of RNA even under strongly

Table 1. Properties of selected nucleases

Name	Source	EC Number [RN]	Specificity	pH optimum	Inhibitors	Refs
Deoxyribonuclease I	Pancreas	3.1.4.5 [9003-98-9]	DNA	5.0	Oxine 2-Nitro-5-thio-cyanobenzoate Actin Aurintricarboxylic acid Daunomycin Adriamycin EDTA	1 2 3 4 5 5
Endodeoxyribonuclease		3.1.4.30 [37211-67-9]	DNA			
Micrococcal nuclease	*Staphylococcus aureus*	3.1.4.7 [9013-53-0]	DNA or RNA	8.8	EGTA Nucleoside Phosphonic acids	6
Alkaline ribonuclease	Swine pancreas	2.7.7.16 [9001-99-4]	RNA	8.0	Vanadyl ribonucleosides Aurintricarboxylic acid RNAsin	7 8 9
Acid ribonucleases						
T2	*Aspergillus oryzae*	3.1.27.1	RNA (3' of A)	4.5		10
M	*Aspergillus saitoi*	3.1.4.23 [37278-25-4]	RNA (3' of A)	4.5		11

Abbreviations: EDTA, ethylenediaminetetraacetic acid; EGTA, ethylene glycol-bis(β-aminoethyl ether) N,N,N',N'-tetraacetic acid.

denaturing conditions and rapid deproteinization is essential. Conversely, when getting rid of RNA in order to examine DNA or proteins, the use of denaturing conditions to inhibit other hydrolases is very useful.

2.3. Proteases

Proteases may have a broad specificity (*Table 3*), act as specific endoproteases (*Table 4*) or act on the N-terminal (*Table 5*) or C-terminal (*Table 6*) ends of proteins.

Proteases are generally less robust than ribonucleases and must be treated with respect. As with the other enzymes, metalloproteases may be inhibited with chelating agents. Serine proteases are frequently inhibited by serine modifying reagents, such as phenylmethylsulfonyl fluoride (PMSF), and thiol proteases are inactivated by thiol-specific reagents, such as *p*-chloromercuribenzoate (PCMB). In addition, a number of specific inhibitors are known (*Table 7*). As with ribonucleases, some proteases are active under denaturing conditions, although many are inactivated by powerful denaturants. When specific inhibitors are used during protein purification be prepared to add fresh inhibitor to the sample at every stage.

2.4. Carbohydrases

Table 8 lists carbohydrases that are frequently encountered. These enzymes can be inhibited both by protein-modifying agents and specific inhibitors.

3. NATURALLY OCCURRING OR PHYSIOLOGICAL INHIBITORS

Many of these enzymes have a cognate physiological inhibitor that regulates activity *in vivo*. Some, such as the soy bean trypsin inhibitors and placental nuclease inhibitor (RNAsin or RNAGuard) are well known, effective and commercially available. There are many of these but only a few are mentioned (*Tables 2* and *7*) because of lack of availability or characterization.

Table 2. Selected nuclease inhibitors

Name	Formula	Mol. wt	Solubility	Inhibits	Refs
Dyes					
Actinomycin D [50-76-0]	$C_{62}H_{86}N_{12}O_{16}$	1255.5	Sol. in H_2O, EtOH glycols	DNase I also inhibits RNA formation *in vivo*	3
Adriamycin [25316-40-9]	$C_{27}H_{29}NO_{11}$	543.5	Sol. in H_2O, MeOH, alcs	DNase I	
Daunomycin [20830-81-3]	$C_{27}H_{29}NO_{10}$	527.5	Sol. in H_2O, MeOH	DNase I	
Proteins					
Actin	—	—	Sol. in H_2O	DNase I	
Placental RNase inhibitor	—	—	Sol. in H_2O	Ribonucleases	
Alkylating agents					
Diethyl pyrocarbonate [1609-47-8]	$C_6H_{10}O_5$	162.1	Immisc. in H_2O	Alkylating agents nonspecifically modify any proteins in the sample	
Iodoacetamide [144-48-9]	C_2H_4INO	233	Sol. in H_2O		
Iodoacetic acid [64-69-7]	$C_2H_3IO_2$	186	Sol. in H_2O		

Table 2. Continued

Name	Formula	Mol. wt	Solubility	Inhibits	Refs
Transition state and base analogs					
Aurintricarboxylic acid [4431-00-9]	$C_{22}H_{14}O_9$	473.4	Sol. in H_2O	DNase I RNase I S1 nuclease Exonuclease III	8
Nucleoside phosphonic acids				Micrococcal nuclease	6
Vanadyl ribonucleosides				Translation ribonucleases A, N1, T1, U2 Micrococcal nuclease Restriction enzymes Polynucleotide phosphorylase	7

Abbreviations: alcs, alcohols; immisc., immiscible; MeOH, methanol; sol., soluble.

Table 3. Broad specificity or nonspecific proteases

Name	Source	EC number [RN]	Optimum pH	Inhibitors	Refs
Alkaline protease	*Streptomyces griseus*	No EC [9075-80-3]		Ovomucoid	17
Dispase	*Bacillus polymyxa*	3.4.24.4 [42613-33-2]	8.5	None	18
Pepsin[a]	Pig stomach	3.4.23.1 [9001-75-6]	1.0	Pepstatin	19
Pronase	*Streptomyces griseus*	No EC [9036-06-0]	7	α1-Antitrypsin Broad bean trypsin inhibitor Trypsin inhibitors	20 21
Protease K	*Tritirachium album*	3.4.21.14 [39450-01-6]	8.0	Chloromethyl ketones Alkyloyl carboxylates Ovomucoid Broad bean trypsin inhibitor	22 23 23 25, 26, 27
Subtilisin	*Bacillus subtilis*	3.4.21.14 [9014-01-1]	8.0	TK-23 SSI	28 29

[a] Pepsin shows some preference for Phe–X, Met–X, Leu–X, Trp–X, where X is hydrophobic.

Table 4. Endoproteases

Name	Source	EC number [RN]	Optimum pH	Specificity	Inhibitors	Refs
Ancrod	Agkistrodon	3.4.21.28 [9046-56-4]	7.5	Arg–X esp. Arg–Gly	Gabexate mesilate Guanidino cinnamates Guanidinophenyl propionates	31 30
Bromelain	Pineapple	3.4.22.3 [9001-00-7]	8.2	Lys–X, Ala–X Tyr–X	Estatin A, B α1-Antitrypsin	32 20
α-Chymotrypsin	Bovine pancreas	3.4.21.1 [9004-07-3]	8.0	Tyr–X, Phe–X, Trp–X, less active against Leu–X, Met–X, Ala–X	PMSF, TPCK Aprotinin	23
Clostripain	*Clostridium histolyticum*	3.4.21.1 [9028-00-6]	7.2	Arg–X	Trypsin inhibitor (weak) α2-Macroglobulin Leupeptins Antipain	19 33 34
Collagenase	*Achromobacter iophagus*	3.4.24.8 [9001-12-1]	7.2	X–Gly of Pro–X–Gly–Pro		35
	Clostridium histolyticum	3.4.24.3 [9001-12-1]	7.0			35
	Leech			Degrades collagen to peptides		36

Table 4. Continued

Name	Source	EC number [RN]	Optimum pH	Specificity	Inhibitors	Refs
Factor Xa	Bovine plasmin	3.4.21.6 [9002-05-5]	8.3	Arg–X of Gly–Arg–X	α2-Plasmin OM-189 Antithrombin III Benzamidine derivs	37 38 39 40
Ficin	*Ficus* (fig tree)	3.4.22.3 [9001-33-6]	6.4	Bonds involving uncharged or arom. amino acids	Cystatin, see also Papain	41 19
Kallikrein	Pig pancreas	3.4.21.8 [9001-01-8]	7.5	Arg–X, pref. of Phe–Arg–X, Leu–Arg–X	Aprotinin	42
Papain	*Carica papaya*	3.4.22.2 [9001-73-4]	6.5	Bonds involving Arg, His, Glu, Lys, Gly, Tyr	Leupeptin Antipain E-64 Chymostatin Cystatin	43 44 45 43 46
Thermolysin	*Bacillus thermoproteolyticus*	3.4.24.4 [9073-78-3]	8.0	X–Leu, X–Phe & other nonpolar residues	Phosphoramidon MK-1	48 49 19
Thrombin	Bovine plasma	3.4.21.5 [9002-04-4]	8.6	Arg–X	Antithrombin	50 19

Table 4. Continued

Name	Source	EC number [RN]	Optimum pH	Specificity	Inhibitors	Refs
Trypsin	Bovine pancreas	3.4.21.4 [9002-07-7]	8.0	Arg–X	Antipain Aprotinin Leupeptins PMSF, TLCK	19
V8 protease	*Staphylococcus aureus*	3.4.21.9 [66676-43-5]	4.0, 7.8	Glu–X, less active against Asp–X	3,4-Dichloro-isocoumarin Isocoumarin derivs	51

Abbreviations: arom., aromatic; derivs, derivatives; esp., especially; PMSF, phenylmethylsulfonyl fluoride; pref., preferably; TPCK, N-tosyl-phenylalanine chloromethyl ketone; TLCK, Nα-p-tosyl-L-lysine chloromethyl ketone.

Table 5. Aminopeptidases (N-terminal specific enzymes)

Name	Source	EC number [RN]	Optimum pH	Specificity	Inhibitors	Refs
Aminopeptidase M	Pig kidney	3.4.11.2 [9054-63-1]	7.2	Cleaves N-terminal residue from di- and tripeptides	Amastatin	51
Cathepsin B		3.4.22.1 [9047-22-7]		Blocked by Lys, Arg or Pro	EACA Leupeptin Cystatin Antipain	52 43 53 46
Cathepsin C	Bovine spleen	3.4.14.1 [9047-22-7]	5.0			
Cathepsin H		3.4.22.16			Cystatin	46
Leucine aminopeptidase	Pig kidney	3.4.11.1 [9001-61-0]	9.2	N-terminal residue	Amastatin Bestatin Chloromethyl ketones t-Butyl-Thr–Phe–Pro	54 55 56
Pyroglutamate aminopeptidase	Calf liver	3.4.11.8 [9075-21-2]	8.0	N-terminal pyroglutamate	5-Oxoprolinal pyroglutamyl diazomethyl ketone	57 58

Abbreviations: EACA, ε-amino caproic acid (6-aminohexanoic acid).

Table 6. Carboxypeptidases (carboxy-terminal specific proteases)

Name	Source	EC number [RN]	Optimum pH	Specificity	Inhibitors	Refs
Carboxypeptidase A	Bovine pancreas	3.4.2.1; 3.4.12.2; 3.4.17.1 [11075-17-5]	8.0	Preference for arom. and aliph.	Potato CPI	59
					Ascaris CPI	60
					Rhein	61
					Phenylalanine phosphonates	62
					Arphamenines	63
					Benzyl malate	64
					FMPI	
					Talopeptin	65
					PLT	
Carboxypeptidase B	Pig pancreas	3.4.2.2; 3.4.12.3; 3.4.17.2 [9025-24-5]	8.0	Arg, Lys, Orn	Arphamenines	63
					Histargin	61
Carboxypeptidase P	*Penicillium janthinellum*	3.4.12.4 [9075-64-3]	3.7	Any C-term. residue		
Carboxypeptidase Y	*Saccharomyces cerevisiae*	3.4.12.1; 3.4.16.1 [9046-67-7]	6.0	Preference for arom. and aliph.	Yeast CPI	67

Abbreviations: arom., aromatic residues; aliph., aliphatic residues; CPI, carboxypeptidase inhibitor; C-term, C-terminal; FMPI, N_2-(*N*-phosphono-L-phenyalanyl)-L-arginine).

Table 7. Inhibitors of proteases

Inhibitor	Mol. wt[a]	Active against	Refs
Amastatin [67655-94-1]	474.3	Aminopeptidases A, M Leucine aminopeptidase	51
ε-Amino caproic acid [60-32-2]	131.1	Cathepsin B, C Plasmin	52
Anticin [79633-33-3]		Ficin Papain Bromelain	68
Antipain [37691-11-5]	604.3	Papain Trypsin	
Antithrombin III [9000-94-6]		Thrombin	49 19
α1-Antitrypsin Broad bean trypsin inhibitor		Pronase Protease K Bromelain	20 21
APMSF [71933-13-6]	216.1	Serine proteases	19
Aprotinin [9087-70-1]		Trypsin Kallikrein	19
Arphamenines A [85713-14-0] B [88465-81-0]	320.2 336.2	Aminopeptidase B Carboxypeptidase A Carboxypeptidase B	64
Benzamidine [618-39-3] and derivs	120.1	Plasmin Thrombin Factor Xa Trypsin Angiotensin II-forming serine protease	40
Benzyl malic acid [51692-93-4]	224.9	Carboxypeptidase A	64
Bestatin [58970-76-6]	308.2	Leucine aminopeptidase	55
Bowman–Birk soya bean trypsin inhibitor [37330-34-0]		Trypsins	
t-Butylthreonyl-phenylalanyl proline [38510-47-3]	419.3	Leucine aminopeptidase	56

Table 7. Continued

Inhibitor	Mol. wt[a]	Active against	Refs
Calpastatin [79079-11-1]		Calpain	69
N-α-Carbobenzoxy pyroglutamyl diazoketone [81910-23-8]	287.1	Pyroglutamate aminopeptidase	58
Chymostatin [9076-44-2]		Chymotrypsin Papain	35
Cystatin [81989-95-9]		Thiol proteases	46
3,4-Dichlorocoumarin [5117-56-6]	215.0	Serine proteases Acetylcholinesterase Kallikrein Factor Xa Trypsin Subtilisin Factor XIa Mast cell neutral protease Plasmin Factor XIIa Complement factor D Thrombin	50
E64 [66701-25-5]	357.2	Papain Cathepsin B	45
Eglin [64886-91-5]		Chymotrypsin Elastase Cathepsin G	70
Elafin [133249-66-8]		Elastase Proteinase 3	71
Estatin A [106455-06-5]	391.2	Bromelain	32
Estatin B [106396-24-1]	407.2	Bromelain	32
FMPI [92116-83-1]	401.2	Carboxypeptidase A *Aspergillus oryzae* metalloprotease Angiotensin-converting enzyme	65
Gabexate mesilate [56974-61-9]	417.3	Ancrod Thrombin Trypsin	30

HYDROLYTIC ENZYMES

Table 7. Continued

Inhibitor	Mol. wt[a]	Active against	Refs
Heparin [9005-49-6]		Factor Xa	
Histargin [93361-66-1]	355.2	Carboxypeptidase B	66
p-Hydroxymercuribenzoate [138-85-2]	360.6	Thiol proteases	
Iodoacetic acid [64-69-7]	185.9	Thiol proteases	
Leupeptin [545-77-99-0] & leupeptins		Cathepsin B, C Clostripain	43
α2-Macroglobulin		Trypsin	
2-Mercaptomethyl-3-guanidino ethyl thio propanoate [77102-28-4]	237.2	Carboxypeptidase N Angiotensin-converting enzyme	72
MK-I [73804-40-7]	543.3	Thermolysin Elastase	48
OM-189 [60182-37-8]	488.3	Plasmin Thrombin Factor Xa Trypsin *Bothrops atrox* serine Proteinase	38
Ovoinhibitor [62449-23-4]			
Ovomucoid		*Streptomyces griseus* alk. protease Trypsin Elastase	17
PCMB [59-85-8]	357.1	Thiol proteases Cathepsins A, B, C, D	73
Pepstatin [39324-30-6]		Pepsin Acid proteases Aspartic proteases Aspergillopepsin F Yeast protease A Cathepsin D	52
Phenylalanine phosphonates		Carboxypeptidase A	62

Table 7. Continued

Inhibitor	Mol. wt[a]	Active against	Refs
[80556-06-5]	392.1		
[80556-07-6]	391.1		
[80556-08-7]	408.2		
Phosphoramidon [36357-77-4]	543.2	Thermolysin Carboxypeptidase A	47
α2-Plasmin inhibitor		Factor Xa Factor XII Thrombin Factor XIa	37
Plasminogen activator inhibitor [105844-41-5]		Serine proteases Plasminogen activators	74
PMSF [329-98-6]	174.1	Serine proteases Chymotrypsin Trypsin	
Rhein [478-43-3]	284.1	Pepsin A Trypsin Elastase Carboxypeptidase A	61
Serpin [96282-35-8]		Serine proteases	
SSI [120433-50-3]		Subtilisin	29
Stefin [88844-95-5]		Cysteine proteases	75
Stefin A [107544-29-6]		Cysteine proteases	76
Stefin B [106178-19-2]		Cysteine proteases	76
Talopeptin [84235-60-9]	543.2	Thermolysin Metalloproteases	77
Thiostatin [114602-40-3]		Cysteine proteases	
TIMP [86102-31-0]		Elastase Collagenase Gelatinase	78
TK-23		Subtilisin	28
TLCK [2364-87-6]		Trypsin	79

Table 7. Continued

Inhibitor	Mol. wt[a]	Active against	Refs
TPCK [402-71-1]		Chymotrypsin	80, 81
Trypsin inhibitor[b] [9035-81-8]		Trypsin Clostripain Plasmin	

[a] Calculated from molecular formula and ignoring water of crystallization. When no molecular weight is given it is because the inhibitor is macromolecular and it is usually measured by activity (units/mg) rather than by weight.

[b] A number of naturally occurring trypsin inhibitors, including basic pancreatic trypsin inhibitor, are meant here.

Abbreviations: APMSF, amidino-phenylmethylsulfonyl fluoride (benzenemethane sulfonyl fluoride, 4-amidino); derivs, derivatives; FMPI, N_2-(N-phosphono-L-phenylalanyl)-L-arginine; Gabexate mesilate; benzoic acid, 4-[[6-[(aminoiminomethyl)amino]-1-oxohexyl]oxy]-, ethyl ester, monomethane sulfonate; PCMB, p-chloromercuribenzoate (mercurate(1-),(4-carboxylatophenyl)chloro-, hydrogen); PMSF, phenyl methylsulfonyl fluoride (benzenemethane sulfonyl fluoride); SSI, *Streptomyces* subtilisin inhibitor; TIMP, tissue inhibitor of metalloproteases; TLCK, $N\alpha$-p-Tosyl-L-lysine chloromethyl ketone (benzenesulfonamide, N-[5-amino-1-(chloroacetyl)pentyl]-4-methyl, (S)-); TPCK, N-Tosyl-phenylalanine chloromethyl ketone (benzenesulfonamide, N-[3-chloro-2-oxo-1-(phenylmethyl)propyl]-4-methyl, (S)-).

Table 8. Carbohydrate-degrading enzymes

Enzyme	Source	EC number [RN]	Optimum pH	Type of enzyme	Specificity	Refs
β-N-Acetyl-D-glucosaminidase	Beef kidney	3.2.1.30 [9012-33-3]	4.5	Exo	Nonspecific	
α-Amylase	Bacillus subtilis	3.2.1.1 [9000-90-2]	5.7	Endo	α-1,4-Glucosides	82
	Swine pancreas		7.0			
β-Amylase	Various plants	3.2.1.2 [9000-91-3]	4.8	Endo	β-1,4-Glucosides	
Amyloglucosidase	Aspergillus niger	3.2.1.3 [9032-08-8]	4.7	Exo	α-1,4 and α-1,6 bonds at nonreducing termini	
Cellulase	Various fungi	3.2.1.4 [9012-54-8]	4.8	Endo	β-1,4 Bonds in cellulose	83
Chitinase	Streptomyces antibioticus	3.2.1.14 [9001-06-3]		Endo	α-1,4 Bonds involving acetamido-2-deoxy-D-glucose	
Endo-α-N-acetyl galactosaminidase	Diplococcus pneumoniae	3.2.1.97 [59793-96-3]	—	Endo	Gal-β-1,3-GalNac links in glycoproteins	84
Endo-β-galactosidase	Bacteroides fragilis	3.2.1.103 [52720-51-1]	5.8	Endo	β-1,4-Galactosidic links in GlcNAc–Gal–GlcNAc	85, 86
Endoglycosidase D	Diplococcus pneumoniae	3.2.1.96	6.5	Endo	High mannose glycans	87, 88
Endoglycosidase F	Flavobacterium meningosepticum	[37278-88-9]	5.7	Endo	High mannose glycans	89, 90

Hydrolases

Table 8. Continued

Enzyme	Source	EC number [RN]	Optimum pH	Type of enzyme	Specificity	Refs
Endoglycosidase H	*Streptomyces*	3.2.1.96 [37278-88-9]	5.5	Endo	High mannose glycans	91, 92
Glusulase	*Helix pomatia*		5.8		Multienzyme prepn for protoplasting, will digest most polysaccharides	
Hyaluronidase	Mammalian testes	3.2.1.35 [37326-33-3]	5.2	Endo	1,4 links between N-acetyl hexosamine and D-glucuronate in hyaluronic acid, chondroitin, etc.	
Hyaluronidase	Leech	3.2.1.35 [37326-33-3]			α-Glucuronidase specific for hyaluronic acid	93
Laminarinase	*Spisula solidissima*	3.2.1.6 [62213-14-3]	—	Endo	β-1,3 and β-1,4 links adjacent to α-1,3 links	
Lysozyme (muraminidase)	Chicken egg white	3.2.1.17 [9001-63-2]	6.5	Endo	β-1,4 Links between GlcNAc and N-acetyl-muramic acid in glycoproteins	94, 95, 97
Neuraminidase	*Arthrobacter ureafaciens* *Clostridium perfringens*	3.2.1.18 [9001-67-6]	5.2	Exo	NeuAc-α-2,6-Gal, NeuAc-α-2,6-GalNAc, or NeuAc-α-2,3-Gal	
Novozyme	*Trichoderma harzianum*	No EC [134214-76-9]			Multienzyme prepn for protoplasting yeast cells	

Table 8. Continued

Enzyme	Source	EC number [RN]	Optimum pH	Type of enzyme	Specificity	Refs
Novozyme 234		No EC [65187-57-7]			Multienzyme prepn for protoplasting fungal mycelium	
Novozyme	*Aspergillus aculeatus*				Multienzyme prepn for protoplasting plant cells	
Peptide N glycosidase F	*Flavobacterium meningosepticum*	[83534-39-8]	7.5	Endo	Glycoproteins between Asn and GlcNAc; activity is a contaminant of endoglycosidase F (EC 3.2.2.18)	96
Pullulanase	*Aerobacter aerogenes*	3.2.1.41 [9075-68-7]	5.0	Endo	α-1,6-Glucans (*pullulans*)	

Abbreviations: endo, endocellular; exo, exocellular; prepn, preparation.

4. REFERENCES

1. Liu, J.-K. and Su, I.-J. (1972) *J. Chinese Biochem. Soc.*, **1**, 89.
2. Liao, T.-H. and McKenzie, L.J. (1979) *J. Biol. Chem.*, **254**, 9598.
3. Lazarides, E. and Lindberg, U. (1974) *Proc. Natl. Acad. Sci. USA*, **71**, 4742.
4. Hallick, R.B., et al. (1977) *Nucleic Acid Res.*, **4**, 3055.
5. Facchinetti, T., et al. (1978) *Chem.-Biol. Interact.*, **20**, 97.
6. Rammler, D.H., Bagdasarian, A. and Morris, F. (1972) *Biochemistry*, **11**, 9.
7. Puskas, R.S., et al. (1982) *Biochemistry*, **21**, 4602.
8. Schulz-Harder, B. and Tata, J.R. (1982) *Biochem. Biophys. Res. Commun.*, **104**, 903.
9. Eichler, D.C., Tatar, T.F. and Lanater, L.S. (1981) *Biochem. Biophys. Res. Commun.*, **101**, 396.
10. Ohgi, K. and Irie, M. (1975) *J. Biochem.*, **77**, 1085.
11. Uchida, T. and Egami, F. (1971) in *Enzymes (3rd edn)* (P.D. Boyer ed.). Academic Press, New York, Vol. 4, p. 205.
12. Ehrenberg, L., Fedorcsak, I. and Solymosy, F. (1976) *Prog. Nucleic Acid Res. Mol. Biol.*, **16**, 189.
13. Furlong, J.C. and Lilley, D.M.J. (1986) *Nucleic Acids Res.*, **14**, 3995.
14. Scholten, P.M. and Nordheim, A. (1986) *Nucleic Acids Res.*, **14**, 3981.
15. Herr, W. (1985) *Proc. Natl. Acad. Sci. USA*, **82**, 8009.
16. Peattie, D.A. and Gilbert, W. (1980) *Proc. Natl. Acad. Sci. USA*, **77**, 4679.
17. Nagata, K. and Yoshida, N. (1983) *J. Biochem.*, **93**, 909.
18. Nakanishi, T. and Yamamoto, T. (1974) *Agric. Biol. Chem.*, **38**, 2391.
19. Carrey, E.A. (1989) in *Protein Structure: A Practical Approach* (T.E. Creighton ed.). IRL Press, Oxford, p. 117.
20. Blatrix, C.C., et al. (1971) *Bibl. Haematol.*, **38**, 504.
21. Warsy, A., Norton, G. and Stein, M. (1974) *Phytochem.*, **13**, 2481.
22. Betzel, C., et al. (1986) *FEBS Lett.*, **197**, 105.
23. Ardelt, W. and Laskowski, Jr., M. (1985) *Biochemistry*, **24**, 5313.
24. Morihara, K. and Tsuzuki, H. (1975) *Agric. Biol. Chem.*, **39**, 1489.
25. Ebeling, W., et al. (1974) *Eur. J. Biochem.*, **47**, 91.
26. Jani, K.D. and Mayer, B. (1985) *Biol. Chem. Hoppe-Seyler*, **366**, 485.
27. Weigers, U. and Hilz, H. (1971) *Biochem. Biophys. Res. Commun.*, **44**, 513.
28. Oda, K., Koyama, T. and Murao, S. (1980) *Biochim. Biophys. Acta*, **571**, 5313.
29. Kojima, S., Kumagai, I. and Miura, K. (1990) *Gendai Kagaku Zokan*, **19**, 74.
30. Okutome, T., et al. (1984) *Chem. Pharm. Bull.*, **32**, 1854.
31. Menegatti, E., et al. (1989) *J. Enzyme Inhib.*, **2**, 249.
32. Yaginuma, S., et al. (1986) Eur. Pat. Appl. EP196189 A2.
33. Siffert, O., Emod, I. and Keil, B. (1976) *FEBS Lett.*, **66**, 114.
34. Giroux, E. and Vargaftig, B.B. (1978) *Biochim. Biophys. Acta*, **525**, 429.
35. Keil, B. (1979) *Mol. Cell Biochem.*, **23**, 87.
36. Rigbi, M., et al. (1987) *Comp. Biochem. Physiol.*, **87B**, 567.
37. Saito, H., et al. (1979) *Proc. Natl. Acad. Sci. USA*, **76**, 2013.
38. Sakuragawa, N., et al. (1978) *Acta Med. Biol.*, **25**, 140.
39. Marciniak, E. (1973) *Brit. J. Haematol.*, **24**, 391.
40. Stuerzebecher, J., Markwardt, F. and Walsmann, P. (1976) *Thromb. Res.*, **9**, 637.
41. Bjoerk, I. and Ylinenjaervi, K. (1990) *Biochemistry*, **29**, 1170.
42. Amouric, M. and Figarella, C. (1980) *Hoppe-Zeylers' Z. Physiol. Chem.*, **361**, 85.
43. Aoyagi, T. and Umezawa, H. (1975) *Cold Spring Harbor Conf. Cell Prolif.*, **2**, 429.
44. Suda, H., et al. (1972) *J. Antibiotics*, **25**, 263.
45. Hanada, K., et al. (1978) *Agric. Biol. Chem.*, **42**, 523.
46. Anastasi, A., et al. (1983) *Biochem. J.*, **211**, 129.
47. Komiyama, T., et al. (1975) *Arch. Biochem. Biophys.*, **171**, 727.
48. Murao, S., et al. (1975) *Agric. Biol. Chem.*, **44**, 701.

49. Markwardt, F. (1972) *Folia. Haematol.*, **98**, 381.

50. Harper, J.W., Hemmi, K. and Powers, J.C. (1985) *Biochemistry*, **24**, 1831.

51. Rich, D.H., Moon, B.-J. and Harbeson, S. (1984) *J. Med. Chem.*, **27**, 417.

52. Kocmierska-Grudzca, D. and Goutier, R. (1972) *Strahlentherapie*, **144**, 245.

53. Watanabe, M., *et al.* (1988) *J. Histochem. Cytochem.*, **36**, 783.

54. Aoyagi, T., *et al.* (1978) *J. Antibiotics*, **31**, 636.

55. Umezawa, H., *et al.* (1976) *J. Antibiotics*, **29**, 97.

56. Akhtar, M. and Birch, P.L. (1972) *Biochem. J.*, **126**, 23.

57. Friedman, T.C., Kline, T.B. and Wilk, S. (1985) *Biochemistry*, **24**, 3907.

58. Fujiwara, K., *et al.* (1982) *Biochim. Biophys. Acta*, **702**, 149.

59. Ryan, C.A., Hass, G.M. and Kuhn, R.W. (1974) *J. Biol. Chem.*, **249**, 5495.

60. Peanasky, R.J., *et al.* (1974) *Bayer-Symp.*, **5**, 649.

61. Raimondi, L., *et al.* (1982) *Pharmacol. Res. Commun.*, **14**, 103.

62. Jacobsen, N.E. and Bartlett, P.A. (1981) *ACS Symp. Ser.*, **171**, 221.

63. Ohuchi, S., *et al.* (1984) *J. Antibiotics*, **37**, 1741.

64. Tanaka, T., *et al.* (1984) *J. Antibiotics*, **37**, 682.

65. Kasai, N., *et al.* (1983) *Agric. Biol. Chem.*, **47**, 2915.

66. Umezawa, H., *et al.* (1984) *J. Antibiotics*, **37**, 1088.

67. Matern, H., Barth, R. and Holzer, H. (1979) *Biochim. Biophys. Acta*, **567**, 503.

68. Sumi, H. and Toki, N. (1981) *Proc. Soc. Exp. Biol. Med.*, **167**, 530.

69. Seemueller, U. *et al.* (1977) *Hoppe-Zeyler's Z. Physiol. Chem.*, **358**, 1105.

70. Dauter, Z., *et al.* (1988) *FEBS Lett.*, **236**, 171.

71. Wiedow, O., Luedemann, J. and Utecht, B. (1991) *Biochim. Biophys. Res. Commun.*, **174**, 6.

72. Plummer, Jr., T.H. and Ryan, T.J. (1981) *Biochem. Biophys. Res. Commun.*, **98**, 448.

73. Salama, Z.B. and Bohley, P. (1977) *Acta Biol. Med Ger.*, **36**, 1939.

74. Pannekoek, H., *et al.* (1986) *EMBO J.*, **5**, 2539.

75. Brzin, J., *et al.* (1983) *Hoppe-Zeyler's Z. Biol. Chem.*, **364**, 1475.

76. Turk, V., *et al.* (1986) *Biomed Biochim. Acta*, **45**, 1375.

77. Kitagishi, K. and Hiromi, K. (1984) *Agric. Biol. Chem.*, **48**, 1827.

78. Morales, T.I., Kuettner, K.E., Howell, D.S. and Woessner, J.F. (1983) *Biochim. Biophys Acta*, **760**, 221.

79. Shaw, E., Mares-Guia, M. and Cohen, W. (1965) *Biochemistry*, **4**, 2219.

80. Kostka, V. and Carpenter, F.H. (1964) *J. Biol. Chem.*, **239**, 1799.

81. Shaw, E. (1972) *Methods Enzymol.*, **25**, 655.

82. Bernfeld, P. (1951) *Adv. Enzymol.*, **12**, 379.

83. Takebe, I., Otsuki, Y. and Aoki, S. (1968) *Plant Cell Physiol.*, **9**, 115.

84. Glasgow, L.R., Paulson, J.C. and Hill, R.L. (1977) *J. Biol. Chem.*, **252**, 8615.

85. Scudder, P., *et al.* (1983) *Biochem. J.*, **213**, 585.

86. Scudder, P., *et al.* (1984) *J. Biol. Chem.*, **259**, 6586.

87. Kobata, A. (1979) *Anal. Biochem.*, **100**, 1.

88. Taniguchi, T., *et al.* (1986) *J. Biol. Chem.*, **261**, 1730.

89. Elder, J.H. and Alexander, S. (1982) *Proc. Natl. Acad. Sci. USA*, **79**, 4540.

90. Tarentino, A.L., Gomez, C.M. and Plummer, T.H. (1985) *Biochemistry*, **24**, 4665.

91. Tarentino, A.L. and Maley, F. (1974) *J. Biol. Chem.*, **249**, 811.

92. Trimble, R.B. and Maley, F. (1984) *Anal. Biochem.*, **141**, 515.

93. Yuki, H. and Fishman, W.H. (1963) *J. Biol. Chem.*, **238**, 1777.

94. Hamlyn, P.F., *et al.* (1981) *Enzyme Microbiol. Tech.*, **3**, 321.

95. Murao, S. (1979) *J. Ferment. Technol.*, **57**, 151.

96. Murao, S. and Sakamoto, R. (1979) *Agric. Biol. Chem.*, **43**, 1791.

97. Plummer, T.H., *et al.* (1984) *J. Biol. Chem.*, **259**, 10700.

CHAPTER 8
CHARACTERISTICS OF SELECTED PROTEINS
J. A. A. Chambers

Several thousand classes of nonenzymic proteins have been identified by criteria that make them chemically, structurally or functionally distinct. The following listing is a brief description of a selection of these classes, giving an outline of their biological roles and properties. The intent is to provide a guide to the names, structures and functions that may be useful for the researcher, rather than a theoretical outline of protein classification by structure.

For many of these protein classes, a clearly identifiable relationship exists between a structure, a name and a function, e.g. keratins or tubulins, but for others, such as nucleotide-binding regulatory proteins or interferons, the definition is primarily by functional criteria with very little that is structurally distinct.

Actin-binding proteins, ABP-120 and ABP-280
Elongated actin-binding proteins that cause *actin* microfilaments to form gels (1). ABP-280 is a 280 kDa tail-to-tail dimer with a domain structure similar to that of *spectrins*, but the repeat sequence is not related to that of spectrin. ABP-120 is a 120 kDa dimer with sequence similarity to ABP-280 and domain structure comparable to α-*actinin* but using ABP-280 repeat sequences. ABP-120 may have the same relationship to ABP-280 that α-*actinin* has to *spectrin*. See also *filamins*.

Actinins
Cytoskeletal proteins, interact with *actins*, *myosins* and profilins. Active α-actinin is a homodimer of 94–103 kDa. Proteins have an N-terminal globular actin-binding domain, a fourfold repeat of a very α-helical ≈120 amino acid domain and a C-terminal calcium-binding domain. (Note: muscle actinins are calcium-insensitive, Ca^{2+} binding by cytoskeletal actinins lowers the affinity for actin.) Several conserved sequences are found in other actin-binding proteins (*spectrin*, *dystrophin*). No particularly hydrophobic regions (1, 2).

Actinogelins
Calcium-responsive actin-binding protein involved in the gelation of *actin* in the cytoskeleton. Dimeric, monomer is 105–115 kDa (mammalian). Binds 10–12 G-actin per dimer with binding at ends of dumbell-shaped dimer (3).

Actins
Component of the contractile mechanism of muscle and cytoskeleton (microfilaments), monomer ≈43 kDa, acid pI (pH 4.3–4.5). The monomer is water soluble and forms filaments with ATP hydrolysis at high calcium concentrations. Interacts with *myosins* in muscle and a large number of microfilament-regulating proteins in cytoskeletal functions. Abundant, highly conserved, contains the unusual amino acid 3-methylhistidine (4).

Adhesins
Proteins of pathogenic bacteria responsible for adhesion to target cells, typically by binding to cell surface carbohydrates. Most are associated with fimbriae but are structurally distinct from *fimbrins*. Structurally these proteins are heterogeneous (5).

Adhesin is also sometimes used as a term for eucaryotic adhesion proteins.

Adseverins
Actin-binding 74 kDa protein, severs and caps barbed ends of actin microfilaments. Binding is regulated by calcium and phospholipids. Structurally similar to *gelsolin* and *vimentin* (6).

Albumins
Generic name for storage proteins that originated with a method for classifying proteins by their solubility properties. Includes ovalbumins, lactalbumins and some *plant storage proteins*. Mammalian serum albumins (68 kDa) act to bind small molecules in serum, e.g. drugs, fatty acids, and controls colloidal osmotic pressure. Usually freely water soluble, slightly acidic (high in glutamate and aspartate), has several hydrophobic pockets for binding small molecules, and may be glycosidated. See also *plant storage proteins, conalbumin*.

Anchorins — see *ankyrin*

Ankyrins
(Anchorins, *syndeins*.) Spectrin-binding 215 kDa membrane protein of erythrocytes and nerve tissue. Also binds tubulin monomers (7).

Annexins
Calcium-binding membrane-binding proteins. All are immunologically cross-reactive with one another. Includes calcimedins, calelectrins, calpactins, chromobindins, endonexins I and II and lipocortins. These proteins have a repeating structure based on a 70 amino acid sequence with the number of repeats a function of the size of the protein. The repeat sequence may be a calcium-binding domain related to that of *calmodulin*. The amino terminal 25–35% of the proteins shows greatest variability. Bind phospholipids and bind to cytoskeletal proteins (8).

Apolipoproteins
The proteins involved in the formation of the plasma lipoproteins (chylomicrons, high-density lipoprotein (HDL), low-density lipoprotein (LDL) and very-low-density lipoprotein (VLDL)) involved in lipid transport in animal serum. Their properties are summarized in *Table 1* (10).

APs
Activator, adaptor or associated proteins. Proteins that interact with clathrin triskelions in the formation of coated pits. The best characterized is brain AP-2, a complex ≈ 350 kDa protein with subunit constitution $\alpha\beta\gamma_2\delta_2$ where α and β are similar 100 kDa peptides, γ is a 60 kDa peptide and δ is a 16 kDa peptide. AP-2 is phosphorylated *in vivo*. AP-1 is similar to AP-2 in subunit composition and is found in non-brain tissue, subunits are 100, 47 and 19 kDa. AP-180 is a single 180 kDa polypeptide found in the brain. The protein is phosphorylated and may be related to the α and β subunits of AP-2. APs are membrane associated and bind to the domains of a number of receptors (9).

Axonins
Secreted proteins of axons involved in cell growth (11).

Bacteriorhodopsins — see *opsins*

Table 1. General properties of apolipoproteins of normal human plasma (reproduced from Scanu 1987 (ref. 12) with permission from Academic Press)

Apolipoprotein	Concentration in plasma (mg dl^{-1})	Mol. wt	Isoelectric point pI	Physiological role	Lipoprotein association
A-I	100–200	28016	A-I$_2$ = 5.85 A-I$_3$ = 5.74 A-I$_4$ = 5.65 A-I$_5$ = 5.52 A-I$_6$ = 5.40	LCAT activator	Chyl, HDL$_2$, HDL$_3$, VHDL
A-II	30–40	17440	4.9	Unknown	Chyl, HDL$_2$, HDL$_3$
A-IV	16–20	46000	5.5	Transport	VLDL, HDL$_2$
B-100	90–110	≈500000	—	Cholesterol carrier; ligand apoB, E receptor	VLDL, IDL, LDL
B-48	0	250000	—	Unknown	Chyl
C-I	4–6	6630	7.5	LCAT activator	VLDL, IDL, HDL$_2$
C-II	3–5	8824	4.9	Lipoprotein lipase activator	VLDL, IDL, HDL$_2$
C-III	12–14	8764	C-III-0 = 5.0[a] C-III-1 = 4.85 C-III-2 = 4.65	Inhibition remnant uptake	VLDL, IDL, HDL$_2$
E	3–6	34145	E-2 = 5.89 E-3 = 6.02[b] E-4 = 6.18	Ligand for apoE and apoB, E receptor	VLDL, IDL, HDL$_c$

[a] C-III-0, C-III-1, and C-III-2 each contain 0, 1 and 2 moles of sialic acid per mole protein, respectively.
[b] Disialylation of E-3 changes the pI of this isoprotein: E-3$_{s-1}$ = 5.89; E-3$_{s-2}$ = 5.78; E-3$_{s-3}$ = 5.68; s is the degree of sialylation in moles per mole protein.
Abbreviations: Chyl, chylomicrons; HDL, high-density lipoproteins; IDL, intermediate-density lipoproteins; LCAT, lecithin: cholesterol acyl transferase; LDL, low-density lipoproteins; VLDL, very-low-density lipoproteins.

Blood-coagulation factors
The protein components of the cascade that converts fibrinogen into an interconnected network of fibrin molecules in the formation of a blood clot fall into a limited number of structural classes. The largest group is the vitamin K-dependent group of factors: factor VII (50 kDa), factor IX (56 kDa), factor X (59 kDa), prothrombin (72 kDa), protein C (62 kDa) and protein S (71 kDa). Factors VII and IX, prothrombin and protein S are single-chain molecules converted to a two-chain form, the remainder are two-chain precursors. All have an N-terminal region containing several γ-carboxylglutamic acid residues followed by repeats of an *epidermal growth factor* domain and a serine protease domain (except protein S), and in prothrombin the epidermal growth factor domain is replaced by kringle domains. Factors V and VIII (both 330 kDa) have two repeats of a 350 amino acid A domain at the N-terminus, a long stalk of a repeated nonapeptide, another A domain and a repeated 150 amino acid C-domain. Proteins are glycosidated. Factor XI is different from all other members of the family as a 143 kDa disulfide-linked homodimer of a protein containing a serine protease domain and four repeats of the 'apple' domain. Hydroxyaspartic acid and hydroxyasparagine are found in factor X, protein C and protein S (13).

Bradykinin — see *kinins*

Brevins
Actin-binding 85 kDa protein, severs and caps barbed ends of actin microfilaments. Binding is regulated by Ca^{2+} and phospholipids. Calcium-binding domains are in the C-terminal region, actin-binding domains in the N-terminal region. Structurally similar to *gelsolin* (14).

Cadherins
Calcium-dependent intercellular adhesion molecules of vertebrates. Large family of transmembrane proteins including L-CAM, uvomorulin, and cell-CAM. This family has three major members E-, N- and P-cadherin, of size 723–748 amino acids (86–90 kDa). The N-terminal extracellular domain defines the specificity of interaction and has two major repeats, including a number of short internal repeats. Intracellularly the protein appears to interact with microfilaments via a 94 kDa protein (15).

Calbindins
Vitamin D-induced calcium-binding proteins of animal organs involved in Ca^{2+} uptake. There are two forms, of 9 or 28 kDa. The 28 kDa form binds four Ca^{2+} ions per molecule. Structurally related to *troponin C* (16).

Calcimedin — see *annexins*

Caldesmon
A major calmodulin and actin-binding protein of smooth muscle and non-muscle cells. There are two isoforms: H- (120–150 kDa of which 89 kDa is peptide), found in smooth muscle, and L- (70–80 kDa, 59 kDa peptide) found in non-muscle cells and stress fibers. It is involved in the regulation of contraction and cell motility, it inhibits superprecipitation and actomyosin Mg-dependent ATPase. Binding to actin and calmodulin is regulated by Ca^{2+} levels (17).

Calelectrins — see *annexins*

Calmodulins

Intracellular mediators of calcium-dependent responses. Upon binding Ca^{2+}, the protein changes conformation and binds to calmodulin-regulated proteins. Properties: 17 kDa, acidic ($pI \approx 4$) containing $\approx 30\%$ (Glu + Asp), no Trp, heat stable, soluble in water. Highly conserved with a conserved calcium-binding site called the E-F hand, which is also found in other calcium-binding proteins. The sequence is related to that of *troponin C* (18).

Calpactins — see *annexins*

Calponins

A ≈ 35 kDa calcium-binding protein of smooth muscle that interacts with *calmodulin*, actin and *tropomyosin*. The protein has structural similarities to *troponins*. The C-terminal region of the protein is based upon a fourfold (calponin α) or threefold (calponin β) repeat, similar to that found in skeletal and cardiac *troponin* T (17).

Caseins

The major group of proteins of milk. They form micelles that act as a nitrogen and phosphate sources (heavily phosphorylated). Caseins of bovine milk are:

α_{S1}		199 amino acids, 7–9 phosphates;
α_{S2}		207 amino acids, 2 Cys, 10–13 phosphates;
β		209 amino acids, 5 phosphates;
γ		degradation products of β caseins;
κ		169 amino acids, 2 Cys, 1 phosphate, some glycosidation; important in the stabilization of caseinate micelles in milk.

Caseins of other animals are essentially similar (19).

CD antigens

Cluster differentiation (CD) antigens, antigenic surface markers for T lymphocytes. Their identification as antigens has resulted in the name CD antigen being given to a number of structurally and functionally distinct proteins. These are further outlined below (20).

CD4	55 kDa monomeric glycoprotein that plays a major role in the immune response by controlling lymphocyte proliferation.
CD8	Two 32 kDa chains, α and β, found as $\alpha\alpha$, $\alpha\beta$ or $\beta\beta$. Functionally similar to CD4.
CD2	50–55 kDa monomeric glycoprotein involved in T-cell activation.
CD11	see *integrins*.
CD18	see *integrins*.
CD44	80 kDa glycoprotein (37 kDa protein) of T cells, granulocytes, macrophages, fibroblasts and erythrocytes. Involved in the interaction between lymphocytes and endothelium and in the interaction of CD2 and CD58.
CD3	A multisubunit protein intimately associated with the T-cell receptor. Subunits are: γ, δ, ε, ν, ζ, with the composition $\gamma\delta\varepsilon\nu\zeta_2$.

Cell-CAM — see *cadherins*

Ceruloplasmins

Copper-binding α_2-glycoprotein of mammalian plasma involved in Cu^{2+} transport; 132 kDa, contains six strongly bound copper atoms that do not exchange. The Cu^{2+} taken up from plasma is more loosely bound (21).

Chaperonins
Chaperonins and *nucleoplasmins* are molecular chaperones that play a role in the folding of nascent proteins to their correct structure. They may also play a role in the formation of oligomeric proteins in the correct stoichiometry and conformation without themselves becoming part of the protein. Their modes of function are not clear, although they may mask surfaces of high charge density and so control ionic interactions (e.g. the role of nucleoplasmin in nucleosome formation). These proteins are structurally heterogeneous and include chaperonins, *nucleoplasmin*, the groEL proteins of bacteria, and some *heat-shock proteins* (22).

Chlorophyll a/b-binding proteins
These are proteins forming the primary light-harvesting antennae of the reaction centers of the chloroplast. These integral membrane proteins contain three amphipathic α-helices. The binding of chlorophyll *a* or *b* to the protein is not covalent. In any plant, several forms of the protein may be associated with the two reaction centers (23).

Chromobindins — see *annexins*

Chromochelatins — see *metallothioneins*

Chromogranins
Catecholamine-binding proteins of storage vesicles — see *granins*.

Cingulins
Extrinsic membrane protein associated with tight junctions and removed by washing with physiological solutions. The isolated protein is a dimer of 108 and 140 kDa proteins; it is slightly acidic and heat stable. Protein *ZO-1* is similar but is more tightly bound to the membrane, requiring chaotropic agents to elute it (24).

Clathrins
Membrane proteins that form triskelions involved in the coated pits of uptake and transport processes. The molecule is a hexamer of $\alpha_3 \beta_y \beta'_{(3-y)}$ where α is a large subunit of 180 kDa and β and β' are small subunits of 36 and 33 kDa, respectively. The small subunits show variability associated with tissue specificity. They interact with assembly proteins and have conserved, very acidic N-terminal regions (25).

Cofilins
Actin-binding phosphoproteins related to *destrin*.

Collagens
Structural proteins of animal connective tissue, ≈ 100 kDa. The structure is based on the repeating unit Gly–Xxx–Yyy–Gly, where Xxx is any amino acid but is most often Pro, and Yyy is any amino acid but often hydroxyproline. Forms a three-stranded helix *in vivo*. Essentially insoluble in physiological buffers. The triple helix is stabilized by covalent bonds between lysine or hydroxylysine and adjacent groups. *Gelatin* is a commercial preparation from collagen-rich tissues (26).

Conalbumins
Non-heme iron-binding albumins from avian blood, involved in iron uptake from the digestive tract. There are several forms of ≈ 77 kDa, glycosidated. This protein carries two Fe(II)/molecule (26).

Connexins

Integral membrane proteins involved in intercellular gap junctions. Liver protein: 32 kDa, very α-helical, forms hexamers in the membrane with a central pore for the passage of inorganic ions. Monomer size is very variable depending upon the source. There are two extracellular domains containing invariant cysteines, with N- and C-termini in the cytoplasm (27).

Crystallins

Major soluble proteins of the lens of the vertebrate eye; marked by extreme chemical stability. The forms: α, β and γ. There are two forms of α, 800 kDa and ⪢ 800 kDa, formed by combinations of two precursors (αA1 and αB2) and their processing products (αA1 and αB1). β-crystallins appear to be encoded by at least seven different genes and are made up of combinations of the gene products. γ-crystallins are oligomers of four or five subunits of ⩽ 28 kDa, and are cysteine rich. Crystallin sequences are very highly conserved (28).

Cyclins — see *proliferating cell nuclear antigen*

Cylindrins

A structural protein of the erythrocyte membrane, found as a pentamer of 22–25 kDa subunits (29).

Cytochromes

Heme-containing proteins involved in redox reactions in respiration, photosynthesis and a number of other energy-conserving reactions. Reversible changes of the oxidation state of the iron atom (between Fe(II) and Fe(III)) bound to the heme moiety of the protein lie at the core of their electron transport properties. These hemes give the proteins, and mitochondria, their distinctive red color and allow them to be assayed spectrophotometrically. There are three major cytochromes involved in the respiratory chain of mitochondria: cytochrome b, cytochrome c and cytochrome $a + a_3$ (also known as cytochrome oxidase). Several variants of these are found in bacterial systems. Cytochrome b has the lowest redox potential and has Fe(II) protoporphyrin IX as a prosthetic group. It is a membrane-bound protein of 60 kDa and is insensitive to CO and cyanide. Cytochrome c is a strongly basic (pI 10.1), ≈ 13 kDa, water-soluble protein, usually with acetylation of the N-terminal alanine or glycine. Cytochrome c_1 is an insoluble, 37 kDa protein that is part of the cytochrome c reductase activity. Cytochrome $a + a_3$ is the last step of the respiratory chain and is that which uses oxygen and so can be inhibited by CO or cyanide. Heart muscle $a + a_3$ is a tetramer of a protein of 4–6 nonidentical subunits, the 440 kDa tetramer containing four hemes and four copper atoms, with ≈ 90 kDa of the protein strongly bound to lipid. Bacterial cytochrome oxidases are structurally less complex, being homodimers of ≈ 60 kDa subunits (26).

Cytochromes P_{450}

Also known as mixed-function oxidases. Relatively nonspecific oxidases associated with microsomal membranes. A heme group is responsible for the distinctive absorption at 450 nm after exposure to CO. Many drugs, insecticides and carcinogens are activated or inactivated by these cytochromes and they perform a number of catabolic and anabolic hydroxylations (26).

Cytokeratins

Also known as prekeratins. A component of cytoskeletal intermediate filaments. They range in size from 40 to 68 kDa, with four main classes of 50, 56.5, 58 and 60 kDa. The proteins have four central conserved α-helical domains with heterogeneity in the N- or C-termini.

They interact with neurofilament triplet proteins, glial fibrillary acidic proteins, *desmins* and *vimentins*. Intermediate filaments are insoluble and resistant to solubilization by nondenaturing detergents such as Triton X-100 (30).

Cytotactins — see *tenascins*

Desmins
Fibrous 52 kDa protein of intermediate filaments, related to α-*keratins* and interact with *cytokeratins* in intermediate filaments (31).

Desmocalmins
A 240 kDa protein of the desmosome, binds calmodulin and keratins but is immunologically and compositionally distinct from *desmoplakins* (32).

Desmogleins
A glycosidated 165 kDa peptide of the desmoglea of the desmosome that shows similarities to *cadherins* (33).

Desmoplakins
Desmosomal structural proteins. There are two forms in bovine snout epithelium: desmoplakin I is 250 kDa, desmoplakin II is 215 kDa, neither is heavily glycosidated but can be heavily phosphorylated. Desmoplakins I and II have a long (850 residue) N-terminal α-helical region with a C-terminal region rich in hydroxy amino acids and with several repeats of a phosphorylation motif Gly–Ser–Arg–Ser (34).

Destrins
A 19 kDa actin-depolymerizing protein of mammals, similar to *cofilin* but the conditions for interaction with actin are slightly different. Interacts with actin at many points on the microfilament rather than at the ends. The protein has a nuclear transport signal and shares features with *gelsolin*, *fragmin* and profilin (35).

Dynamins
A 100 kDa nucleotide-sensitive microtubule-binding protein. May be an ATPase (36).

Dyneins
Axonemal dyneins are the ATPases of cilia and flagella responsible for motility. Cytoplasmic dynein is a 1200 kDa microtubule-associated protein (MAP1C) and is a microtubule-dependent ATPase involved in retrograde motion structures along the microtubule. The protein is made up of four subunits of 53–59 kDa and two of 440 kDa (36).

Dystrophins
Actin-binding protein similar to *spectrins* but much larger (427 kDa) (37).

Elastins
Structural proteins of mammalian elastic tissue. Rich in Gly, Ala, Val (17%) and Leu and Ile (12%). They are cross-linked in a three-dimensional network by the unusual amino acid desmosine. Elastins are insoluble in water and NaOH solutions and are resistant to denaturants and all proteases except elastase (EC 3.4.21.11) (26).

Elastonectins
A 120 kDa protein involved in the adhesion of cells to elastin fibers but separable from elastin receptors (38).

Endonexins — see *annexins*

Entactins
An integral basement membrane glycoprotein, shows tyrosine *O*-sulfation, 158 kDa. Globular N- and C-terminal regions are connected by a rigid cysteine-rich stalk that has four repeats of an *epidermal growth factor*-like domain and a thyroglobulin-related domain. Entactin binds Ca^{2+} and also interacts with a short arm of *laminin* and with *collagen IV*. Nidogen is a proteolytic fragment of entactin (39).

Epidermal growth factor (EGF)
First member of a family of related proteins that are not endocrine in origin but are mitogens and affect several physiological functions. EGF is 50–60 amino acids long with six conserved cysteines that are essential for biological activity. The structure is essentially β-sheet and devoid of α-helix. The protein is derived from a ≈ 150 kDa precursor by proteolytic cleavage. The protein only affects cells carrying the EGF receptor (40).

Erythrocruorins
A hemoglobin-like protein of invertebrates. Large oligomers of a subunit of 18.5 kDa are circulatory oxygen-carriers in some organisms, and intracellular monomers or oligomers are found in others (26).

Extensins
The glycoprotein structural component of plant cell walls. Repeated sequence of Ser-(Hyp)$_4$ with each of the hydroxyproline residues carrying an arabinotetraose and the Ser galactosidated. This protein also contains Lys and Tyr, with the Tyr involved in inter- and intra-molecular cross-links involving the formation of isodityrosine. The carrot cell precursor is 86 kDa, of which 66% is carbohydrate and the only amino acids are Hyp (45%), Ser, His, Tyr, Lys and Val. The protein is also cross-linked to other components of the cell wall with the extent of cross-linking increasing as the cell ages (41).

Ferredoxins
These are Fe–S proteins involved in redox reactions in plants and bacteria. The iron is non-heme. These proteins have eight cysteines out of 55 amino acids. Bacterial (4Fe-4S) ferredoxins have a single Fe–S cluster and a redox potential of -420 mV. The related high potential iron–sulfur protein (HiPIP) has a redox potential of $+350$ mV. Blue-green algae, green algae and higher plants have a 96–98 amino acid ferredoxin containing 4–6 cysteines in a structure called the 2Fe-Fd model (26).

Ferritins
Iron storage proteins of mammals. The monomer is 18.5 kDa; mature ferritin contains 24 monomers and 16 iron atoms as 8 $FeO(OH) \cdot FeO(PO_3H_3)$. Iron is transferred from *transferrin* that has resorbed iron from the intestine to ferritin in the bone marrow, liver, spleen and reticulocytes (26).

α-Fetoproteins
The first globulin formed by embryos, ≈ 70 kDa, replaced by serum albumin in adults. Used as a diagnostic marker for spina bifida or for liver cancer in adults (26).

Fibrins and fibrinogens
Fibrinogen is the precursor of fibrin, the protein that forms the structural base of a blood clot. Fibrinogen is a 350 kDa $\alpha_2\beta_2\gamma_2$ hexamer held together by disulfide bridges. Fibrin is

formed from fibrinogen by removal of the internal fibrinopeptides by blood-coagulation factors and cross-linking by the formation of isopeptide bonds to generate an insoluble complex (26).

Fibronectins
Glycoproteins of 200–250 kDa that interact strongly with proteins such as *collagen* and *fibrin*. They play important roles in connective tissue matrices and cell adhesion and spreading. Related proteins include chondronectins (26).

Filaggrins
A keratin-bundling protein of the mammalian epidermis. Protein is rich in polar amino acids, especially Arg, Glu, Gly, Ser and His. The mature protein is 26–48 kDa (depending upon the source) arising from a phosphorylated precursor of ≈ 600 kDa. The precursor contains several copies of the mature protein that are released by an endoprotease (42).

Filamins
Actin-crosslinking and bundling proteins, found in cytoplasmic microfilaments and the Z-line of skeletal and cardiac muscle. They comprise homodimers of a 250 kDa subunit, with each subunit binding at least one actin monomer. These proteins are phosphorylated predominantly at Ser *in vivo*. Their properties and composition are quite distinct from those of *spectrin* and *myosin*. They are also known as *actin-binding proteins* or HMWPs (43).

Fimbrins
There are two very different classes of proteins called fimbrins:

(i) The structural protein of bacterial fimbriae, related to the pilins.
(ii) A monomeric 68 kDa actin-bundling protein first found in intestinal microvilli. The protein has two actin-binding domains similar to those of other actin-binding proteins but the effect of bundling with fimbrin is to produce tight filaments rather than gels. Binding to actin is modulated by Ca^{2+} binding at two N-terminal domains. Binding of Ca^{2+} modulates the interaction with actin. The structure of fimbrin is very similar to that of villin, although there is no sequence similarity (1, 2).

Flagellins
The primary structural component of bacterial flagella. The protein has an α-keratin structure that is converted to a β-keratin upon stretching; 33–40 kDa; no Cys or Trp; low in His and Pro (26).

Fodrins — see *spectrins*

Fragmin
42 kDa Actin capping and severing protein from *Physarum polycephalum*. Similar to *severin* and to *gelsolin* (44).

Gelactins
An actin-gelling protein similar to *actinin*. A 190 kDa dimer with a pI of 5.45 (45).

Gelsolins
Also known as Band 4.1. 90 kDa Actin capping and severing protein, caps the barbed ends of microfilaments. It binds two molecules of actin/mole gelsolin. Its effects are modulated by Ca^{2+} possibly also by inositol phosphates. The protein sequence shows six repeats of a domain. All functions are found in the N-terminal half of the protein. The structure is related

to those of villins, *fragmins* and *severins* and may have arisen from a duplication of the ancestral fragmin–severin gene (46).

Globulins
A term for the soluble proteins of plasma that survive from the separation of proteins into albumins and globulins by solubility properties (26).

Glutaredoxin — see *thioredoxins*

G-proteins — see *nucleotide-binding proteins*

Granins
Include chromogranins and secretogranins; a group of acidic proteins associated with storage and secretory processes that are widely distributed in endocrine and neural tissues. Sizes from 24 kDa (secretogranin V) to 120 kDa (chromogranin B); pI (denatured) ≈ 5.1; rich in acidic amino acids, especially Glu. Proteins are modified by phosphorylation, glycosidation and sulfation, and there may be some proteolytic processing to yield biologically active peptides such as pancreastatin (47).

Halorhodopsins — see *opsins*

Haptoglobins
Acidic plasma glycoproteins involved in the degradation of free hemoglobin in blood. An $\alpha_2\beta_2$ tetramer of 310 kDa, it can remove heme and proteolytically digests heme-free hemoglobin (26).

Heat-shock proteins
Proteins whose synthesis is strongly induced upon sudden heat shock or other physiological stress that shuts down the synthesis of most other proteins. The response and the proteins are conserved from bacteria to mammals. There are two groups. The first is the hsp70 group (the bacterial homologs are DnaK proteins) of ≈ 70 kDa with weak ATP-binding and ATPase activities. Related constitutive proteins are hsc70 and BiP of the endoplasmic reticulum. The second group is hsp60 (the bacterial homolog is groEL) of ≈ 60 kDa. The function of these proteins in both groups appears to be related to unfolding and refolding of proteins in various pathways, e.g. the refolding of partially heat-denatured proteins and the assembly of immunoglobulins (see also *chaperonins*) (48).

Hemocyanins
Copper-containing oxygen-transporting proteins of mollusc and arthropod hemolymph. Mollusc hemocyanins have a subunit of 400 kDa and contain 16 copper atoms but no porphyrins and can bind 8 oxygen molecules. Arthropod hemocyanin is a hexamer of 75 kDa subunits each containing two copper atoms (26).

Hemoglobins
Oxygen-transporting, heme-containing protein of blood. Tetrameric $\alpha_2\beta_2$ or $\alpha_2\delta_2$ protein; the δ and β subunits are very similar but distinct. Subunits ≈ 16.3 kDa, slightly acid pI. *Myoglobins* are very similar but monomeric (26).

Hemopexins
Heme-binding glycoprotein of plasma, 57 kDa, 22% glycoprotein. N- and C-terminal halves are essentially identical, with each made up of four similar repeats with the first and fourth

connected by disulfide bonds. This structure is also found in a somewhat modified form in *vitronectins* (26).

High-mobility group (HMG) proteins
Acidic proteins of the eucaryotic nucleus. There are four major HMGs characterized by their solubility in dilute TCA and $HClO_4$. Slightly acidic, freely water soluble (49).

Histones
Primary structural proteins of the mammalian nucleus. Five major classes: H1, H2A, H2B, H3 and H4. Molecular weights 10 000–12 000. All are very basic (pI 11–12) and rich in arginine and/or lysine. *In vivo* H2A, H2B, H3 and H4 form an octameric complex called the nucleosome, which is involved in the first layer of packaging of chromosomal DNA. Histones show considerable function-related post-translational modification (acetylation, phosphorylation, ubiquitinylation) (26).

HLA antigens
Human leukocyte antigens: integral membrane glycoproteins that control histocompatibility and some aspects of the immune response in man. The mouse equivalent is the H-2 locus. Structurally, the antigens are similar to the heavy chains of immunoglobulins, showing a high content of β-pleated sheet. Class I antigens have a 40–45 kDa chain of class A, B or C (there are multiple forms of each) that are associated with a β2-microglobulin. Class II antigens are dimers of allelic forms of the class II antigen polypeptides. Class III antigens include some components of the complement system (50).

HMG proteins — see *high-mobility group proteins*

Immunoglobulins
The antigen-binding proteins of the vertebrate immune system. The proteins are tetramers. Two large subunits ('heavy chains') and two small subunits ('light chains') form a Y-shaped molecule (*Figure 1*). The C-terminal region of subunits is constant, but the N-terminal region shows extensive variability as a result of the somatic generation of variants. Major classes of mammalian immunoglobulins (Igs) are IgA, IgD, IgE, IgG and IgM (see *Table 2*) (26).

Integrins
Integral membrane glycoproteins involved in cell adhesion. Related to VLA antigens and receptors for *fibronectin* and *vitronectin*. Two monomers involved: α (11 forms known) and β (six forms known), combinations of subunits are tissue-related. The tripeptide RGD is diagnostic and responsible for the strong adhesion of the protein to its ligands. Interacts with the cytoskeleton (52, 53).

Intercellular adhesion molecules (ICAMs)
Integral membrane-spanning glycoproteins involved in the adhesion of leukocytes to the carrier cell. Two forms, ICAM-1 (55 kDa) and ICAM-2 (29 kDa). Related to immunoglobulins, and ICAM-1 has five copies of a domain similar to the disulfide bond-delimited domains of immunoglobulins; ICAM-2 has two copies (54).

Figure 1. Structure of an IgG molecule. Variable (VL, VH) and constant (CL, CH) parts (domains) of the light and heavy chains, respectively. The molecule shown represents an IgG or IgA secreted Ig. Other classes have different numbers of heavy-chain domains (2–4), and the hinge region may not be present (IgM or IgE). The C-terminal sequence of membrane-bound Ig is different (and slightly larger) than that of the secreted form. CHO = carbohydrate chain. VL and VH form the antigen-binding site. Papain cleavage produces two F_{ab} and one F_c fragment; pepsin cleavage forms a $(F_{ab'})_2$ fragment and a smaller $F_{c'}$ fragment. PC = pyrrolidone carboxylic acid. (Reproduced from Scott and Eagleson 1988 (ref. 51) with permission from Walter de Gruyter.)

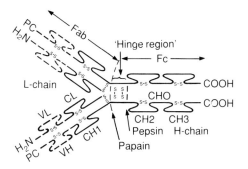

Interferons
A protein that exerts a virus nonspecific, antiviral activity, at least in homologous cells, through cellular metabolic processes involving the synthesis of both RNA and protein. Human interferons and their mouse homologs are classified by their sites of synthesis: α = leukocyte interferon; β = fibroblast interferon, γ = immune system. Human interferons-α are 16–23 kDa proteins that may be glycosidated, although this does not appear necessary for function. They are rich in Asx, Glx and Lys and are very α-helical. Interferon-β is about 20 kDa and similar to interferon α but is definitely glycosidated. Interferon-γ is about the same size and more basic than the others (55, 56).

Interleukins
A term unifying a group of proteins by function rather than structure. They are defined as polypeptides produced by leukocytes and involved in inflammatory responses regardless of their cellular targets. Interleukins stimulate the growth of T lymphocytes and an interleukin may act on the subset of T cells that produces it, or more usually, on one or more subsets. Interleukins are generally about 20 kDa and frequently glycosidated. Proteins defined as interleukins include interleukins 1 to 8 (IL-1 to IL-8), interferon γ, granulocyte–macrophage colony-stimulating factor (GM-CSF), tumor necrosis factors α and β (TNFα, TNFβ) and transforming growth factor β (TGFβ) (57).

Involucrins
A 92 kDa acid (pI = 4.5) protein that is the precursor of the cornified envelope of epithelium. Maturation to form the envelope involves transglutaminase-mediated cross-linking of the protein. The protein contains ≈ 40 repeats of the peptide Glu–Gln–Gln–Gly–Leu–Lys–His–Leu and ≈ 20 copies of peptides loosely related to this sequence (58).

Table 2. Properties of human immunoglobulins (Ig) (reproduced from Scott and Eagleson 1988 (ref. 51) with permission from Walter de Gruyter)

	IgG	IgM	IgA Serum	IgA Secretions	IgD	IgE
Sedimentation constant	6.5...7S	19S	7S	11S	6.8...7.9S	8.2S
M_r, of which the L-chain represents 23 000	155 000	940 000 (pentamer)	170 000	380 000 (dimer)	185 000	196 000
H-chain type and M_r	$\gamma 1...4$ 50 000 to 60 000	μ 71 000	α 64 000		δ 60 000 to 70 000	ε 75 000
Chain formula (L = κ or λ)	$L_2\gamma_2$	$(L_2\mu_2)_5$	$L_2\alpha_2$	$(L_2\alpha_2)_2$	$L_2\delta_2$	$L_2\varepsilon_2$
Carbohydrate fraction of the serum Ig	2...3% 70...75%	10...12% 7...10%	8...10% 10...22%		12.7% 0.03...1%	10...12% 0.05%
Serum concentration mg/100 ml	1 300 (800...1800)	140 (60...280)	210 (100–450)		3 (1...40)	0.03 (0.01 to 0.14)
Valence of binding	2	5(10)	1	2	?	2
Biological half-life (days)	8(IgG3) or 21	5.1	5.8		2.8	2...3
Complement binding	Yes	Yes	No		No	No

Keratins

Structural proteins of mammalian hair, nails, etc.; 40–68 kDa, with four main classes of 50, 56.5, 58 and 60 kDa. These proteins have four central conserved α-helical domains with heterogeneity in the N- or C-termini (see also *cytokeratins*) (26).

Kinesins

Microtubule-binding protein of neural tissue (squid axon, bovine brain). May be involved in ATP-dependent transport of organelles along axons. In squid this is a 600 kDa protein with 110 kDa, 70 kDa and 65 kDa subunits; in cattle it is an $\alpha_2\beta_2$ tetramer of 120 and 62 kDa subunits. The N-terminal head contains microtubule- and ATP-binding sites. The C-terminal domain binds membrane vesicles. The central domain is a fibrous rod (36).

Kininogens

Two forms, H-kininogen (110 kDa) and L-kininogen (68 kDa). Kinins are released from kininogens by the action of kallikreins. Kininogens also inhibit some classes of cysteine proteases. The amino acid sequence shows similarities to cystatins and stefins (see Chapter 7) (59).

Kinins

Nonapeptides derived from *kininogens* by specific cleavage with kallikreins. Kinins are vasodilators that also increase vascular permeability, stimulate the contraction of smooth muscle, and provoke pain. Bradykinin is Arg–Pro–Hyp–Pro–Gly–Phe–Ser–Pro–Phe–Arg. Typically Pro_3 is converted to Hyp (59).

Lactoferrins

Iron-binding protein of milk; see *transferrins*.

Laminins

Large multidomain proteins of the extracellular matrix of 900 kDa, three subunits: A, B1 and B2, all about the same size. Readily extracted as a complex with *nidogen* and *entactin* from cell cultures. The molecule is cruciform with a number of α-helical and globular domains with individual domains, showing internal repeats. Individual domains have distinctive amino acid compositions. Glycosidation is very heterogeneous. Aggregation is calcium-responsive and also binds several other proteins of the extracellular matrix (60).

Lamins

Major structural proteins of the eucaryotic nuclear lamina. There are two to four such peptides in animals. The structure is similar to that of intermediate filament proteins. Mammalian lamins have molecular weights of 70 000 (A), 68 000 (B) and 60 000 (C). A and C are very similar, with lamin A being a C-terminal extended form of lamin C. The central 310 amino acid region is very α-helical with a heptad repeat similar to that of *cytokeratins* and *vimentin*. There is a diagnostic histidine tetrapeptide in the fibrous region. The head and tail regions are rich in Ser and Thr and Cys is found in the C-terminal region. Proteins form oligomers and *in vivo* complexes are broken down upon phosphorylation (61, 62).

L-CAM — see *cadherins*

Legumins

The 11S storage proteins of legumes. Have up to 12 subunits of six types of 22–37 kDa. Unlike *vicilins* they are not glycosidated (63, 64).

Leu-CAMs
Heterodimeric membrane-spanning glycoproteins involved in leukocyte adhesion. The subunits are CD antigens. One subunit is the calcium-binding CD11a, b, or c antigen, with the other subunit the CD18 antigen (54).

Lipocortins — see *annexins*

Lipoproteins — see *apolipoproteins*

MAP — see *microtubule-associated proteins*

Matrix γ-carboxyl glutamic acid protein (MGP)
γ-Carboxyl glutamic acid-containing bone protein is marked by its insolubility. It is 79 amino acids long, containing γ-carboxyl glutamic acids. It contains the chemoattractant pentapeptide Arg-Arg-Gly-Ala-Lys found in *osteocalcin* (67).

Merosins
Laminin-like proteins of basement membranes of some mammalian tissues (65).

Metallothioneins
Sulfur-rich heavy-metal-binding proteins distributed throughout plants, animals and microorganisms. Size, 6–7 kDa; high in Cys (20 of 61 residues) with Cys positions strongly conserved but lacking Trp (66).

β_2-Microglobulins
An extrinsic 12 kDa membrane glycoprotein that is a subunit of class I HLA-A antigens, also found free in the serum. This protein is acidic to neutral, depending upon the source ($pI = 5.0$–7.5) but always has a single internal disulfide bond. The sequence shows similarities to those of other *HLA* antigens (50).

Microtubule-associated proteins (MAPs)
A name given to the high molecular weight proteins (70 000– >250 000) that bind *tubulin* or microtubules. The major groups are MAP1 and MAP2, both of which are very large (>250 000) and structurally related to *tau proteins*. The binding of these proteins to microtubules appears to play an as yet undefined role in microtubule structure or morphology. Other groups of these proteins include *dyneins*, *kinesins*, STOPs and chartins. These proteins have usually been isolated from nervous tissue but they, or their homologs, are probably found in other tissues also (68).

Milk proteins — see *caseins*

Mucins
The glycoproteins of mammalian mucus. Sizes are 10^5– >10^7 daltons. They show very heavy *O*-glycosidation with carbohydrate moieties rich in *N*-acetyl glucosamine. The peptide is <100 kDa and rich in Thr, Ser, Pro, Gly and Ala. Glycosidation is at one end of this peptide with the unglycosidated region somewhat hydrophobic (69).

Myoglobins
Oxygen storage protein of mammalian muscle. Resembles one subunit of *hemoglobin* (26).

Myosins

Along with *actin*, the major protein of the force-generating mechanism of muscle. Muscle myosin is a hexamer of two $\alpha\beta\gamma$ half-molecules. The α-subunit is a very large (≈ 250 kDa) monomer made up of an N-terminal head of ≈ 100 kDa with the remainder being an α-helical tail. The tail is made up of ≈ 40 repeats of a 28 amino acid repeat that is itself a modified repeat of a 7 amino acid motif. The N-terminal head is made up of three domains that include the actin-binding site, an ATP-binding site and an ATPase. There are two forms of the alkali light chains (β-subunit) derived from a single gene from different transcription starts, and one form of the γ-subunit MLC-2. The small subunits have calcium-binding domains related to those of *calmodulin*. MLC-2 is a calcium regulated protein kinase that regulates the affinity of myosin for actin by phosphorylation and dephosphorylation of the large subunit. Myosin contains the unusual amino acid 3-methylhistidine (70, 71).

Nidogens

A basement membrane protein that is a proteolytic fragment of *entactins*.

Nucleoplasmins — see *chaperonins*

Nucleotide-binding (N-) proteins

Proteins involved in transmembrane signalling, e.g. in hormone-mediated systems and in vision. Best known are the G (guanine nucleotide-binding) proteins. Binding to the hormone receptor is mediated by binding and hydrolysis of the nucleotide triphosphate. Proteins may act either to increase or reduce the affected activities, depending upon the hormone and the tissue. G proteins are heterotrimers with the guanine nucleotide-binding 45–52 kDa α-subunit showing the greatest variability and also great similarity to the bacterial elongation factor EFTu. The 35 kDa β- and 8.4 kDa γ-subunits are less variable (72).

Opsins

Retinal-binding 40 kDa integral membrane protein of the retina. The primary photoreceptor of the eye, this protein is monomeric and has seven α-helical domains. Retinal is covalently bound to lysine and the protein is phosphorylated at a cluster of hydroxyamino acids. Bacteriorhodopsins and halorhodopsins are structurally and functionally similar proteins of the purple membranes of halophilic bacteria. Bacteriorhodopsins form massive aggregates with a homotrimer as the minimal functional unit. In all cases, interaction of the retinal moiety with light results in the movement of cations (H^+, Ca^{2+}, Na^+) across the membrane (73).

Osteocalcins

Major protein of bone matrix, 5.2–5.9 kDa, 46–50 amino acids, contains 2–3 γ-carboxyl glutamic acid residues. This protein is anionic, binds hydroxyapatite and Ca^{2+}. Upon binding calcium, the α-helix content reaches 40%. The pentapeptide Arg–Arg–Gly–Ala–Lys of the protein acts as a chemoattractant for chondrocytes (67).

Osteonectins

Also known as SPARC; an acidic, cysteine-rich extracellular protein that binds type I collagen, Ca^{2+} and hydroxyapatite. Involved in the modulation of binding of animal cells to the extracellular matrix in proliferating systems (74).

Osteopontins
Also known as 44 kDa bone phosphoprotein, bone sialoprotein I and 2ar; a 41 kDa protein from osteoblasts, osteocytes and preosteoblasts that plays a role in the early stages of osteogenesis. The protein is rich in Asp, Ser and Glu and contains 12 phosphoserines and one phosphothreonine, the RGD adhesion motif is also found. Osteopontin is glycosidated, with sialic acid being the predominant carbohydrate (75).

Ovotransferrins — see *transferrins*

Patatins
The major protein of potato tubers. About 50 species are found in tubers; they arise from a multigene family. They appear to be storage glycoproteins that accumulate in vacuoles but are unusual in having an esterase activity (76).

Plakoglobins
A structural protein of desmosomes and other adherent cell–cell junctions. An unglycosidated homodimer of an 83 kDa subunit. The free and bound forms are in equilibrium, with 20–30% of the protein free (77).

Plant storage proteins
Proteins laid down in plant seeds that act as a store of nitrogen for the developing seedling. In legumes the two major classes are *legumins* and *vicilins*, and in cereals they are *prolamins* and *glutelins* (63, 64).

Plastins
There are two forms, L- and T-, of 65 kDa, and they are phosphorylated in interleukin-2-stimulated cells. No function has been assigned to these proteins but their sequences are very similar to that of eucaryotic *fimbrin* (78).

Plectins
A widely distributed protein associated with intermediate filaments and *vimentin* but not with *actins*. Plectin is a 300 kDa, hydrophobic, acidic protein (denatured $pI = 4.7–5.0$), found in the sub-plasma-membrane region and associated with, but not restricted to, desmosomes. Monomers are 30% α-helical and form a dumbell-shaped tetramer in solution. The protein is phosphorylated by *calmodulin* and cAMP-dependent mechanisms (79).

Ponticulins
A transmembrane 17 kDa glycoprotein of *Dictyostelium discoideum* involved in nucleation of *actin* microfilaments. A similar protein has been found in human leukocytes (80).

Porins
Proteins responsible for the formation of generally permeable pores of bacterial and outer mitochondrial membranes. In bacteria they are major proteins of the cell membrane, including the ompF and ompC proteins of *Escherichia coli* and are associated with the peptidoglycan of the cell wall. Porins form a homotrimer in the membrane with a central pore that allows solutes of up to several hundred daltons to pass through the membrane. There may be some voltage control of the channel. Bacterial porins are 30–50 kDa, mitochondrial porins are 30–35 kDa. All porins are very α-helical and rich in polar amino acids ($\approx 45\%$ Arg, Asp, Glu, Lys). Mitochondrial porins are slightly basic with a pI of 7–8 (81).

Prolamines — see *prolamins*

Prolamins
Acidic, ethanol-soluble storage proteins of plants. High in Glu (30–45%) and proline (15%) but low in essential amino acids. All plants studied have distinct prolamins including: avenins (oats), coixins (*Coix*), hordeins (barley), kaffirins (sorghum), pennisetins (pearl millet), secalins (rye) and zeins (maize). Zein peptides are ≈ 19 kDa and ≈ 22.5 kDa. The N- and C-termini of heavy and light chains are variable, but the main body of the protein is based upon repeats of an eicosapeptide with variations between adjacent repeats (26, 82).

Proliferating cell nuclear antigen (PCNA)
A nuclear antigen associated with DNA polymerase α and showing cell cycle related behavior; accumulates in the nucleus during the S-phase. The mature protein is 33–36 kDa, of which 28.7 kDa is peptide (83).

Protamines
Small, ≈ 4–4.5 kDa, strongly basic DNA-binding proteins of avian and fish sperm, where they replace histones during spermatogenesis. Very high in Arg (80–85%), the balance is of Ala, Gly, Pro, Ser, and Val or Ile. Similar proteins, such as illexins, have been found in invertebrate sperm (26).

Protein C
A name to cause confusion, seven very distinct groups of proteins are called protein C or C protein.

Protein synthesis initiation and elongation factors
Proteins associated with initiation and elongation stages of protein synthesis. Can be stripped from ribosomes with low-salt washes. Three initiation factors in bacteria: IF1 of 8 kDa; IF2 of 75 kDa and IF3 of 30 kDa (see *Table 3*) (26).

Ribosomal proteins
The protein components of ribosomes. Functions have not been definitively assigned to individual proteins. In *Escherichia coli* there are 31 proteins (5.4–30 kDa) associated with the large subunit, and 21 (8.4–61.2 kDa) with the small. All are basic and rich in Lys, Arg, Ala and Val, with some methylated and acetylated amino acids. Eubacterial ribosomal proteins are conserved between species but show a remarkable lack of similarity to one another in the same species. The proteins are very α-helical and appear globular in solution but are rod-like when associated with the ribosome. Mitochondrial and chloroplast ribosomal proteins are similar to bacterial proteins. Eucaryotic cytoplasmic ribosomes contain ≈ 80 proteins, the exact number is unclear. There is some similarity to bacterial proteins but it is not strong (84).

Secretogranins — see *granins*

Serum albumin — see *albumins*

Severins
A 45 kDa actin capping and severing protein from *Dictyostelium discoideum*. Similar to *fragmins* and to *gelsolin* (85).

Spectrins
Actin-crosslinking protein, elongated antiparallel $\alpha_2\beta_2$ tetramer. N-terminal actin-binding domain; the remainder is a very α-helical repeated domain, similar to that of *actinins* and *dystrophins*. Human α-spectrin is 280 kDa, β-spectrin is 246 000 (*Drosophila* β-spectrin is 430 000). The β-isoform shows greater variability. Actin binding is probably calcium regulated (86).

Table 3. Procaryotic and eucaryotic protein synthesis factors (reproduced from Scott and Eagleson 1988 (ref. 51) with permission from Walter de Gruyter)

Factor	M_r	Function
IF-1*		Equilibration of 70S ⇌ 50S + 30S units; Stabilization of initiation complex
IF-2*		Binding of fMet-tRNA$_f$ to 30S subunit; process may or may not require mRNA
IF-3*		Prevents association of 30S and 50S subunits
eIF-1	15 000	Stabilization of initiation complex
eIF-2	α subunit, 32 000–38 000; β subunit, 47 000–52 000; γ subunit, 50 000–54 000	Binding of Met-tRNA$_f$ to 40S subunit; process requires GTP and occurs before mRNA is bound
eIF-2A	50 000–96 000	Binding of Met-tRNA$_f$ to 40S subunit; process requires mRNA but not GTP
eIF-3	500 000–750 000 (complex of 7–11 polypeptides)	Prevents association of ribosomal subunits, stabilizes initiation complex
eIF-4A	48 000–53 000	Binding of mRNA to 40S initiation complex
eIF-4B	80 000–82 000	Binding of mRNA to 40S initiation complex
eIF-4C	19 000	Stabilization of initiation complex
eIF-4D	17 000	Stabilization of initiation complex
eIF-5	125 000–160 000	Release of eIF-2 and eIF-3 from initiation complex; binding of the 60S subunit to the 40S complex
Cap recognition protein	24 000	Not clear; binds to mRNA cap
EF-TU*	43 000	GTP-EF-TU binds aminoacyl-tRNA to ribosomal A site
EF-TS*	35 000	Displaces GDP from EF-TU-GDP which has been released from the ribosome; EF-TU:EF-TS complex reacts with GTP to regenerate EF-TU-GTP
EF-G*	80 000	Involved in translocation of pepidyl-tRNA from the A to the P site; GTPase

Table 3. Continued

Factor	M_r	Function
eEF-TU (EF-1α)	53 000	GTP-eEF-TU binds aminoacyl-tRNA to ribosomal A site
eEF-TS (EF-1β)	30 000	Displaces GDP from eEF-TU-GDP which has been released from the ribosome; eEF-TU:eEF-TS complex reacts with GTP to regenerate eEF-TU-GTP
eEF-G (EF-2)		Involved in translocation of peptidyl-tRNA from the A to the P site
RF-1*	47 000	Recognizes UAA and UAG termination codons; releases peptide from ribosome-bound tRNA
RF-2*	35 000–48 000	Recognizes UAA and UGA termination codons; releases peptide from ribosome-bound tRNA
RF-3*	46 000	Stimulated RF-1 and RF-2 activities
eRF	56 000–105 000	Recognizes all three termination codons; has ribosome-dependent GTPase activity

*Procaryotic factors.

Synapsins
Abundant nervous tissue proteins associated with synapses and neurotransmitter release. There are four synapsins: Ia, Ib, IIa and IIb. The a and b forms are derived from the same gene by differential splicing of the transcript. Synapsins I and II have similar domain structures in the N-terminal two-thirds of the protein, with Ia and Ib having a proline-rich C-terminal domain. Ia and Ib are 86 and 80 kDa respectively, have a pI > 10.5, and are prolate ellipsoids having some surfactant properties. Synapsins IIa and IIb are 74 and 55 kDa and neutral, with a pI of 6.6–7.1. These proteins are subject to cAMP or calcium-mediated phosphorylation and interact with microfilaments, microtubules and *spectrin* (87).

Syndeins
Spectrin-binding proteins of the erythrocyte membrane (88).

Talins
A 225–235 kDa protein associated with adhesion plaques. It is involved in cell–cell and cell–matrix interactions and shows limited cell and tissue distribution. It is an elongated molecule at physiological ionic strengths, usually monomeric but may form oligomers. Talin interacts with *F-actin*, *vinculin* and *β-integrins*. It shows post-translational phosphorylation at Ser, Thr and Tyr and specific proteolysis by calcium-dependent protease (89).

τ(tau) proteins
One of the major groups of microtubule-associated proteins. Sizes in the range 40–60 kDa. The N-terminal domains show heterogeneity that is probably due to alternative splicing of messengers. The C-terminal domains are conserved, responsible for tubulin binding and appear to be based upon a repeated element of 31 amino acids (90).

Tenascins
A protein of the extracellular matrix that mediates interaction between the cell and the matrix. *In vivo* an $(\alpha\beta\gamma)_2$ dimer of 1200 kDa. Subunits are 250, 225 and 215 kDa and are held together by disulfide bonds. One of the subunits of chicken tenascin has several tandem repeats of an *epidermal growth factor*-like sequence. Also known as cytotactin or hexabrachion (91).

Thioredoxins
Peptide cofactor in some hydrogen-transfer reactions (e.g. ribonucleotide reduction with ribonucleotide reductase). Hydrogen from NADPH is transferred via a reducible disulfide bridge by the enzyme thioredoxin reductase. The peptide is ≈ 12 kDa in all organisms. The active site peptide Cys–Gly–Pro–Cys is conserved in all organisms. Thioredoxin may have a regulatory role in some systems, e.g. photosynthesis. Glutaredoxin is analogous (92).

Torins
A structural protein of the erythrocyte membrane, an acidic (pI 4.8) 20 kDa multimer (93).

Transcupreins
A copper-binding protein of serum involved in Cu^{2+} uptake and transport from ceruloplasmin to the organs (21).

Transducins
A class of G-proteins specifically associated with signal transduction mechanisms in vision (94).

Transferrins
Iron-binding protein of serum also known as siderophilin; avian egg ovotransferrin and milk *lactoferrin* are structurally similar. Human transferrin is a glycoprotein of 74–80 kDa, with a pI of 5.4–5.9, depending upon iron content. The protein has a bilobate structure, reflected in an amino acid sequence in which the C-terminal half is an imperfect repeat of the N-terminal half. The protein has 19 disulfide bridges. It binds two Fe^{3+} per molecule with a stability constant slightly higher than that of citrate (26).

Tropoelastins
A 70 kDa protein associated with elastin in *elastic* fibers. It contains some hydroxyproline; but is not glycosidated and cross-linked by the same mechanism as *elastin* (95).

Tropomyosins
Tropomyosins and *troponins* form the mechanism for regulation of contraction of striated muscle. *Troponin* is a fibrous protein that is a homo- or heterodimer of 33 kDa subunits. The structure is very α-helical with the α-helices based upon charge-balanced heptad repeats that encourage strong interaction of the monomers. Tropomysin binds actin, with one tropomyosin molecule binding seven actins (4).

Troponins
Proteins involved in the regulation of contraction of striated muscle. Troponin C (17.8 kDa, pI = 4.1–4.4) is noted for its similarity to *calmodulin* and is the calcium-binding subunit. In striated muscle, troponin I (20.9 kDa, pI 9.3) binds to *actin* to prevent interaction with *myosin*, and troponin T (30.5 kDa, pI = 9.1) interacts with *tropomyosin* (4).

Tubulins

The major structural proteins of microtubules. They are typical soluble heterodimeric proteins when not organized into microtubules; there are multiple forms of each subunit in all but the simplest eucaryotes. Both subunits are ≈ 50 kDa. They bind guanine nucleotides, and their hydrolysis and Mg^{2+} are essential for the formation of microtubules. These proteins undergo acetylation, and phosphorylation and the α-subunit shows a reversible C-terminal addition of tyrosine that appears to affect the long-term stability of the microtubules (96).

Ubiquitins

A conserved neutral 76 amino acid peptide (8500 Da, $pI = 6.7$) found throughout procaryotes and eucaryotes. It lacks cysteine and tryptophan and is chemically very stable. It plays a role in protein degradation, chromatin structure, the heat-shock reponse and other phenomena. In most roles it is reversibly attached to another protein by an isopeptide bond in an ATP-dependent enzymic process (97).

Uvomorulin — see *cadherins*

Vicilins

The 7S storage glycoprotein of legumes. This protein has three subunits of 43–46, 47–48 and 51–53 kDa. It is noted for its low content of sulfur amino acids and lack of disulfide bonds. It shows post-translational heterogeneity (59, 60).

Vimentins

An intermediate filament protein of ≈ 52 kDa that is specific to mesenchymal tissue. The arrangement of α-helices and β-sheets is similar to that of *desmins* and prekeratins. The terminal peptides have opposite net charges, with the N-terminal region showing no similarity to the corresponding region of other intermediate filament proteins (98).

Vinculins

A 116 kDa protein associated with adhesion plaques involved in cell–cell and cell–matrix adhesion in animals. Has a very acidic N-terminal domain of 90 kDa that includes two or three repeats of a 110 amino acid domain, a proline-rich central domain of about 50 amino acids and a very acidic C-terminal domain. It may be phosphorylated on tyrosine and fatty acylated (99).

Vitronectins

An adhesion protein of mammalian tissue, also known as an α_1-glycoprotein or serum spreading factor, that is functionally similar to, but structurally distinct from, *fibronectins*. The protein is synthesized predominantly in liver as a 75–80 kDa glycoprotein (52.5 kDa protein) with several independent domains and contains the sequence Arg–Gly–Asp (RGD) found in adhesive proteins. The N-terminal domain includes a 44 amino acid region identical to the hormone somatomedin B; distal to this is a set of repeats similar to those found in *hemopexin* and including a cysteine repeat unit that is distinct from that of *epidermal growth factor*. The N-terminal region is similar to a heparin-binding domain. Biological functions of vitronectins are dominated by their interactions with *integrins* (100).

REFERENCES

1. Dubreuil, R.R. (1991) *BioEssays*, **13**, 219.
2. Blanchard, A., Ohanian, V. and Critchley, D. (1989) *J. Muscle Res. Cell Motility*, **10**, 280.
3. Asano, A., Kohno, K. and Mimura, N. (1983) *Tumor Res.*, **18**, 1.
4. Leavis, P.C. and Gergely, J. (1984) *CRC Crit. Rev. Biochem.*, **16**, 235.
5. Nimmich, W. (1990) *Acta Biotechnol.*, **10**, 151.
6. Sakurai, H., et al. (1991) *J. Biol. Chem.*, **266**, 15979.
7. Haigler, H.T., Fitch, J.M., Jones, J.M. and Schlaepfer, D.D. (1989) *Trends in Biochem. Sci.*, **14**, 48.
8. Burgoyne, R.D. and Geisow, M. (1989) *Cell Calcium*, **10**, 1.
9. Keen, J.H. (1990) *Ann. Rev. Biochem.*, **59**, 415.
10. Scanu, A.M. (1987) in *The Plasma Proteins* (F.W. Putnam ed.). Academic Press, New York, Vol. V, p. 141.
11. Stoeckli, E.T., et al. (1991) *J. Cell Biol.*, **112**, 449.
12. Scanu, A.M. (1987) in *The Plasma Proteins (2nd edn)* (F.W. Putnam ed.). Academic Press, New York, Vol. V, p. 141.
13. Davie, E.W., Fujikawa, K. and Kisiel, W. (1991) *Biochemistry*, **30**, 10363.
14. Bryan, J. and Hwo, S. (1986) *J. Cell Biol.*, **103**, 1439.
15. Takeichi, M. (1990) *Ann. Rev. Biochem.*, **59**, 237.
16. Gross, M. and Kumar, R. (1990) *Am. J. Physiol.*, **259**, F195.
17. Lehman, W. (1991) *J. Muscle Res. Cell Motility*, **12**, 221.
18. Persechini, A., et al. (1989) *Trends in Neurosci.*, **12**, 462.
19. Jenness, R. (1982) *Dev. Dairy Chem.*, **1**, 87.
20. Zaleski, M.B. (1991) *Immunol. Invest.*, **20**, 103.
21. Goode, C.A., Dinh, C.T. and Linder, M.C. (1989) *Adv. Exp. Med. Biol.*, **258**, 131.
22. Ellis, R.J. (1991) *Ann. Rev. Biochem.*, **60**, 321.
23. Bassi, R., Rigoni, F. and Giacommeti, F. (1990) *Photochem. Photobiol.*, **52**, 1187.
24. Stevenson, B.R., Heintzelmann, M.B., Anderson, J.M., Citi, S. and Mooseker, R. (1989) *Am. J. Physiol.*, **257**, C621.
25. Keen, J.H. (1990) *Ann. Rev. Biochem.*, **59**, 415.
26. Scott, T. and Eagleson, M. (1988) *Concise Encyclopedia of Biochemistry (2nd edn)*. Walter de Gruyter, New York.
27. Beyer, E.C., Paul, D.L. and Goodenough, D.A. (1990) *J. membr. Biol.*, **116**, 187.
28. Slingsby, C., et al. (1991) *Biochem. Soc. Trans.*, **19**, 853.
29. Harris, J.R. and Naeem, I. (1981) *Biochim. Biophys. Acta*, **670**, 285.
30. Moll, R., Franke, W.W. and Schiller, D.L. (1982) *Cell*, **31**, 1.
31. Ip, W., et al. (1985) *Ann. NY Acad. Sci.*, **455**, 185.
32. Tsukita, S. and Tsukita, S. (1985) *J. Cell Biol.*, **101**, 2070.
33. Schmelz, M., et al. (1986) *Eur. J. Cell Biol.*, **42**, 177.
34. Osborn, M. (1984) *J. Invest. Dermatol.*, **82**, 443.
35. Moriyama, K., et al. (1990) *J. Biol. Chem.*, **265**, 5768.
36. Vallee, R.B. and Shpetner, H.S. (1990) *Ann. Rev. Biochem.*, **59**, 909.
37. Monaco, A.P. (1989) *Trends in Biochem. Sci.*, **14**, 412.
38. Robert, L., et al. (1989) *Path. Biol.*, **37**, 736.
39. Chung, A.E. and Durkin, M.E. (1990) *Am. J. Respir. Cell Mol. Biol.*, **3**, 275.
40. Carpenter, G. and Cohen, S. (1990) *J. Biol. Chem.*, **265**, 7709.
41. Wilson, L.G. and Fry, J.C. (1986) *Plant Cell Environ.*, **9**, 239.
42. Dale, B.A., et al. (1985) *Ann. NY Acad. Sci.*, **455**, 330.
43. Weihing, R.R. (1985) *Can. J. Biochem. Cell Biol.*, **63**, 397.
44. Hasegawa, T.S., et al. (1980) *Biochemistry*, **19**, 2677.
45. Gruda, J. and Therien, H.M. (1985) *Can. J. Biochem. Cell Biol.*, **62**, 1072.
46. Way, M., et al. (1988) *Colloq. INSERM Structure and Functions of the Cytoskeleton* (B.A.F. Rousset ed.). INSERM, Paris, p. 401.
47. Huttner, W.B., Gerdes, H.-H. and Rosa, P. (1991) *Trends in Biochem. Sci.*, **16**, 27.
48. Langer, T. and Neupert, W. (1991) *Curr. Top. Microbiol. Immunol.*, **167**, 3.

49. Goodwin, G.H., *et al.* (1985) *Chromosomal proteins and gene expression*, NATO ASI Ser., Ser A 101, p. 221.

50. Strominger, J.L. (1987) *Brit. Med. Bull.,* **43**, 81.

51. Scott, T. and Eagleson, M. (1988) *Concise Encyclopedia of Biochemistry (2nd edn)*. Walter de Gruyter, New York.

52. Ruoslahti, E. (1991) *J. Clin. Invest.,* **87**, 1.

53. Hemmler, M.E. (1991) *Ann. Rev. Immunol.,* **8**, 365.

54. Gahmberg, C.G., *et al.* (1990) *Cell. Differ. Dev.,* **32**, 239.

55. Knight, Jr., E.J. (1984) *Interferon* (A. Billau ed.). Elsevier, Amsterdam, Vol. 1, p. 61.

56. Zoon, K.C. and Wetzel, R. (1984) *Hand. Exp. Pharmacol.,* **71**, 79.

57. Schrader, J.W. (1988) *Immunol. Cell Biol.,* **66**, 111.

58. Eckert, R.L. and Rorke, E.A. (1989) *Environ. Health Perspect.,* **80**, 109.

59. Müller-Esterl, W. (1989) *Thrombosis and Haemostasis,* **61**, 2.

60. Beck, K., Hunter, I. and Engel, J. (1990) *FASEB J.,* **4**, 148.

61. Franke, W.W. (1987) *Cell,* **48**, 3.

62. McKeon, F.D. (1987) *BioEssays,* **7**, 169.

63. Pernollet, J.-C. and Mossé, J. (1983) *Ann. Proc. Phytochem. Soc. Eur.,* **20**, 155.

64. Slightom, J.L. and Chee, P.P. (1987) *Biotechnol. Adv.,* **5**, 29.

65. Ehrig, K., *et al.* (1990) *Proc. Natl. Acad. Sci. USA,* **87**, 3264.

66. Hunziker, P.C. and Kägi, J.H.R. (1985) *Topics Mol. Struct. Biol.,* **7**, 149.

67. Hauschka, P.V., *et al.* (1989) *Physiol. Rev.,* **69**, 990.

68. Wiche, G. (1989) *Biochem. J.,* **259**, 1.

69. Roussel, P., *et al.* (1988) *Biochemie,* **70**, 1471.

70. Emerson, Jr., C.P. and Bernstein, S.I. (1987) *Ann. Rev. Biochem.,* **56**, 695.

71. Pollard, T.D., Doberstein, S.K. and Zot, H.G. (1991) *Ann. Rev. Physiol.,* **53**, 653.

72. Gilman, A.G. (1987) *Ann. Rev. Biochem.,* **56**, 615.

73. Findlay, J.B.C. and Pappin, D.J.C. (1986) *Biochem. J.,* **238**, 625.

74. Sage, E.H. and Bormstein, P. (1991) *J. Biol. Chem.,* **266**, 14831.

75. Butler, W.T., *et al.* (1988) in *Chemical Aspects of Regulation of Mineralization* (C.S. Sikes and A.P. Wheeler eds). University of South Alabama Publishing Services, Mobile, AL.

76. Frommer, W., *et al.* (1991) Comm. Eur. Communities [Rep.] Eur. Eur. 13415, 49.

77. Schwarz, M.A., *et al.* (1990) *Ann. Rev. Cell Biol.,* **6**, 461.

78. Zu, Y., *et al.* (1990) *Biochemistry,* **29**, 8139.

79. Wiche, G. (1989) *CRC Crit. Rev. Biochem. Mol. Biol.,* **24**, 41.

80. Luna, E.J. (1990) *Dev. Genet.,* **11**, 354.

81. Benz, R. (1985) *CRC Crit. Rev. Biochem.,* **19**, 145.

82. Spena, A., Viotti, A. and Pirrotta, A. (1982) *EMBO J.,* **1**, 1589.

83. Moriuchi, T. (1990) *Med. Sci. Res.,* **18**, 911.

84. Wittmann-Liebold, B. (1986) in *Structure, Function, and Genetics of Ribosomes* (B. Hardesty and G. Kramer eds). Springer Verlag, Berlin, p. 326.

85. Eichinger, L., Noegel, A.A. and Schleicher, M. (1991) *J. Cell Biol.,* **112**, 665.

86. Bubreuil, R.R. (1991) *BioEssays,* **13**, 219.

87. De Camili, P., *et al.* (1990) *Ann. Rev. Cell. Biol.,* **6**, 433.

88. Yu, J. and Goodman, S.R. (1991) *Proc. Natl. Acad. Sci. USA,* **76**, 2340.

89. Beckerle, M.C. and Yeh, R.K. (1990) *Cell Motil. Cytoskeleton,* **16**, 7.

90. Himmler, A., *et al.* (1989) *Mol. Cell. Biol.,* **4**, 1381.

91. Pearson, C.A., *et al.* (1988) *EMBO J.,* **7**, 2977.

92. Holmgren, A. (1989) *J. Biol. Chem.,* **264**, 13963.

93. Harris, J.R. and Naeem, I. (1981) *Biochim. Biophys. Acta,* **670**, 285.

94. Hwok-Keung Fung, B. (1986) *Prog. Retinal Res.,* **6**, 151.

95. Prosser, I.W. and Mecham, R.P. (1988) in *Self-Assembling Architecture* (J.E. Varner, ed.). Alan R. Liss, New York, p. 1.

96. Macioni, R.B., Serrano, S. and Avila, J. (1985) *BioEssays,* **2**, 165.

97. Wilkinson, K.D. (1988) in *Ubiquitin* (M. Reichsteiner, ed.). Plenum Press, New York, p. 27.

98. Bloemendal, H., *et al.* (1985) *Ann. NY Acad. Sci.,* **455**, 95.

99. Otto, J.J. (1990) *Cell Motil. Cytoskeleton,* **16**, 1.

100. Preissner, K.T. (1991) *Ann. Rev. Cell Biol.,* **7**, 275.

CHAPTER 9
GLYCOPROTEINS AND PROTEIN GLYCOSYLATION
A. P. Corfield and J. M. Graham

1. INTRODUCTION

The surface membranes of cells are characterized by the presence of carbohydrate conjugated to many different molecules including proteins and lipids (1, 2). The functions of these molecules vary considerably and concern many vital aspects of cell structure, communication and defence. The asymmetry observed in the plasma membrane is due significantly to the localization of the carbohydrate, largely on the extracellular surfaces. This chapter considers the glycosylation patterns found in proteins at the surface of cells. A great number of different molecules in many different organisms come into this category. The first part of the chapter shows examples of glycosylated proteins from a variety of organisms in nature. The review of structural features, enzymic glycosylation and functional roles is limited to eucaryotic systems. These molecules are therefore divided up into glycoproteins and proteoglycans, the two main families of glycosylated membrane proteins at the eucaryote cell surface (1–9).

2. OCCURRENCE OF GLYCOSYLATED PROTEINS IN NATURE

2.1. Range and types of glycosylated proteins

Glycosylated proteins are present in the form of glycoproteins (1, 3–5) and proteoglycans (6–9) associated with cell surface membranes through transmembrane peptide sequences, by specific phospholipid membrane anchor structures, inserted partially into the lipid bilayer, or bound at the surface to other proteins or the lipid membrane, as shown in *Figure 1* (2). Examples of these molecules for a range of organisms throughout nature are given in *Table 1*. The selection presented gives an indication of the enormous variety encountered and serves to underline the importance of protein glycosylation in general.

Many organisms have a carbohydrate-rich cell surface coat, which has been termed the glycocalyx (2). This shows considerable variation in composition depending on the organism and the origin of the cells. The glycocalyx may also be organized into discrete structural components, as found in plant and Gram-negative bacterial cells. In contrast, animal cells show tissue-specific variation of glycocalyx, with local functional roles related to, for example, membrane enzymes. Some examples are given in *Table 2*. The presence of the glycocalyx allows a number of extracellular glycoproteins to interact non-covalently through the abundant oligosaccharide. These interactions include those with molecules of the extracellular matrix, such as the collagens, proteoglycans, fibronectin, vitronectin, laminin and thrombospondin.

Animal membranes depend on the cytoskeleton for stability and generally for maintaining cell form and shape. The cytoskeleton contains a number of proteins and glycoproteins

▶ p. 196

Figure 1. Occurrence of proteins in membranes. The five categories of membrane proteins are shown: A, the polypeptide crosses the bilayer once; B, the polypeptide crosses the bilayer more than once; C, extrinsic proteins bound ionically to other membrane components; D, lipid link to the bilayer through a fatty acid linked by the carboxyl to a protein amino acid side-chain; E, glycosyl phosphatidylinositol (GPI) membrane anchor. Abbreviations: Eth, ethanolamine; Ins, myo-inositol, GlcN, D-glucosamine; P, phosphate; 〜〜〜 , fatty acid.

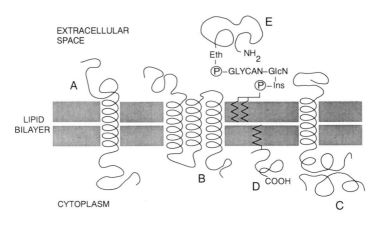

Table 1. Examples of glycosylated proteins in nature

Group/name	Carbohydrate type	No. of chains	Mol. wt (kDa)	Location
Viruses and micro-organisms				
Influenza virus hemagglutinin HA1	N-	6	36	Viral coat
Sindbis virus E1-glycoprotein	N-	2	50	Viral coat
HIV-1 virus gp-120	N-	24	120	Viral coat
Halobacteria Cell surface glycoprotein	O-Gal-Thr N-Glc-Asn N-GalNAc-Asn	15 10 1	120	Surface membrane
Trypanosomal Variant surface glycoprotein	N-	2–3	60	Cell coat
Leishmania Zn-protease	N-	2	63	Cell surface ecto-enzyme

Table 1. Continued

Group/name	Carbohydrate type	No. of chains	Mol. wt (kDa)	Location
Animals				
Type A				
N-CAM	N-	6	130–160	Neural cells
HLA type II α-chain	N-	2	34	Plasma membrane
EGF receptor	N-	12	170	Plasma membrane
Insulin receptor β-subunit	N-	4	95	Plasma membrane
Transferrin receptor (subunit)	N- O-	3 1	90	Plasma membrane
LDL receptor	N- O-	2 18	160	Plasma membrane
Type B				
Erythrocyte membrane band 3	N-	1	90–100	Plasma membrane anion exchange
Glucose transporter GLUT 1	N-	1	55	Erythrocyte hexose transport
Rhodopsin	N-	2	40	Vertebrate rod cell membranes
Type C				
Sodium channel β-subunit	N-	2	36	Plasma membrane widespread
Insulin receptor α-subunit	N-	11	130	Plasma membrane protein kinase activity
Type E				
Thy-1 receptor	N-	3	23	Thymocyte plasma membrane
Glypican	PG–HS	3–4	Variable	Fibroblasts and epithelia
NCAM	N-	6	120	Neural cells

Abbreviations: Asn, L-asparagine; EGF, epidermal growth factor; Gal, D-galactose; GalNAc, N-acetyl-D-galactosamine; Glc, D-glucose; HIV, human immunodeficiency virus; LDL, low-density lipoprotein; N-CAM, neural cell adhesion molecule; PG-HS, proteoglycan-heparansulfate; Thr, L-threonine.

Table 2. Typical glycosylated proteins in the animal glycocalyx

Name	Carbohydrate type	No. of chains	Mol. wt (kDa)	Location
Alkaline phosphatase	N-	2	67	Widespread
Glycophorin A	O- N-	15 1	62	Erythrocytes
Leukosialin	O-	80	90–150	Leukocytes
Versican	PG-CS	12–15	Variable	Fibroblasts
Decorin	PG-CS/DS	1	70	Connective tissue cells
Thrombomodulin	PG-CS	1	90	Endothelial cells
CD 44	PG-CS N-	0–4 6	Variable	Lymphoid and epithelial cells

Abbreviations: PG-CS, proteoglycan-chondroitin sulfate; PG-CS/DS, proteoglycan-chondroitin sulfate/dermatan sulfate.

which can interact reversibly to achieve the shape and structure necessary for the cell. Glycosylation of these molecules is also found as an integral feature. Intercellular junctions are further important features associated with cellular membranes; these complexes are important structural features concerned with cell communication and polarity. Many of the components of these complexes are also structural glycoproteins, relying on carbohydrate for the assembly and/or function of the intact complex *in situ*. Examples of such structural glycoproteins are given in *Table 3*.

In animal cells the membrane proteins have been divided up into five categories, A–E, depending upon the interaction of the molecules with the membrane (*Figure 1*). Examples of glycoproteins found in the various classes are given in *Table 1*. In spite of the diversity of structure and function for glycoproteins from different organisms, common features are apparent in *Tables 1–3*. The types of glycosylated proteins found at the cell surface membranes reflect the many functions carried out at this location. This topic is further considered in Section 5.

3. STRUCTURAL FEATURES OF GLYCOSYLATED PROTEINS

3.1. Linkage of carbohydrate to proteins

There are very few different types of linkages between carbohydrate and proteins (1, 3, 4, 6, 9). The common N- and O-glycosidic linkages found in eucaryotic glycosylated proteins are shown in *Table 4*. The small number of glycosidic linkages found reflects the preservation of this important feature of protein carbohydrate structure in evolution.

Table 3. Typical structural glycosylated proteins in animal membranes

Name	Carbohydrate chains	Mol. wt (kDa)	Location/role
Erythrocyte cytoskeleton			
Band 4.1	O-GlcNAc	78	Cross-bridges with spectrin
Nuclear pore glycoproteins			
gp190	N-	190	Nuclear pore complexes
gp62	O-GlcNAc	62	Nuclear pore complexes
Cell junction/adhesion molecules			
Integrins			
e.g. platelet glycoproteins			
gpIIb/IIIa	N-	α: 125 β: 105	Platelet membranes
Cadherins			
e.g. N-CAM	N-	130–160	Neural cells

Abbreviations: GlcNAc, N-acetyl-D-glucosamine.

Glycoproteins

Two types of glycosidic bond are found in glycoproteins (*Table 4*). First, those attached through N-acetylglucosamine to the amide nitrogen of asparagine, the N-glycosidic bond (1, 3, 4). The second type of linkage is that containing oxygen between the monosaccharides N-acetylgalactosamine, N-acetylglucosamine or galactose and either of the hydroxy amino acids serine or threonine (1, 5, 6, 9–11). The GalNAc-Ser/Thr linkage is the most common in eucaryotic membrane glycoproteins. The GlcNAc-Ser linkage is unique in representing a family of glycoproteins substituted only by this single monosaccharide in O-glycosidic linkage (12). Together with the novel O-glycosidic bond found in collagens between galactose and hydroxy L-lysine, this group constitutes the O-glycosidic linkages.

Proteoglycans

The glycosaminoglycan chains of chondroitin, dermatan and heparan sulfates have a common linkage (1, 6, 9). These chains have an O-glycosidic β linkage to serine residues in the core protein through xylose (*Table 5*). In addition, the glycosaminoglycan chains of keratan sulfate are attached through two types of linkage dependent on the origin of the proteoglycan. These are the two glycosidic linkages most common in glycoproteins, the N-glycosidic link of GlcNAc with asparagine and the O-glycosidic link of GalNAc to serine. Some chondroitin and heparan sulfate chains may also have an N-glycosidic linkage. Most membrane linked proteoglycans discovered thus far belong to the heparan sulfate type, although chondroitin, dermatan and keratan sulfates are all cell surface located as shown in *Tables 1* and *2* (6–9).

Table 4. Carbohydrate–protein linkages

Amino acid	Linkage monosaccharide	Type	Structure	Example
L-Asparagine	N-Acetyl-D-glucosamine	N-	GlcNAcβ1-Asn	N-Glycosyl proteins
L-Serine L-Threonine	N-Acetyl-D-galactosamine	O-	GalNAcα1-Ser	O-Glycosyl proteins
L-Serine	N-Acetyl-D-glucosamine	O-	GlcNAc1-Ser	O-GlcNAc proteins
L-Serine	D-Xylose	O-	Xylβ1-Ser	Proteoglycans
Hydroxy-L-lysine	D-Galactose	O-	Galβ1-Hyl	Collagen

Abbreviations: Asn, L-asparagine; Gal, D-galactose; GalNAc, N-acetyl-D-galactosamine; GlcNAc, N-acetyl-D-glucosamine; Hyl, hydroxy-L-lysine; Ser, L-serine; Xyl, D-xylose.

Table 5. Linkage and core structures in proteoglycans

Proteoglycan	Linkage to core protein
Chondroitin sulfate Dermatan sulfate Heparan sulfate Heparin	-β1-4GlcUAβ1-3Galβ1-3Galβ1-4Xylβ1-O-Ser

Keratan sulfate:

Cornea:
```
  -β1-3Galβ1-4GlcNAcβ1-2Manα
                            1
                             ↘6
                              Manβ1-4GlcNAcβ1-4AGlcNAcβ1-N-Asn
                             3                    |
  -β1-3Galβ1-4GlcNAcβ1-2Manα 1↗                   |α1-6
                                                  Fuc
```

Cartilage:
```
  -β1-3Galβ1-4GlcNAcβ
                    1
                     ↘6
                      GalNAcα1-O-Ser(Thr)
                     3
  Neu5Acα2-3Galβ    1↗
```

Abbreviations: Asn, L-asparagine; Fuc, fucose; Gal, D-galactose; GlcNAc, N-acetyl-D-glucosamine; GlcUA, D-glucuronic acid; Man, D-mannose; Neu5Ac, N-acetylneuraminic acid; Ser, L-serine; Thr, L-threonine; Xyl, D-xylose.

Peptide recognition sites for glycosylation

In the case of N-linked oligosaccharide attachment to peptides, a characteristic amino acid sequence has been identified, asparagine-X-serine or threonine, where X is any other amino acid as a potential glycosylation site. These sites are 'potential' glycosylation sites as sequence analysis of glycoprotein peptide has shown that such sequences are not always glycosylated (13). This demonstrates that other factors are also involved in the recognition of N-linked glycosylation sites in proteins.

There does not appear to be an equivalent sequence for sites of O-glycosylation. The two amino acids serine and threonine are commonly found in proximity to proline residues in GalNAc-Ser/Thr linkages (5, 11). This may result in topographical arrangements favorable for glycosylation. No consensus sequence has been identified. Similarly, there are no consensus sequences for GlcNAc-Ser linkages, but they occur in serine/threonine rich areas of peptide with the presence of proline within one to three amino acids of the linkage (12).

Neither do proteoglycans contain a universal peptide sequence acting as a recognition site for attachment of the initial xylose residue, although in some cases the sequence serine-glycine-X-glycine, preceded by several acidic amino acids, acts in this way (6, 9). In general, it appears that the peptide sequence is only one factor affecting the glycosylation of proteins.

3.2. Oligosaccharide structures

The common monosaccharides found in eucaryotic proteins occur in a number of glycosidic linkages with their partners in oligosaccharides, as listed in *Table 6* (1, 4). In eucaryotes a pattern of glycosylation has been identified and used to characterize the vast number of oligosaccharide structures described for membrane proteins (1, 3–5). This identifies the three important regions which may be present in any oligosaccharide and simplifies the description of the large structural variations encountered in glycosylated proteins.

Core structures

The *N*-linked oligosaccharides are commonly built up from a pentasaccharide core comprised of two *N*-acetylglucosamine and three mannose units as shown in *Figure 2* (1, 3, 4). This is carried on a dolichol-phosphate lipid prior to transfer to the peptide (3). They serve as precursors for the many different *N*-linked oligosaccharide structures found and indicated in *Figure 2*.

The GalNAc-Ser/Thr, *O*-linked oligosaccharides in glycoproteins and keratan sulfate exhibit a small number of common 'core' oligosaccharides (*Table 7*) in the linkage region of oligosaccharides (1, 4, 5, 10, 11).

Proteoglycan carbohydrate core structures have been identified which are typical for the different types of proteoglycan and are shown in *Table 5* (6, 9). In general the sequence of glucuronic acid–galactose–galactose–xylose–serine is found for the glycosaminoglycan chains, while keratan sulfate chains show core structures in common with the *N*- and *O*-linkages found in glycoproteins.

Table 6. Linkages between monosaccharides in eucaryotic protein oligosaccharides

Monosaccharide	Anomeric linkage	Link position
D-Xylose	β	2
L-Fucose	α	2, 3, 4, 6
D-Galactose	α	3
	β	3, 4, 6
D-Mannose	α	2, 3, 6
	β	4
D-Glucuronic acid	β	3, 4
L-Iduronic acid	α	3, 4
N-Acetyl-D-glucosamine	β	2, 3, 4, 6
N-Acetyl-D-galactosamine	α	3
	β	4
N-Acetyl neuraminic acid[a]	α	3, 6, 8

[a] Also occurs as *N*-glycolyl neuraminic acid.

Figure 2. Examples of *N*-linked oligosaccharide structure. The core structure common to most *N*-linked oligosaccharide chains is shown, with arrows indicating the positions of further addition. Representative oligosaccharide structures for the four main groups of *N*-linked chains are shown. Those substituents which may also be present are shown by an arrow and ±.

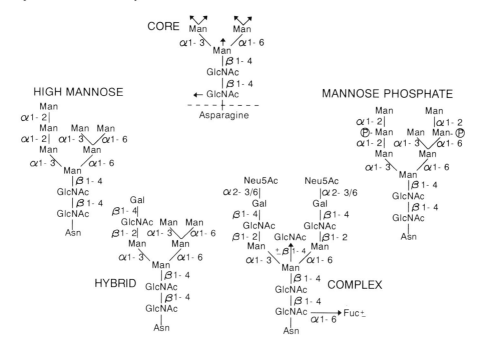

Table 7. Core structures in *O*-linked oligosaccharides

Core type	Core structure
1	Galβ1-3GalNAcα-*O*-Ser(Thr)
2	GlcNAcβ1-6(Galβ1-3)GalNAcα-*O*-Ser(Thr)
3	GlcNAcβ1-3GalNAcα-*O*-Ser(Thr)
4	GlcNAcβ1-6(GlcNAcβ1-3)GalNAcα-*O*-Ser(Thr)
5	GalNAcα1-3GalNAcα-*O*-Ser(Thr)
6	GlcNAcβ1-6GalNAcα-*O*-Ser(Thr)

Abbreviations: Gal, D-galactose; GalNAc, *N*-acetyl-D-galactosamine; GlcNAc, *N*-acetyl-D-glucosamine; Ser, L-serine; Thr, L-threonine.

Backbone, repeat structures
Glycoproteins show a range of repeating sequences which act as backbones in glycoprotein oligosaccharides (*Table 8*). Many of these structures are common to both *N*- and *O*-linked oligosaccharides and give characteristic size and antigenic properties to the glycoproteins (1, 4, 10).

Proteoglycans are characterized essentially by the nature of the repeating structures found in their oligosaccharides, together with the sulfation patterns in these repeat units (6, 9). The most common features of these repeat units are shown in *Table 9*.

Peripheral structures
The glycoproteins contain an array of chain-terminating or peripheral structures. These terminal additions may contain information or donate physicochemical and antigenic properties (1, 4). The most important of these additions are the blood group antigens, sialylation and fucosylation. These additions are found in both *N*- and *O*-linked oligosaccharides (*Table 10*), each type of chain resulting in specific structures.

The proteoglycans contain a number of chain-terminating structures which arise during biosynthesis and serve to limit the chain length of the glycosaminoglycan chains (*Table 10*). The formation of these terminating units correlates closely with the activity of the repeat structure biosynthetic enzyme complexes (6, 7, 9).

Additional modifications
In addition to the structural features of the carbohydrates described above there are a number of additional modifications to oligosaccharide structure with significant physiological roles (*Table 11*). These include sulfation, phosphorylation, acetylation, fatty acid acylation and GPI-membrane anchoring (1, 4, 10, 12–15).

Table 8. Repeating backbone carbohydrate structures in glycoproteins

Nomenclature	Structure	Oligosaccharide occurrence
Type 1	-Galβ1-3GlcNAc	*N*- and *O*-
Type 2	-Galβ1-4GlcNAc	*N*- and *O*-
Polylactosaminoglycans		
Blood group i	-(Galβ1-4GlcNAcβ1-3-)$_n$	*N*- and *O*-
Blood group I	-Galβ1-4GlcNAcβ1↘6 Galβ- -Galβ1-4GlcNAcβ1↗3	*N*- and *O*-
Human glycoproteins	-GlcNAcβ1-6Galβ-	*N*- and *O*-

Abbreviations: Gal, D-galactose; GlcNAc, *N*-acetyl-D-glucosamine.

Table 9. Common repeating carbohydrate units in proteoglycans

Proteoglycan type	Repeat structure and sulfation
Chrondroitin sulfate	β1-4GlcUAβ1-3GalNAcβ1-4GlcUAβ1-3 $\qquad\qquad\qquad\qquad\;\;$ | $\qquad\qquad\qquad\qquad$ 4 or 6-O-SO$_3^-$
Heparan sulfate and heparin	$\qquad\qquad\qquad$ 6-O-SO$_3^-$ $\qquad\qquad\qquad\quad$ | α1-4IdUAα1-4GlcNβ1-4GlcUAβ1-4 $\qquad\quad$ |$\qquad\quad$ | $\quad$ 2-O-SO$_3^-\;$ N-SO$_3^-$ $\qquad\qquad\quad$ or $\qquad\qquad\;$ COCH$_3$[a]
Dermatan sulfate	β1-4IdUAα1-3GalNAcβ1-4GlcUAβ1-3 $\quad$ |$\qquad\qquad\;$ | $\;\pm$2-O-SO$_3^-\;$ 4 or 6-O-SO$_3^-$
Keratan sulfate	β1-4GlcNAcβ1-3Galβ1-4GlcNAc $\qquad\qquad\qquad\qquad\;\;$ | $\qquad\qquad\qquad\quad$ 6-O-SO$_3^-$

[a] N-sulfation is extensive in heparin and less frequent in heparan sulfate.
Abbreviations: Gal, D-galactose; GalNAc, N-acetyl-D-galactosamine; GlcN, D-glucosamine; GlcNAc, N-acetyl-D-glucosamine; GlcUA, D-glucuronic acid; IdUA, L-iduronic acid.

4. ENZYMIC GLYCOSYLATION — SYNTHESIS OF SEQUENCE

4.1. Monosaccharide transport, interconversion and activation

The assembly of oligosaccharides in cells throughout the organism depends on a supply of monosaccharides which have to be transported into the cell, or into and out of subcellular compartments such as lysosomes, the endoplasmic reticulum and the Golgi apparatus before they can be utilized (16, 17). In addition, the transfer of nucleotides and nucleotide sugars across membranes within the cell is closely linked with the successful achievement of glycosylation (18). This is carried out by a number of membrane-located transport systems specific for monosaccharides, nucleotides and nucleotide sugars (*Table 12*).

The monosaccharides required for glycosylation processes are derived from the diet and endogenous degradation. There is often a need to interconvert monosaccharides, and the established pathways include both *de novo* and salvage routes (*Figure 3*). The end-products of these pathways are the nucleotide sugars (10), the activated forms of the individual sugars, which are required as substrates in the glycosyltransferase reactions in oligosaccharide biosynthesis for glycoproteins and proteoglycans.

Table 10. Peripheral structures in protein-linked oligosaccharides

Structure	Example
Neu5Acα2-3Galβ1-3GalNAcα-	O-linked glycoproteins[a]

-3GalNAcα-
   | α2-6
   Neu5Ac
	O-linked glycoproteins[a]

Neu5Acα2-3/6Galβ1-4GlcNAcβ1-	N-linked glycoproteins[a]
Neu5Acα2-8(Neu5Acα2-8)$_n$- Polysialyl oligosaccharide	Embryonic human N-CAM
SO_3^- GalNAcβ1-4GlcNAcβ1-2Manα1- sulfated N-linked oligosaccharide	Human lutropin

GalNAcα1-3Galβ1-3GlcNAcβ1-
       | α1-2
       Fuc
Blood group A — type 1 chain
	N- and O-linked glycoproteins[a]

Galα1-3Galβ1-4GlcNAcβ1-
     | α1-2
     Fuc
Blood group B — type 2 chain
	N- and O-linked glycoproteins[a]

Galβ1-4GlcNAcβ1-
  | α1-2
  Fuc
Blood group H — type 2 chain
	N- and O-linked glycoproteins[a]

Neu5Acα2-3Galβ1-4GlcNAcβ1-
        | α1-3
        Fuc
Sialyl-Lewis X
	Neutrophil and monocyte membrane glycoproteins

GalNAcβ1-4GlcUAβ1- | 4-O-SO_3^-	Non-reducing GalNAc-4-O-SO_3^- acts as terminator structure in chondroitin sulfate chain biosynthesis

[a] Widespread occurrence.
Abbreviations: Fuc, fructose; Gal, D-galactose; GalNAc, N-acetyl-D-galactosamine; GlcNAc, N-acetyl-D-glucosamine; GlcUA, D-glucuronic acid; Man, D-mannose; N-CAM, neural cell adhesion molecule; Neu5Ac, N-acetylneuraminic acid.

Table 11. Post-translational modifications to glycoslyated proteins

Modification	Site	Example
Sulfation	Gal, GlcNAc GalNAc GalNAc, IdUA, GlcNAc	Mucus glycoproteins Glycoprotein hormones Proteoglycans
Phosphorylation	Peptide, serine	Erythrocyte Band 4.1 EGF receptor
O-Acetylation	Neu5Ac/Neu5Gc	Mucus glycoproteins B-cell glycoproteins
Fatty acid acylation	Peptide	Insulin receptor Erythrocyte Bands 3, 4.1 Glycophorin Mucus glycoproteins
GPI-anchor	Peptide C-terminus	Alkaline phosphatase Glypican N-CAM

Abbreviations: EGF, epidermal growth factor; Gal, D-galactose; GalNAc, N-acetyl-D-galactosamine; GlcNAc, N-acetyl-D-glucosamine; IdUA, L-iduronic acid; N-CAM, neural cell adhesion molecule; Neu5Ac, N-acetylneuraminic acid; Neu5Gc, N-glycolylneuraminic acid.

Table 12. Carbohydrate transport systems in animal cell membranes

Carbohydrates transported			Transporter location
Ligand	From	To	
Glc, Gal, Man	Extracellular environment	Cytoplasm	Plasma membranes
Glc-P-Dol Man-P-Dol	Cytoplasm	Lumen of ER	Endoplasmic reticulum
Sialic acids	Lysosomal vesicle	Cytoplasm	Lysosomal membranes
UDP-Gal UDP-GlcNAc UDP-GlcUA GDP-Man GDP-Fuc CMP-Neu5Ac	Cytoplasm Cytoplasm Cytoplasm Cytoplasm Cytoplasm Nucleus/cytoplasm	Golgi/ER lumen	Golgi/ER membranes

Abbreviations: CMP, cytosine monophosphate; Dol, dolichol; Fuc, fructose; Gal, D-galactose; GDP, guanosine diphosphate; Glc, D-glucose; GlcNAc, N-acetyl-D-glucosamine; GlcUA, D-glucuronic acid; Man, D-mannose; Neu5Ac, N-acetylneuraminic acid; UDP, uridine diphosphate.

Figure 3. Metabolic routes in glycoconjugate biochemistry. The general interconversion pathways of monosaccharides and the formation of activated monosaccharides (nucleotide sugars). The nucleotide sugars (shown in boxes) are the substrates in glycosyltransferase reactions, forming all oligosaccharides on glycosylated proteins. The activated form of sulfate, phospho-adenosine phosphosulfate (PAPSO$_3^-$) is also shown. *De novo* pathways are indicated in heavy arrows. Monosaccharides arising from glycoconjugate degradation and feeding into salvage pathways are shown in bold type.

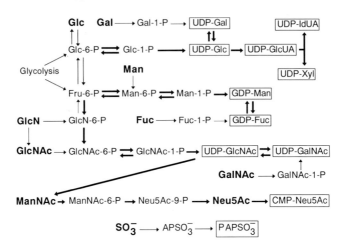

4.2. Glycosyltransferases

A large family of enzymes

The glycosyltransferases are the group of enzymes responsible for sequential transfer of monosaccharides on to acceptors to form oligosaccharide structures (1, 3, 5, 19). They catalyze reactions of the general type:

$$\text{monosaccharide–XDP} + \text{acceptor} \rightarrow \text{acceptor–monosaccharide} + \text{XDP}.$$

A large number of glycosyltransferases have been described and some examples are given in *Table 13* to illustrate the range of monosaccharides transferred and the variation in acceptor. The prediction that glycosyltransferases function as multienzyme complexes in order to optimize the specificity and fidelity of oligosaccharide sequence formation was made by Roseman (20).

Substrate specificity

The specificity of the glycosyltransferases depends on the nature of the glycoconjugate acceptor, e.g. glycoprotein or proteoglycan, the non-reducing terminal sugar and the oligosaccharide sequence (1, 5, 19). The result of this specificity is the formation of certain structures, the exclusion of others and the existence of ordered pathways in the formation of defined oligosaccharide structures. An example of a pathway illustrating such features is shown in *Figure 4*. Not all of the glycosyltransferases show such narrow substrate specificity, and some of the heterogeneity found in the glycosylation of proteins may arise as a result of this type of glycosyl transfer.

Table 13. Glycosyltransferases — examples of transfer

Glycosyltransferase EC number	Reaction	Nucleotide sugar	Glycosylated protein
Galactosyl-			
EC 2.4.1.38/90	β1-4Gal→GlcNAc in GlcNAcβ1-2Manα1- or Glc/GlcNAc as monosaccharides	UDP-Gal	N–linked (lactose or lactosamine)
EC 2.4.1.37	α1-3Gal→Gal in Fucα1-2Galβ1-	UDP-Gal	N-/O-linked
EC 2.4.1.122	β1-3Gal→GalNAc in GalNAcα-O-Ser(Thr)	UDP-Gal	O-linked
Mannosyl-			
EC 2.4.1.83	αMan→dolichol Manβ-O-dolichol	GDPαMan	Lipid linked (N-linked)
EC 2.4.1.x	β1-4Man→GlcNAc in GlcNAcβ1-4GlcNAcα-	GDPαMan	Lipid linked (N-linked)
EC 2.4.1.132	α1-3Man→βMan in Manβ1-4GlcNAcβ1-	Manβ1-O-Ⓟ-Dol	Lipid linked (N-linked)
Sialyl-			
EC 2.4.99.1	α2-6Neu5Ac→Gal in Galβ1-4GlcNAc-	CMP-Neu5Ac	N-linked
EC 2.4.99.6	α2-3Neu5Ac→Gal in Galβ1-4GlcNAc-	CMP-Neu5Ac	N-linked
EC 2.4.99.4	α2-3Neu5Ac→Gal β1-3Gal→GalNAcα	CMP-Neu5Ac	O-linked
EC 2.4.99.x	α2-8Neu5Ac→Neu5Ac -(α2-8Neu5Ac)$_n$-	CMP-Neu5Ac	N-(O-?) linked polysialic acid
Fucosyl-			
EC 2.4.99.68	α1-6Fuc→GlcNAc -4GlcNAcβ1-N-Asn	GDP-Fuc	N-linked
EC 2.4.99.69	α1-2Fuc→Gal Galβ1-3/4-	GDP-Fuc	N-/O-linked
EC 2.4.99.152	α1-3Fuc→GlcNAc in Galβ1-4GlcNAc	GDP-Fuc	N-/O-linked

Table 13. Continued

Glycosyltransferase EC number	Reaction	Nucleotide sugar	Glycosylated protein
N-acetylglucosaminyl-			
EC 2.4.1.101	β1-2GlcNAc→Man in Manα1-3Manβ1-4 (GlcNAc transferase I see *Figure 7*)	UDP-GlcNAc	*N*-linked
EC 2.4.1.144	β1-4GlcNAc→Man in Manβ1-4GlcNAc (GlcNAc transferase III, bisecting GlcNAc)	UDP-GlcNAc	*N*-linked
EC 2.4.1.146	β1-3GlcNAc→Gal in Galβ1-3GalNAcα- (Core 1 elongation enzyme)	UDP-GlcNAc	*O*-linked
EC 2.4.1.x	α1-4GlcNAc→GlcUAβ1 or IdUAα1 in GlcUAβ1-4GlcNAcα1-4IdUAα1-	UDP-GlcNAc	Proteoglycan (heparan sulfate)
N-acetylgalactosaminyl-			
EC 2.4.1.41	αGalNAc→Ser(Thr)-peptide	UDP-GalNAc	*O*-linked
EC 2.4.1.40	αGalNAc→Gal in Fucα1-2Galβ1- Blood group A enzyme	UDP-GalNAc	*N*-/*O*-linked
EC 2.4.1.x	β1-4GalNAc→GlcUA in GlcUAβ1-3GalNAcβ1-	UDP-GalNAc	Proteoglycan (chondroitin sulfate)
Glucuronyl-			
EC 2.4.1.x	β1-3GlcUA→GalNAc in GalNAcβ1-4GlcUAβ1-	UDP-GlcUA	Proteoglycan (dermatan sulfate)

Abbreviations: Asn, L-asparagine; CMP, cytosine monophosphate; Dol, dolichol; Fuc, fructose; Gal, D-galactose; GalNAc, *N*-acetyl-D-galactosamine; GDP, guanosine diphosphate; Glc, D-glucose; GlcNAc, *N*-acetyl-D-glucosamine; GlcUA, D-glucuronic acid; IdUA, L-iduronic acid; Man, D-mannose; Neu5Ac, *N*-acetylneuraminic acid; Ⓟ, phosphate; Ser, L-serine; Thr, L-threonine, UDP, uridine diphosphate. EC numbers ending in 'x' have not yet been allocated.

Figure 4. Biosynthesis of an oligosaccharide. The biosynthetic pathway for the formation of a sialylated tetrasaccharide common to many membrane glycoproteins shows how glycosyltransferase specificity can govern the final product on a pathway. The bars ▬ show 'dead end' products or reactions that do not occur, or are slower ⊏⊐ due to glycosyltransferase specificity. The glycosyltransferases in the above scheme are: 1, polypeptide α-O-N-acetylgalactosaminyl transferase; 2, α-N-acetylgalactosamide β1-3 galactosyltransferase; 3, galactosyl β1-3 N-acetylgalactosamide α2-3 sialyltransferase; 4, sialyl α2-3 galactosyl β1-3 N-acetylgalactosamide α2-6 sialyltransferase; 5, α-N-acetylgalactosamide α2-6 sialyltransferase. E, pathways leading to extended structures through other glycosyltransferases.

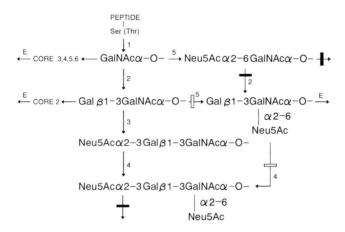

Peripheral additions, chain terminations
A variety of oligosaccharide structures can be identified as common terminator sequences (see *Tables 10, 11* and *Figure 4*). The completion of oligosaccharide chains is dependent on the type of chain, N-, O-linked or glycosaminoglycan (1, 5, 19). It is achieved through the specificity of the oligosaccharide multienzyme, biosynthetic complexes 'designed' for each oligosaccharide (20). Many of the structures resulting in chain termination also have a role in the biodegradation of oligosaccharides on proteins, through their influence on glycosidase activity. This is also true for other post-translational modifications, such as sulfation and acetylation (see Section 3.2).

Regulation and fidelity of sequence reproduction
The production of an oligosaccharide sequence with high fidelity is of great importance where information is being expressed. This is the case, for example, in the production of the ABO blood group antigens. The integrity of these transfer reactions is dependent on the specificity of the glycosyltransferases, the subcellular localization and compartmentalization of these enzymes as functional multienzyme complexes (20) and the intermediate metabolic regulation of acceptor and nucleotide sugar substrates.

Glycosyltransferases — subcellular distribution
The membrane location of the glycosyltransferases is of great significance for the biosynthesis of glycoconjugates in the cell (1, 3, 5, 19). The organization of N- and O-linked biosynthesis in glycoproteins and proteoglycans shows significant differences which correlate with activity located in specific subcellular membranes.

The biosynthesis of N-linked oligosaccharides proceeds via the stepwise addition of monosaccharides from nucleotide sugars and lipid carriers, the dolichol phosphates (*Figure 5*), in the membrane-mediated process illustrated in *Figure 6* (3, 18). The process initially involves the transfer of sugars to the dolichol carrier on the cytoplasmic side of the endoplasmic reticulum membrane, and subsequent completion of the structure after a 'flip' to the endoplasmic reticulum lumen mediated by a flippase (see *Figure 6*). This results in the formation of an oligosaccharide containing 14 monosaccharides linked to the dolichol pyrophosphate carrier. The entire oligosaccharide is transferred *en bloc* to peptide acceptors and then processed or trimmed to give high-mannose intermediate structures (*Figure 7*), which are then further modified in the Golgi apparatus to yield the complex and hybrid types of N-linked oligosaccharides shown in *Figure 2* (1, 18, 19). Inhibitors for specific steps on the pathway have been discovered which act at different stages in the biosynthesis and processing of the N-linked oligosaccharide chains (*Figures 6* and *7*). These are based on naturally occurring products and have used as tools to determine the pathways themselves and, more recently, as therapeutic agents (21).

O-linked oligosaccharides do not show this type of *en bloc* transfer and all of the glycosyl-transfer takes place in a step-by-step process on the acceptor polypeptide at sites throughout the rough endoplasmic reticulum and Golgi apparatus (1, 5).

Figure 5. Structure of a dolichol-phosphate-linked oligosaccharide. The diagram shows the structure of the complete oligosaccharide linked to the lipid carrier dolichol through a pyrophosphate bridge. △, glucose; □, mannose; ●, N-acetyl-glucosamine.

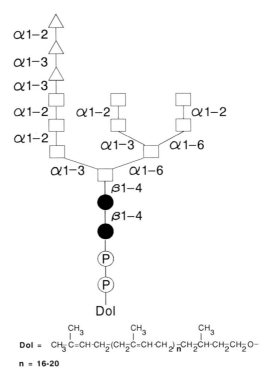

Proteoglycans are biosynthesized in an integrated membrane-mediated system, which includes the enzymes required for the initiation, elongation and repeat structures found in the glycosaminoglycan chains together with those required for N- and O-linked oligosaccharide chains (6, 9, 22). In addition, there are also those enzymes necessary for O- and N-sulfation, epimerization and deacetylation (22). The arrangement of these enzymes in the membranes is believed to allow the simultaneous synthesis of multisite oligosaccharides.

5. EXAMPLES OF FUNCTIONAL ROLES FOR PROTEIN GLYCOSYLATION

There have been many studies documenting the chemical nature of protein glycosylation and the enzyme processes leading to it. Far fewer studies have been successful in demonstrating

Figure 6. Biosynthesis of a dolichol-linked oligosaccharide. Steps in the formation of the initial peptide-bound oligosaccharide through the stages of lipid-linked glycosyltransfer. Note that some transfers take place through nucleotide sugar donors and others from dolichol phosphate monosaccharides. The site of action of the N-linked oligosaccharide inhibitor, tunicamycin, is shown by T. For explanation of other symbols see *Figure 5*.

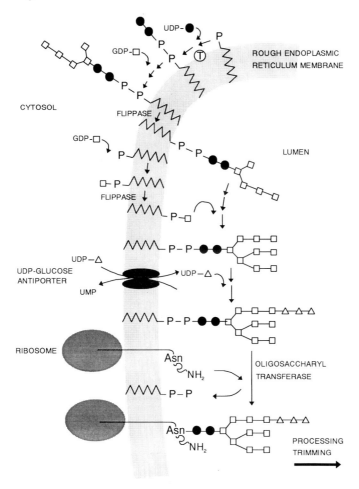

Figure 7. Processing pathways for *N*-linked oligosaccharides. The pathways of endoplasmic reticulum and Golgi membrane mediated processing of *N*-linked oligosaccharides are shown. The sites of action of the glycoprotein processing inhibitors are given by: C, castanospermine; S, swainsonine; D, deoxynojirimycin; DM, deoxymannonojirimycin. For explanation of other symbols see *Figure 5*.

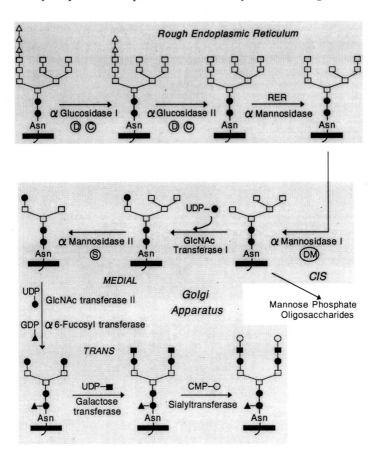

structure–function relationships for membrane-linked glycosyl-proteins. The roles of the carbohydrate chains are varied (23) and many of the examples given below belong in more than one group. A few examples have been chosen to illustrate the main groups identified so far and are shown in *Table 14*.

Structural and antigenic components.   Roles in the preservation of normal membrane structure, including its flexibility and surface antigenic properties, have been identified for many glycosylated proteins both at the cytoskeleton level and at the cytoplasmic (e.g. glycocalyx) and extracellular surfaces (e.g. intercellular junction components, microvillar enzymes).

Cell adhesion.   Many cell surface glycoproteins and proteoglycans (7–9, 22, 24, 25) have been shown to play a role at the plasma membrane, involving attachment to other cells and to other surfaces. These interactions are subject to modification by other cell-bound and extracellular matrix components.

Table 14. Functions of some glycosylated proteins

Group/example	Type	Location (example)
Structural and antigenic roles		
Epiglycanin	Mucus glycoprotein	Mouse TA3 cells
Versican	CS Proteoglycan	Fibroblasts
Glycophorin	O-/N-Glycoprotein	Erythrocytes
Band 4.1	O-GlcNAc Glycoprotein	Erythrocytes
Nuclear pore 62 kDa	O-GlcNAc Glycoprotein	Widespread
Cell adhesion		
CD44	CS/HS Proteoglycan N-Glycoprotein	Lymphocytes and epithelial cells
Decorin	CS/DS Proteoglycan	Connective tissue cells
Fibroglycan	HS Proteoglycan	Fibroblasts
Leukosialin	O-/N-Glycoprotein	Leukocytes
Transport		
H^+-ATPase	Type B glycoprotein	Widespread
GLUT 1	Type B glycoprotein	Intestinal cells
Sodium channel	Type B glycoprotein	Nerve cells
Growth factor binding		
Decorin	CS/DS Proteoglycan	Connective tissue cells (TGF-β)
Betaglycan	CS/HS Proteoglycan	Fibroblasts (TGF-β)
Targeting		
Selectin E ligand	Glycoprotein	Endothelial cells
Sialoadhesin ligand	Glycoprotein	Bone marrow precursor cells
CD75	Glycoprotein	Activated B-cells
Receptors		
Insulin	Glycoprotein	Adipocytes
LDL	Glycoprotein	Fibroblasts
Thy-1	Glycoprotein	Mouse lymphocytes
Transferrin	Glycoprotein	Hepatocytes
EGF	Glycoprotein	Fibroblasts

Abbreviations: CS, chondroitin sulfate; DS, dermatan sulfate; EGF, epidermal growth factor; GlcNAc, N-acetyl-D-glucosamine; HS, heparan sulfate; LDL, low-density lipoprotein; TGF, transforming growth factor.

Receptors. A common feature of the many receptors detected at specific cell surfaces is the presence of oligosaccharide chains. These have a range of functions including stabilization, specific ligand recognition and receptor cycling phenomena within the cell (26).

Targeting. The recent discovery of leukocyte and endothelial cell carbohydrate ligands for the selectins, a family of cell surface lectins functioning to direct leukocytes to sites of injury or to direct recirculation into the lymph nodes, has highlighted a highly specific series of interactions (24, 25). Similar carbohydrate-mediated interactions are known in other cells.

Transport. Many integral membrane-transport proteins contain N-linked oligosaccharides. The role of the carbohydrate in these molecules is most likely in directing the proteins to their plasma membrane location (27).

Growth factor binding. Cell surface proteoglycans have been found to act as specific receptors for several important growth factors, including TGF-β and fibroblast growth factor (7–9, 22). In addition, a number of specific growth-factor-binding glycoproteins exist (2, 28).

6. REFERENCES

1. Allen, H.J. and Kisailus, E.C. (1992) *Glycoconjugates, Composition, Structure and Function.* Marcel Dekker, Basel.

2. Evans, W.H. and Graham, J.M. (1989) *Membrane Structure and Function.* IRL Press at Oxford University Press, Oxford.

3. Kornfeld, R and Kornfeld, S. (1985) *Ann. Rev. Biochem.,* **54**, 631.

4. Montreuil, J. (1982) *Comp. Biochem.,* **19B** (II), 1.

5. Sadler, J.E. (1984) in *Biology of Carbohydrates* (V. Ginsberg and P.W. Robbins, eds). Wiley, New York, p. 199.

6. Poole, A.R. (1986) *Biochem. J.,* **236**, 1.

7. Esko, J.D. (1991) *Curr. Opin. Cell Biol.,* **3**, 805.

8. Wight, T.N., Kinsella, M.G. and Qwarnstrom, E.E. (1992) *Curr. Opin. Cell Biol.,* **4**, 793.

9. Gallagher, J.T. (1989) *Curr. Opin. Cell Biol.,* **1**, 1201.

10. Carlstedt, I., Sheehan, J.K., Corfield, A.P. and Gallagher, J.T. (1985) *Essays Biochem.,* **20**, 40.

11. Jentoft, N. (1990) *Trends in Biochem. Sci.,* **15**, 291.

12. Haltiwanger, R.S., Kelly, W.G., Roquemore, E.P., Blomberg, M.A., Dennis Dong, L.-Y., Kreppel, L., Chou, T.-Y. and Hart, G.W. (1991) *Biochem. Soc. Trans.,* **20**, 264.

13. Rademacher, T.W., Parekh, R.B. and Dwek, R.A. (1988) *Ann. Rev. Biochem.,* **57**, 785.

14. Cross, G.A.M. (1990) *Ann. Rev. Cell Biol.,* **6**, 1.

15. Ferguson, M.A.J. (1992) *Biochem. Soc. Trans.,* **20**, 243.

16. Silverman, M. (1991) *Ann. Rev. Biochem.,* **60**, 757.

17. Gould, G.W. and Bell, G.I. (1990) *Trends Biochem. Sci.,* **15**, 18.

18. Hirschberg, C.B. and Snider, M.D. (1987) *Ann. Rev. Biochem.,* **56**, 63.

19. Schachter, H. (1991) *Curr. Opin. Struct. Biol.,* **1**, 755.

20. Roseman, S. (1970) *Chem. Phys. Lipids,* **5**, 270.

21. Winchester, B. and Fleet, G.W.J. (1992) *Glycobiology,* **2**, 199.

22. Kjellen, L. and Lindahl, U. (1991) *Ann. Rev. Biochem.,* **60**, 443.

23. Paulson, J.C. (1989) *Trends in Biochem. Sci.,* **14**, 272.

24. Varki, A. (1992) *Curr. Opin. Cell Biol.,* **4**, 257.

25. Springer, T.A. (1990) *Nature,* **346**, 425.

26. Dohlman, H.G., Caron, M.G. and Lefkowitz, R.J. (1987) *Biochemistry,* **26**, 2657.

27. Catterall, W.A. (1986) *Ann. Rev. Biochem.,* **55**, 953.

28. Carpenter, G. (1987) *Ann. Rev. Biochem.,* **56**, 881.

CHAPTER 10
CHEMICAL AND POST-TRANSLATIONAL MODIFICATION OF PROTEINS

Once synthesized, proteins adopt a specific three-dimensional structure. Protein structures have been analyzed by studying their chemical modifications; this is covered in the first part of the chapter. Proteins are also modified in nature as one of the mechanisms for modifying their activities; the second part of the chapter focuses on this topic.

1. CHEMICAL MODIFICATION OF PROTEINS
G. E. Means and R. E. Feeney

The chemical modification of proteins has been an important part of the methodology for determining the structure and function of proteins. A number of reagents that modify the amino acid residues of proteins has been characterized. Lists of these reagents and their specificities are given in *Tables 1* and *2*. It is important to note that few of the reagents are absolutely specific for a particular group and the reaction of groups of an amino acid may be affected by their environment.

2. ENZYME-CATALYZED COVALENT MODIFICATION REACTIONS: AN OVERVIEW
B. L. Martin

2.1. Introduction
The alteration of protein structure by enzyme-catalyzed post-translational modification reactions provides a sensitive system for tightly regulating enzyme activity (144–146). These reactions are crucial for understanding cellular responses to essentially all external signals and the regulation of every biological system. *Table 3* lists the more commonly identified modifications and some of the events with which they are involved. This list, is designed to be illustrative and not comprehensive, as is this discussion. Good sources for additional information are volumes 106 and 107 of the series *Methods in Enzymology*.

Post-translational modification reactions provide an amplification mechanism for biological signals. In blood clotting (165) the initial activation of the first enzyme initiates a series of reactions called a cascade. Activation of the first enzyme leads to a subsequent increase in reaction rate at each step of the reaction, resulting in a significant increase in the reaction rate of each step in the cascade. This is an example of an irreversible, noncyclic cascade. More important for cellular regulation are cyclic cascade processes, including protein phosphorylation–dephosphorylation, acetylation–deacetylation, adenylylation–deadenylylation and (ADP)-ribosylation–de(ADP)-ribosylation. These cyclic cascades provide a sensitive regulatory mechanism because they:

(i) allow amplification of the primary signal;
(ii) involve multiple enzymes with a potential for an increase in the number of sites for allosteric control;

▶ p. 234

Table 1. Reagents for selective chemical modification of proteins[a]

Amino acid side-chain	Reaction/reagents	Competing reactions[b]	Comments advantages/disadvantages
Lysine (ε-NH_2)	Amidination/methyl or ethyl acetimidate (.HCl)	α-NH_2*	White, hygroscopic, water-sol. solid; aq. solns. undergo rapid hydrolysis; specific reaction with -NH_2 at pH $\approx$ 9; some cross-linking at lower pH; relatively small substituent retains cationic charge; analysis by TNBS (1, 2)
	Dimethyl suberimidate, dimethyl adipimate, and many other bis-imido esters (usually as HCl salts)	αNH_2*	Very similar to above; gives inter- and intramol. cross-link of various lengths; to determine distances between reactive -NH_2 groups, subunit organization, etc. (3, 4)
	2-Iminothiolate.HCl (Traut's reagent)	c	Very similar to above; cationic substituent with -SH moiety; introduced -SH can be used to prepare secondary derivs to effect cross-linking, etc. (5, 6)
Lysine (ε-NH_2)	Reductive methylation/ formaldehyde + $NaBH_4$, $NaBH_3CN$ or pyridine borane	α-NH_2*	Specific for NH_2; optimal reaction at pH $\approx$ 9 with $NaBH_4$ and $\approx$ 7 with $NaBH_3CN$; small cationic methyl- and dimethylamino-derivs retain cationic charge; dimethyl amino derivs usually predominate except at low levels of modification; used to introduce ^3H, ^{13}C and ^{14}C labels; other aldehydes/ketones (e.g. pyridoxal phosphate, aldoses, etc.) can be used to give corresponding monoalkylamine moieties; anal. by TNBS or AA anal. (7, 8)
	Acetylation/acetic anhydride	Tyr*	Colorless liquid; hydrolyzes rapidly in water; reaction at $\approx$ pH 8–10, also modifies Tyr; eliminates cationic charge; anal. by TNBS (9, 10)

Reagent	Target	Description
N-Hydroxysuccinimidyl esters (NHS esters)	α-NH_2	Used to introduce p-hydroxyphenyl propionyl (Bolton–Hunter reagent), biotin (NHS biotin), fluorescent probes (e.g. NHS 1-pyrenebutyrate), various cross-linking (NHS p-azidosalicylate, diNHS suberate, NHS 3-(2-pyridylthio)–propionate, etc.) and other special-purpose groups; reaction at $\approx$ pH 7 and above; high selectivity for -NH_2 groups; eliminate cationic charge; Na salts of sulfo-NHS esters are more water sol. (11, 12)
Maleic and citraconic anhydrides	α-NH_2* -SH	White cryst. solids; reaction at $\approx$ pH 8 and above; selective for NH_2 groups; undergo rapid hydrolysis; introduce anionic charge in place of original cationic charge; reversible upon incubation at pH 3–4; used for temporary blocking of -NH_2 groups to prevent cleavage by trypsin; to increase solubility and effect subunit dissocn; determine extent of reaction from absorption at $\approx$ 250 nm or by reaction with TNBS (13, 14)
2,4,6-Trinitrobenzenesulfonate (TNBS)	-SH*	Pale yellow cryst. solid, water-sol, reaction at $\approx$ pH 8–10; very specific for -NH_2 groups; hydrolyzes slowly under alk. conditions; large hydrophobic substituent eliminates cationic charge; forms Meisenheimer-type complexes with sulfite ion; strong visible absorption ($\varepsilon_{345} \approx 14\,500$ $M^{-1}\,cm^{-1}$ or $\varepsilon_{420} \approx 19\,200\,M^{-1}\,cm^{-1}$ for the sulfite complex) used for the quantitative detn of -NH_2 groups (15, 16)
2,3-Butanedione	-NH_2	Yellow water-sol. liq.; reaction at $\approx$ pH 7–8, eliminates cationic charge; reaction promoted by borate buffer; more than one product; partially reversible upon dialysis or dilution; detn from decreased Arg upon AA anal. (17, 18)
Phenylglyoxal (hydrate)	-NH_2	White, water-sol., cryst. solid; reaction at $\approx$ pH 7.5–8.5 with 2 equivs; hydrophobic deriv. eliminates cationic charge; reaction promoted by carbonate-bicarbonate buffer; detn from decreased Arg upon AA anal. (19, 20)

Arginine (guanidino)

Table 1. Continued

Amino acid side-chain	Reaction/reagents	Competing reactions[b]	Comments advantages/disadvantages
Histidine (imidazole)	Diethyl pyrocarbonate (ethoxyformic anhydride)	-NH_2* Tyr*	Colorless liq.; slightly sol. and hydrolyzes rapidly in water; reaction at ≈ pH 6–7 (low pH suppresses reaction of other nucleophilic groups); extent of reaction can be detd from increase in absorbance at ≈ 240 nm ($\varepsilon \approx 3200$ M^{-1} cm^{-1}); reversible upon addn of NH_2OH (21, 22)
Aspartic and glutamic acids	Water-sol. carbodiimide + amine (e.g. ethyldimethylaminopropyl-carbodiimide + glycine ethyl ester)	α-COOH* Tyr -SH	White, water-sol. solids; carbodiimides undergo slow hydrolysis in aq. soln.; reaction at ≈ pH 4.5–5; eliminates cationic charge; different amines may be used to introduce substituents of different sizes, charges, etc., also used without amine to introduce so-called zero-length cross-links between closely spaced carboxyl and amino groups (especially promoted by the presence of N-hydroxy succinimide); products of reaction with Tyr can be eliminated by treatment with NH_2OH (23–25)
Cysteine (-SH)	Alkylation/iodoacetate, iodoacetamide, bromoacetate bromoacetamide	His -NH_2 Met	White, cryst. solids; sol. but slowly hydrolyze in water; reaction at ≈ pH 7; very high reactivity with -SH groups; much slower reaction with other nucleophilic groups — may be significant at high reagent concns, with long reaction times or in the case of site-directed reagents; detn by DTNB or AA anal; many other halocarbonyl compds have been used to introduce spectroscopic probes (e.g. N-(1-pyrene)iodoacetamide, 5-iodoacetylaminoethyl)amino-1-naphthalene sulfonic acid, 4-acetamido-TEMPO, etc.), biotin (e.g. iodoacetylbiotin), to effect cross-linking (e.g. NHS p-iodoacetamidobenzoate, p-iodoacetamidobenzo-

Reagent	Target	Description
(continued)		phenone, etc.), other special-purpose groups and affinity labels (e.g. TPCK, bromoacetylthyroxine, bromopyruvate, 8-[(4-bromo-2,3-dioxobutyl)thio]-adenosine triphosphate, etc. (26, 27))
N-Ethylmaleimide	$-NH_2$ His	White, cryst. solid; slightly sol. and slowly decomposes in water; high reactivity with -SH groups at $\approx$ pH 6–7; slower reaction with $-NH_2$ groups may be significant at higher pH; detn by DTNB or AA anal.; many other N-alkyl and N-arylmaleimides can be used to introduce spectroscopic probes (e.g. N-(1-pyrene)maleimide, N-(4-TEMPO)-maleimide, etc.), other special-purpose groups and to effect cross-linking (e.g. bismaleimidohexane, m-maleimidobenzoic acid NHS ester, etc. (28, 29))
5,5'-Dithiobis (2-nitrobenzoic acid) (DTNB or Ellman's reagent)		Pale yellow, cryst. solid; water-sol. and stable; specific reaction with -SH groups; detn by absorption increase at $\approx$ 412 nm (ε = 13 600 M^{-1} cm^{-1}); large anionic substituent ($\varepsilon_{310} \approx$ 2500 M^{-1} cm^{-1}) may be displaced by addn of excess thiol; subsequent reaction with cyanide gives thiocyanate derivs (30–32)
p-Hydroxymercuribenzoate		White, cryst. solids; water-sol. and stable; specific reaction with -SH groups at $\approx$ pH 5–8; large hydrophobic substituent; detn from absorbance at $\approx$ 225 nm ($\varepsilon \approx$ 7600 M^{-1} cm^{-1}) potentially reversible upon addn of excess thiol (33, 34)
Hydrogen peroxide	-SH* Trp	Usually as $\approx$ 30% aq. soln, reaction usually below pH 5; AA anal. after aryl or alkylsulfonic acid hydrolysis; reversible upon addn of excess thiol (35, 36)
Methionine (thioether)		

Table 1. Continued

Amino acid side-chain	Reaction/reagents	Competing reactions[b]	Comments advantages/disadvantages
Tyrosine	Iodination/I_2, I_3^-, ICl (I^- + chloramine T or one of many other Cl^+ donors) or lactoperoxidase + I^- and H_2O_2	His* -SH* Trp	Metallic, dark gray crysts; water-sol. in the presence of I^- (as I_3^-); used for radiolabeling, usually with $Na^{131}I$ ($t_{1/2}$ = 8 days) or $Na^{125}I$ ($t_{1/2}$ = 56 days) plus oxidant; reaction at ≈ pH 8–9; gives mono- and diiodo derivs of both Tyr and His; large substituents decrease pK_a and increase UV absorbance of the modified Tyr and His residues (37–39)
	Nitration/tetranitromethane	-SH* Trp	Yellow. liq. low sol. in water; stable in ethanol; reaction at ≈ pH 8 or above; decreases pK_a of modified Tyr to ≈ 6.8–7.0 and increases absorbance at 360 nm (ε = 2790 M^{-1} cm^{-1}); detn from increased absorbance at 360 or 428 nm (ε = 4100 M^{-1} cm^{-1} above pH 9) or by AA anal. (40)
	Acetylation/acetyl imidazole	-NH_2*	White cryst. solid; sol. but slowly hydrolyzes in water; reaction at ≈ pH 7–8; detn from decreased absorption at 280 nm (Δ = 1200–1400 M^{-1} cm^{-1}); deacetylation by NH_2OH (41)
Tryptophan (indole)	N-Bromosuccinimide	-SH* Tyr Met His	White cryst. solid; water-sol.; reaction usually at ≈ pH 4.5 or lower; detn from decreased absorbance at 280 nm (Δ ≈ 4000 M^{-1} cm^{-1} (42))

[a] Selected on the basis of their widespread use and apparent usefulness. A more extensive list of reagents and the amino acid side-chains with which they usually react is presented in Table 2.
[b] Listed in the approximate order of their reactivity. An asterisk (*) is used to indicate those other groups that will usually react under normal conditions. probably also reacts with α-NH_2.
[c] Complete information not available; probably also reacts with α-NH_2.
Abbreviations: AA, amino acid; addn, addition; alk, alkaline; anal., analysis; aq., aqueous; compds, compounds; concns, concentrations; cryst, crystalline; derivs, derivatives; detd, detected; detn, detection; dissocn, dissociation; DTNB, 5,5′-dithiobis(2-nitrobenzoic acid); equivs, equivalents; intermol., intermolecular; intramol, intramolecular; liq, liquid; NHS, N-hydroxysuccinimidyl; sol, soluble; solns, solutions; soly, solubility; TNBS, 2,4,6-trinitro-benzenesulfonate; TPCK, N-tosyl-phenylalanine chloromethyl ketone.

Table 2. The specificity of reagents used to chemically modify proteins[a,b]

Reagent	Amino acid and side-chain										Ref.
	Amino[c] (Lys)	Guanidino (Arg)	Imidazole (His)	Carboxyl (Glu, Asp)	Thiol (Cys)	Thioether (Met)	Disulfide (CysS)	Phenol (Tyr)	Indole (Trp)		
Acetic anhydride [108-24-7]	+++	—	±	—	±	—	—	++[e]	—		43
N-Acetylhomocysteine thiolactone [1195-16-0]	+++	—	±	—	—[d]	—	—	—	—		44
Acetylimidazole [2466-76-4]	++	—	±	—	±	—	—	+++[e]	—		41
(S-Acetylthio)succinic anhydride [6953-60-2]	+++	—	±	—	±	—	—	±	—		45
Acrylodan [86636-92-2]	—	—	+	—	+++	—	—	—	—		46
Acrylonitrile[f] [107-13-1]	+	—	+	—	+++	—	—	—	—		47
Aldehydes + NaBH$_4$, NaBH$_3$(CN), or pyridine borane	+++	—	—	—	—	—	—	—	—		7, 8
4-(Aminosulfonyl)-7-fluorobenz-2-oxa-1,3-diazole [91366-65-3]	—	—	—	—	+++	—	—	—	—		48

Table 2. Continued

Reagent	Amino acid and side-chain									Ref.
	Amino[c] (Lys)	Guanidino (Arg)	Imidazole (His)	Carboxyl (Glu, Asp)	Thiol (Cys)	Thioether (Met)	Disulfide (CysS)	Phenol (Tyr)	Indole (Trp)	
Ammonium 4-chloro-7-sulfobenz-2-oxa-1,3-diazole [81377-14-2]	+	—	—	—	+++	—	—	—	—	49
Arsenite, sodium $(NaAsO_2)$[f] [7784-46-5]	—	—	—	—	+++[e,g]	—	—	—	—	50
Aryl azides[h]	+	+	+	+	+	+	—	+	++	51
Azobenzene-2-sulfenyl bromide [2849-62-9]	—	—	—	—	+++	—	—	—	—	52
4,4'-Bis(dimethylamino)diphenylcarbinol [119-58-4]	—	—	—	—	+++	—	—	—	—	53
Bismaleimidyl hexane [4856-87-5]	+	—	+	—	+++	—	—	—	—	54
Bromobimane (mono and dibromo) [71418-44-5]	+	—	+	—	+++	+	—	+	—	55
Bromoethylamine [107-09-5]	—	—	+	—	+++	—	—	—	—	56

Reagent											Ref.
p-Bromophenacyl bromide [99-73-0]	+	—	++	+[i]	+++	+	—	—	—	—	57
N-Bromosuccinimide [128-08-5]	—	—	+	—	+++	++[e]	—	++	+++	—	58
Butanedione[k] [431-03-8]	—	+++	—	—	—	—	—	—	—	—	59
Carbodiimide (dicyclohexyl) [538-75-0]	+[j]	—	—	+++	+	—	—	+[e]	—	—	60
Carbodiimide (water sol.) + RNH$_2$	+[j]	—	—	+++	+	—	—	+[e]	—	—	61
N-carboxyanhydrides	+++	—	±	—	±	—	—	±	—	—	62
Chloramine T [127-65-1]	—	—	—	—	+++	++[e]	—	—	—	—	63
7-Chloro-4-nitrobenz-2-oxa-1,3-diazole (NBD chloride) [10199-89-0]	+++	—	—	—	+++	—	—	+	—	—	64
N-Chlorosuccinimide [128-09-6]	—	—	—	—	+++	+++[e]	—	—	—	++	65
Citraconic anhydride [616-02-4]	+++[e]	—	±	—	+	—	—	±	—	—	13,66
Cyanate, potassium (KCNO) [590-28-3]	+++	—	±	±	±±±	—	—	±	—	—	67

Post-translational Modification

Table 2. Continued

Reagent	Amino acid and side-chain									Ref.
	Amino[c] (Lys)	Guanidino (Arg)	Imidazole (His)	Carboxyl (Glu, Asp)	Thiol (Cys)	Thioether (Met)	Disulfide (CysS)	Phenol (Tyr)	Indole (Trp)	
Cyanogen bromide[f] [506-68-3]	–	–	+	–	+++	+++	–	–	–	67
Cyanuric fluoride [675-14-9]	+	–	+	–	+	–	–	+++	–	69
1,2-Cyclohexanedione[k] [765-87-7]	–	+++	–	–	–	–	–	–	–	70
Dansylaziridine [51908-46-4]	–	–	–	–	+++	+	–	–	–	71
Diamide (diazenedicarboxylic acid bis(N,N-dimethyl amide) [10465-78-8]	–	–	–	–	+++	–	–	–	–	72
Diazoacetates	–	–	–	+++	++	–	–	–	–	73
Diazomethyl-N,N-dimethyl-benzene sulfonamide [71398-36-2]	–	–	–	+++	+	–	–	–	–	74
Diazonium ions	+++	+	+++	–	++	–	–	+++	+	40, 75
Dichlorotriazines	+++	–	+	–	++	–	–	+++	–	76
Didansylcystine [18468-46-7]	–	–	–	–	+++[e]	–	–	–	–	77

Reagent											Ref.
Diethyl pyrocarbonate [1609-47-8]	+	—	—	+++e	±	+e	—	—	+e	+p	21
Dimethylaminonaphthalene-sulfonyl chloride [605-65-2]	+++	—	—	++	—	++	—	—	+++	+	78
Dimethyl(2-hydroxy-5-nitrobenzyl) sulfonium bromide [28611-73-6]	—	—	—	—	—	+++	±	—	—	++	79
Dimethyl suberimidate (bisimido esters) [29878-26-0]	+++	—	—	—	—	—	—	—	—	—	3,4
Dinitrofluorobenzene [70-34-8]	+++	—	—	++	—	+++	—	—	++	—	80
Diphenyldiazomethane [883-40-9]	—	—	—	—	+++	—	—	—	—	—	81
5,5'-Dithiobis(2-nitrobenzoic acid) [67-78-3]	—	—	—	—	—	+++e	—	—	—	—	30
Dithiobis(succinimidyl propionate) [57757-57-0]	+++	—	±	—	—	±+l	—	—	+	—	11,82
2,2'-Dithiodipyridine [2127-03-9]	—	—	—	—	—	+++e	—	—	—	—	83
4,4'-Dithiopyridine [2645-2-9]	—	—	—	—	—	+++e	—	—	—	—	83
Dithiothreitol [3483-12-3]	—	—	—	—	—	—	—	+++e	—	—	84

Post-translational Modification

Table 2. Continued

Reagent	Amino acid and side-chain									Ref.
	Amino[c] (Lys)	Guanidino (Arg)	Imidazole (His)	Carboxyl (Glu, Asp)	Thiol (Cys)	Thioether (Met)	Disulfide (CysS)	Phenol (Tyr)	Indole (Trp)	
Ethylenimine[f] [151-56-4]	—	—	—	+	+++	—	—	—	—	85
N-Ethylmaleimide[m] [128-53-0]	+	—	+	—	+++	—	—	—	—	86
N-Ethyl-5-phenyl isoxazolium-3-sulfonate [4156-16-5]	+[j]	—	—	+++	—	—	—	—	—	87
Ethyl thiotrifluoroacetate [383-64-2]	+++	—	±	—	±	—	—	±	—	88
Fentons reagent (Fe(II)/O$_2$/ ± ascorbate)	+++	+++	+++	—	—	—	—	—	—	89
Ferrate, potassium (K$_2$FeO$_4$) [13718-66-6]	—	—	++	—	++	++	—	++	++	90
Fluorescein isothiocyanate [27072-45-3]	+++	—	++	—	++	—	—	+	—	91
Formaldehyde [50-00-0]	±±±	+	±±±	—	±±±	—	—	++	++	92
Glutaraldehyde [111-30-8]	+++	+	++	—	+	—	—	—	—	93
Glyoxal [107-22-2]	+[n]	+++	—	—	+	—	—	—	—	94

Reagent									Ref.
Guanyl-3,5-dimethyl pyrazole [2906-75-8]	+++	—	—	—	—	—	—	—	95
α-Haloacetates, amides and ketones^m	+	—	+	+^i	+++	—	+	—	96
Hydrogen peroxide [7722-84-1]	—	—	—	—	+++	++^e	—	—	35
2-Hydroxy-5-nitrobenzyl bromide [2664-60-9]	—	—	—	—	+++	—	—	+++	97
p-Hydroxyphenyl glyoxal [81346-69-2]	+^o	+++	—	—	+	—	—	—	98
N-Hydroxysuccinimidyl esters^m	+++	—	±	—	±	—	+^e	—	99
Iminothiolane [6539-14-6]	+++	—	—	—	—^d	—	—	—	6, 100
Iodine (I_3^-, ICl, etc.)	—	—	++	—	+++	—	+++	+^i	37, 38
N-(Iodoethyl) trifluoroacetamide [6780-56-2]	—	—	—	—	+++	—	—	—	101
o-Iodosobenzoate [304-91-6]	—	—	—	—	+++	+	+	++	102
Isocyanates (RNCO)^f	+++	—	++	—	++	—	++	—	103
Isothiocyanates (RNCS & ArNCS)	+++	—	++	—	++	—	++	—	104
Maleic anhydride [108-31-6]	+++^e	—	±	—	+	—	±	—	105
Mercurials^f	—	—	—	—	+++	—	—	—	106

Table 2. Continued

Reagent	Amino acid and side-chain									Ref.
	Amino[c] (Lys)	Guanidino (Arg)	Imidazole (His)	Carboxyl (Glu, Asp)	Thiol (Cys)	Thioether (Met)	Disulfide (CysS)	Phenol (Tyr)	Indole (Trp)	
Methanol/HCl	—	—	—	+++	—	—	—	—	—	107
2-Methoxy-5-nitrotropone [14628-90-1]	+++[e]	—	—	—	—	—	—	—	—	108
Methyl acetimidate [14777-29-8]	+++	—	—	—	—	—	—	—	—	1, 2
O-Methylisourea [2440-60-0]	+++	—	—	—	+	—	—	—	—	109
Methyl methane thiosulfonate [2949-92-0]	—	—	—	—	+++[e]	—	—	—	—	110
Methyl-p-nitrobenzene-sulfonate [6214-20-6]	—	—	+++	—	+++	—	—	—	—	111
Monochlorotriazines	+++	—	+	—	++	—	—	+++	—	112
o-Nitrophenyl sulfenyl chloride [7669-54-7]	—	—	—	—	+++	—	—	—	+++	113
2-(2-Nitrophenyl sulfenyl)-3-methyl-3-bromoidolene (BNPS skatole) [27933-36-4]	—	—	—	—	+++	—	—	—	+++	114

Reagent											Ref.
p-Nitrophenylsulfonyl fluoride[h] [349-96-2]	—	—	—	—	—	—	—	—	++	—	115
Nitroprusside, sodium (Na$_2$Fe(CN)$_5$NO·2H$_2$O) [14402-89-2]	+[n]	—	—	—	±±±	—	—	—	—	—	116
2-Nitro-5(sulfothio) benzoate + Na$_2$SO$_3$ [95754-92-0]	—	—	—	—	+++[e]	—	—	+++[e]	—	—	117
2-Nitro-5-thiocyanato benzoate [30211-77-9]	—	—	—	—	+++	—	—	—	—	—	118
Nitrous acid [7782-77-6]	+++[o]	—	—	—	+++	—	—	—	+	—	119
4-(Oxoacetyl) phenoxyacetic acid [39270-55-8]	—	+++	—	—	—	—	—	—	—	—	120
2,4-Pentanedione [123-54-6]	+++	++	—	—	—	—	—	—	—	—	121
Performic acid [107-32-4]	—	—	—	—	+++	+++	+++	+++	+++	—	122
Periodate, sodium (NaIO$_4$)[q] [7790-28-5]	—	—	+	—	+++	++[e]	—	—	+	—	123
Phenylarsine oxide [637-03-6]	—	—	—	—	+++[e,g]	—	—	—	—	—	124
Phenylglyoxal [1074-12-0]	+[o]	+++	—	—	+	—	—	—	—	—	19, 20

Post-translational Modification

Table 2. Continued

Reagent	Amino acid and side-chain									Ref.
	Amino[c] (Lys)	Guanidino (Arg)	Imidazole (His)	Carboxyl (Glu, Asp)	Thiol (Cys)	Thioether (Met)	Disulfide (CysS)	Phenol (Tyr)	Indole (Trp)	
Phosphorothioate, sodium (Na$_3$SPO$_3$) [10101-88-9]	—	—	—	—	—[r]	—	+++[e]	—	—	125
Photo-oxidation	—	—	+++	—	+++	+++	—	+	+++	126
o-Phthalaldehyde + RSH [643-79-8]	+++	—	—	—	+[g]	—	—	—	—	127
Salicyldehyde [90-02-8]	+++[e]	—	—	—	—	—	—	—	—	128
Sodium borohydride (NaBH$_4$) [16940-66-2]	—	—	—	—	—	—	+++[e]	—	—	129
Sodium sulfite (Na$_2$SO$_3$) [7757-83-7]	—	—	—	—	—[r]	—	+++[e]	—	—	130
Succinic anhydride [108-30-5]	+++	—	±	—	±	—	—	±±	—	131
N-Succinimidyl acetate [14464-29-0]	+++	—	±	—	±	—	—	+	—	132
N-Succinimidyl 3-(4-hydroxy-phenyl) propionate [34071-95-9]	+++	—	±	—	±	—	—	+	—	133

Reagent										Ref.
N-Succinimidyl 3-(2-pyridyl-thio)propionate [68181-17-9]	+++	—	—	—	—	±±[1]	—	±	—	134
Tetranitromethane[f] [509-14-8]	—	—	—	—	—	+++	—	+++	+	135
Tetrathionate, sodium (Na$_2$S$_4$O$_8$)	—	—	—	—	—	+++	—	—	—	136
Thioureadioxide [4189-44-0]	+++	—	—	—	—	—	—	—	—	137
2,4,6-Tribromo-4-methyl-cyclohexanedione [39953-10-1]	—	+	—	—	+[e]	+++	+	+++	+++	138
Tributylphosphine [998-40-3]	—	—	—	+++[e]	—	—	—	—	—	139
Trimethyloxonium tetrafluoroborate [420-37-1]	—	—	+++	—	—	—	—	—	—	140
Trinitrobenzenesulfonic acid [2508-19-2]	+++	—	—	—	—	+	—	—	—	141
Vanadate (V$_2$O$_5$) [1314-62-1]	—	—	—	—	—	+++	—	—	—	142
4-Vinylpyridine [100-43-6]	+	—	—	—	—	+++	—	—	—	142

[a] This rather long list includes most of the commonly used reagents for chemically modifying biologically active proteins. We have undoubtedly overlooked some useful but less widely used reagents. We have purposely excluded others that we consider to be close analogs, with little or no difference in specificity of reaction, as compared to one or more of those listed. We have, for example, included O-methylisourea but not S-methylisourea; N-ethylmaleimide but none of the many other useful N-alkyl and N-aryl maleimides that have been used, for example, to introduce spectroscopic probes and other special-purpose moieties into proteins. Affinity labels and other reagents that are site-directed have not been included. Heterobifunctional cross-linking reagents are also not included, but their reaction characteristics can be deduced, for the most part, from related agents that are listed.

Table 2. Continued

[b] The indicated reactivities are as follows: +++, highly reactive, extensive reaction under typical conditions; ++, significant reaction under typical conditions. ±±±, ±± and ± are used similarly to indicate the some reaction is possible, but is not usually extensive; −, no reaction expected under typical conditions. The reactivities indicated are those expected reactivities in cases where the resulting derivatives are unstable and spontaneously revert under normal conditions or upon dilution or dialysis, to the unmodified side-chain. Some unambiguities exist due to differences in the effects of conditions on reactivities. The reactivities indicated are those expected under the most commonly used conditions. Where differences of opinion appear to exist, the authors exercised their best judgement. Criticisms and suggested changes will be appreciated and should be directed to one of the authors.
[c] Most reagents that affect ε-amino groups also affect α-amino groups in a similar manner, exceptions are noted.
[d] May react with the initial reaction products.
[e] Readily converted back to the unmodified side-chain under relatively mild conditions.
[f] May be hazardous if handled improperly — check literature before using.
[g] Reaction requires two closely spaced thiol groups.
[h] Other side-chain groups in addition to those shown may also react.
[i] Reaction only in a few special or unusual cases.
[j] In the absence of an added amine may give rise to cross-links with nearby carboxyl moieties.
[k] At pH 7–9 in the presence of borate.
[l] Possible reaction with the disulfide moiety.
[m] Many related reagents are used to effect cross-linking, introduce spectroscopic probes, etc.
[n] Primarily affects ε-amino groups.
[o] Primarily affects α-amino groups.
[p] Not normally observed upon reaction in aqueous solution.
[q] Also reacts with amino-terminal Ser and Thr.
[r] Reacts in the presence of a mild oxidant.
[s] Reaction requires both an amino and a thiol group.

Table 3. Methods for identification of modified amino acids (the methods provided are for situations in which the derivatized amino acid is not stable to standard amino acid analysis)

Modification	Amino acid	Stable to HCl hydrolysis	Alternative procedure for identification	Ref.
Acetylation	Amino terminus	N	Protease digestion	147
	Lys	N	Protease digestion	148
ADP-ribosylation	Arg	N	NH_2OH sensitivity	149
	Cys	N	NH_2OH sensitivity	150
	Glu	N	NH_2OH sensitivity	151
Amidation	Carboxyl terminus	N	Protease digestion	152
Carboxylation	Asp	N	Base hydrolysis	153
	Glu	N	KOH hydrolysis	154
Glycosylation	Asp	N	Trifluoromethane sulfonic acid	155
	Thr	N	Enzymatic digestion	156
Hydroxylation	Pro	Y		
Iodination	Tyr	Y		
Isoprenylation	Cys	N	Methyl iodide reduction	157
Methylation	Arg	Y		
	Asp (isopeptide)	N	Enzymatic digestion	158
	Glu	N	Base hydrolysis	159
	His	Y		
	Lys	Y		
Myristoylation	NH_2 terminal Gly	N	Protease digestion; fatty acid analysis	160
Nucleotidylation	Tyr	N	Enzymatic digestion	161
Palmitoylation	Cys	N	Methanolic KOH	162
Phosphorylation	Ser	N	Base hydrolysis	163
	Thr	N	Base hydrolysis	163
	Tyr	N	Base hydrolysis	163
Sulfation	Tyr	N	$Ba(OH)_2$ hydrolysis	164

Abbreviations: N, no; Y, yes.

(iii) allow rate amplification because of the flux across the cyclic reaction scheme; and
(iv) respond sigmoidally to changing metabolite concentration, particularly if a metabolite has opposing effects on the two reactions of the cyclic process.

Stadtman, Chock and their associates have pioneered the theoretical explanation of these principles (144, 145, 166).

2.2. Influence of modification

Modification reactions change the structures of amino acid residues and potentially change their chemical properties. For example, the ADP-ribosylation of arginyl residues (150, 167) converts a positively charged site on the protein to one with a negative charge. Phosphorylation or sulfation of hydroxyamino acids introduces a negative charge into the protein (168-170) with accompanying changes in properties. Many of the modification reactions identified in proteins are known to occur at many amino acids. Proteins can be phosphorylated at a number of hydroxy amino acid residues including serine, threonine and tyrosine (171-173); ADP-ribosylation occurs at glutamic acid, cysteine and arginine residues (167, 174, 175). Other modifications can have a dramatic influence on the hydrophobicity of the amino acid (176). The most significant differences in hydrophobicity occur after glycosidation (Asn residues) or fatty acylation (Cys and the amino terminus), often found in proteins bound to or associated with membranes. A specific modification may cause definite and unique alterations of the properties of the target site.

Post-translational modification of proteins may also cause distinct changes in structure that are reflected in the physical properties of the protein. Phosphorylation and glycosidation of proteins are known to cause altered migration of proteins during SDS-polyacrylamide gel electrophoresis (177). If the target protein has multiple target sites for the modifying enzyme, a series of bands may be apparent, yielding a 'ladder' effect that may indicate multiple modification sites. On two-dimensional gels this is indicated by a charge-train.

The diverse effects of modification reactions makes it crucial to find out whether a modification is present and, if so, the type of modification. The choice of technique is strongly dependent on the acid-base stability of the modified amino acid. Acid-stable amino acids can be identified by the standard analytical techniques used in amino acid analysis. Acid-labile derivatives require different approaches. Common techniques include high-performance liquid chromatography (178), mass spectrometry (179) or gas chromatography (180) of derivatives generated by base hydrolysis or enzymic digestion. *Table 3* indicates the appropriate approach for the determination of common modifications.

2.3. Characteristics of modification reactions

Post-translational modifications often occur at a specific recognition sequence. Predicting the site of modification in a protein of known sequence is, therefore, possible for a number of modification reactions. For isoprenylation (181, 182), the acceptor specificity of the farnesylation and geranylgeranylation reactions have been defined. Both reactions occur on cysteine residues contained in a sequence, CAAX-COOH, where A is an aliphatic amino acid and X is the carboxyl terminal residue (181, 183) of the protein. Differences are evident; farnesylation requires a leucine as the carboxyl terminal residue whereas geranylgeranylation generally requires serine (184, 185). Some substitutions are allowed, but activity is diminished by any change in the carboxyl terminal residue. Many of the proteins isoprenylated are subjected to further modification; proteolytic trimming and then carboxyl methylation (186-188). Proteolysis removes their terminal three residues, resulting in a carboxyl terminal cysteine which is isoprenylated. The carboxyl terminal cysteine is methylated. These processing reactions absolutely require that isoprenylation is the initial event (186).

Specific recognition sequences serve as determinants for many different modification reactions. Various protein kinases have been studied extensively to define their recognition sites (189–195). Some of these kinases, including the cAMP-dependent protein kinase and phosphorylase kinase, require basic amino acids near the site phosphorylated (189–191). Others, casein kinase II and tyrosine kinases, seem to require acidic residues near the phosphorylation site (193–195). Still other modifications require very specific sequences in the site modified. As stated, the site for isoprenylation is specific. N-linked glycosylation reactions (196) occur at asparagine residues in the sequence Asn-X-(Thr/Ser). The recognition sequences for many of the commonly observed modifications are given in *Table 4*.

2.4. Physiological role of the modification

Establishing the physiological role of a particular reaction requires that certain criteria are met. At a minimum, the modifying enzyme, its target substrate and the appropriate donor molecule must share the same cellular localization. Availability of the donor molecule is also an important mechanism for the regulation of enzyme activity. Protein isoprenylation is a useful example. Introduction of mevalonate into a cell culture can result in either farnesylation or geranylgeranylation; mevalonate serves as the precursor for formation of the donor of both processes (208–210). Obviously, the relative distribution of the two isoprenylation events is dependent on the regulation of specific donor metabolism, substrate availability and the presence of the appropriate enzyme. The two isoprenylation reactions are catalyzed by distinct heterodimeric enzymes (211–213). Interestingly, farnesyltransferase and one geranylgeranyl-transferase share a common α-subunit (214). The other subunit, β, is distinct and seems to define the acceptor specificity of the enzyme (215). The α-subunits may be responsible for recognition of the isoprenyl group. The mechanism for the recognition of the

Table 4. Recognition sequences and donors for protein modification

Modification	Recognition region	Group donor	Ref.
Acetylation	Amino terminal residues	Acetyl CoA	197
Amidation	Carboxy terminal Gly	Conversion of terminal Gly	198
Carboxylation	Glu-X-X-X-Glu-X-Cys	CO_2	199, 200
Glycosylation	Asn-X-Thr/Ser	Sugar nucleotide diphosphates	201, 202
Hydroxylation		Oxygen	203
4-Pro	X-Pro-Gly		
3-Pro	Pro-(4-Hyd)Pro-Gly		
Isoprenylation	Cys-Ali-Ali-X-OOH	Mevalonyl CoA	181, 183
Methylation		S-Adenosyl methionine	204
Myristoylation	NH_2 terminal Gly	Myristoyl CoA	205, 206
Phosphorylation	Many	ATP	189, 194
Sulfation	Acidic residues	PAPS	207

Abbreviations: Ali, aliphatic residues; PAPS, adenosine-3'-phospho-5'-phosphosulfate.

different isoprenyl groups is unclear. If the shared subunit binds the donor molecule, how do the enzymes discriminate between possible donors? Does the acceptor molecule influence the recognition of the donor? The acceptor sequences for the two classes for isoprenylation are similar but distinct. It is possible that prior binding of the acceptor substrate modulates the subsequent binding of the donor molecule. If so, another mechanism is plausible for modulating the level of modification. The ratio of farnesylation to geranylgeranylation may be influenced by the availability of the acceptor, which is controlled by the compartmentalization of the enzyme and substrate.

To be physiologically relevant, a protein modification site identified *in vitro* must be confirmed to occur *in vivo*. The location of the modification within the protein sequence must be accomplished. Typically, the modified protein is enzymatically digested, the modified peptide isolated, and the modified amino acid identified by peptide sequencing or amino acid hydrolysis. If available, a radiolabeled analog of the donor should be utilized in the labeling reaction. The radioisotope serves as a convenient reporter for isolating the labeled site.

The precise role of each covalent modification reaction can be investigated with many different tools; inhibitors (216–220), genetics and molecular biology (221, 222), and microinjection (223), for example. Again, consider protein isoprenylation; the reaction is regulated by the level of mevalonate metabolites. The inhibition of mevalonate synthesis by inhibitors of HMG-CoA reductase, such as lovastatin (217), causes a net decrease in protein isoprenylation. Coupled to this effect are the multiple cellular effects of lovastatin, including the shutdown of cholesterol biosynthesis. Use of lovastatin *in vivo*, therefore, is complicated by the pleiotropic effects in cellular systems, and the results may be difficult to interpret. Similarly, the role of histone acetylation has long been investigated using *n*-butyrate as an inhibitor of deacetylation (219). Unfortunately, the effect of butyrate is not specific for acetylation, but has a variety of cellular effects. Application of an inhibitor with more stringent specificity enables more detailed examination of the modification. For example, the deacetylation of histones is specifically inhibited *in vitro* and *in vivo* by the fungal component, trichostatin A (224). Trichostatin A was initially characterized as a fungal compound with antibiotic activity and shown to induce cell differentiation. Subsequently, trichostatin was found to inhibit histone deacetylation, but the relationship between this effect and the induction of cell differentiation is unclear. Isolated natural products have been used to examine the role of modification reactions. A variety of natural compounds have been used to probe the cellular function of protein phosphorylation, directed at both serine/threonine and tyrosine kinases (225–228).

The information acquired from the use of an inhibitor is potentially increased as the specificity of the inhibitor for the reaction is increased. Benzamide and other compounds are inhibitors of ADP-ribosylation, but are not specific for a particular class of the reactions (220). These inhibitors are not, therefore, adequate for investigating the cellular function of arginine ADP-ribosylation. Other compounds must be found which are more specific for a certain reaction. Substrate analogs are particularly useful because of their structural similarity to the substrate. Examples include spermidine amide (229) and oxaproline peptides (230) for acetylation and hydroxylation, respectively. Spermidine amide is a multisubstrate analog that mimics both the acceptor and donor portions. This compound is a potent inhibitor of histone acetylation; the inhibition constant is less than 10^{-8} M. This class of inhibitors is increasingly being examined to probe the entire span of the substrate binding regions on the enzyme. Appropriate design of a multisubstrate structural inhibitor can provide a view of the interaction of the substrate with the enzyme. Many of the more common inhibitors are given in *Table 5*.

Table 5. Inhibitors of post-translation modifications

Modification	Inhibitor	Ref.
Acetylation	n-Butyrate [107-92-6]	219
	Trichostatins A	224
ADP-Ribosylation	Benzamide [55-21-0]	220
	meta-Iodobenzylguanidine [80663-95-2]	231
Amidation	4-Phenyl-3-butenoic acid [2243-53-0]	232
Carboxylation	Warfarin [81-81-2]	233
	Tetrachloropyridinol [2322-38-5]	234
Glycosylation	Tunicamycins	235
	Hydroxynorvaline [2280-42-4]	236
Hydroxylation	Poly(ADP-ribose)	237
Isoprenylation	Fluoromevalonate [2822-77-7]	238
Methylation	S-Adenosylhomocysteine [979-92-0]	239
Myristoylation	Oxo-myristic acid	240
Phosphorylation	Protein kinase inhibitor	241
	Staursporine [62996-74-1]	242
	Tyrphostins	243
	β-Nitrostyrenes	244
Sulfation	Chlorate [14866-68-3]	245
Ubiquitination	Arsenite [15502-74-6]	246

Although few examples are available (247–249), the microinjection of a purified enzyme, a specific inhibitor or a specific antibody is another mechanism for modulating the amount of a modifying enzyme activity. This technique has been useful in the study of protein phosphorylation. Both the cAMP-dependent protein kinase (248) and the cell cycle p34^{cdc2} kinase (249) have been examined. Specifically, it was shown that microinjection of protein kinase inhibitor (PKI) caused chromatin condensation (248). These studies have implicated that a role for the cAMP-dependent protein kinase in cell cycle regulation has been discovered. Similarly, a diverse cellular role for type-1 protein phosphatase has been elucidated with this technique (250–252). In fact, a role for phosphatase-1 has been demonstrated when purified enzyme, polyclonal antibody or two different inhibitors have been introduced into cells by microinjection. These examples portend a future for the beneficial application of microinjection to a variety of covalent modifications.

2.5. Reversibility of reactions

A less stringent requirement for assigning a role for covalent modification reactions in cellular regulation is the matter of reversibility. Typically, it is expected that for a modification to be significant in cellular regulation, the event must be reversible, with the cyclic process yielding the starting protein. Indeed, for most events, enzymes which reverse the modification are known. *Table 6* illustrates common reversible modifications. The reverse reaction is

Table 6. Examples of reversible post-translational modifications

Modification	Donor	Reversing enzyme	Ref.
Acetylation	Acetyl CoA	Acyl-peptide hydrolase [95567-87-6]	253
ADP-ribosylation	NAD$^+$	Mono ADP-ribosyl hydrolase [89190-48-7]	254 255
		Poly ADP-ribosyl glycosylhydrolase [9068-16-0]	256 257
		ADP-ribosyl pyrophosphatase [9024-83-3]	258
		ADP-ribosyl protein lyase [9068-16-0]	259 260
Glycosylation	NDP-glucose	Protein-N-Amidase [85334-39-8]	261
Methylation	S-adenosylmethionine	Methylesterase [69552-31-4]	262 263
Phosphorylation	ATP	Protein phosphatases	264 265
Ubiquitination	Ubiquitin	Ubiquitin hydrolase [123175-79-1]	266

crucial for returning the system to a 'de-stimulated' condition. Inability to reverse the modification will result in a system that is unable to respond to future external signals.

Although most modification reactions are reversible, there are some which are not readily reversible. Typically, irreversible reactions do not serve regulatory functions, but do alter the properties of the target protein. Such irreversible modifications include the proteolytic reactions involved in complement fixation (267), the coagulation cascade (165, 268) and the processing of protein during biosynthesis (269), including polypeptide hormones (270). Renewal of these systems requires protein synthesis. The removal of protein linked tyrosine sulfate groups also seems to occur after the degradation of protein. No reversing enzyme, a sulfatase, has apparently been identified. Instead, free tyrosine sulfate has been found in urine. Other examples are known for which the modification may be removed, but the final product does not correspond to the starting material.

Glutaminyl cyclase catalyzes the formation of an amino terminal pyroglutamate from glutamine in some peptide hormones, such as thyrotropin releasing hormone and gonadotropin releasing hormone (271). An obvious effect of this modification is effectively to remove the negative charge at the amino terminus and may be a determinant in the targeting of the hormones. In contrast to other proteins with modifications at the amino terminus, pyroglutamyl-containing proteins are not recovered as the free protein. Instead, pyroglutamyl residues are hydrolyzed by pyroglutamate aminopeptidase (272), leaving a truncated protein.

2.6. Summary

Enzyme catalyzed post-translational modifications of proteins serve many diverse functions in the regulation of biological processes. Suspicion that a protein is a substrate for some modification directs the researchers to establish the type and influence of the involved modification. Hopefully, this overview will serve as a basis on which investigators can add subsequent research.

3. REFERENCES

1. Hunter, M.J. and Ludwig, M.L. (1962) *J. Am. Chem. Soc.*, **34**, 3491.

2. Makoff, A.J. and Malcolm, D.B. (1981) *Biochem. J.*, **193**, 245.

3. Davies, G.E. and Stark, G.R. (1970) *Proc. Natl. Acad. Sci. USA* **66**, 651.

4. Dombradi, V., Hajdu, J., Bot. G. and Freidrich, P. (1980) *Biochemistry*, **19**, 2295.

5. Ju, R., Lambert, J.M., Pierce, L.R. and Traut, R.R. (1978) *Biochemistry*, **17**, 5399.

6. Cover, J.R., Lambert, J.M., Norman, C.M. and Traut, R.R. (1981) *Biochemistry*, **20**, 2843; McCall, M.J., Diril, H. and Meares, F. (1990) *Bioconjugate Chem.*, **1**, 222.

7. Means, G.E. and Feeney, R.E. (1968) *Biochemistry*, **7**, 2192.

8. Jentoft, N. and Dearborn, D.G. (1979) *J. Biol. Chem.*, **254**, 4359; Wong, W.S.D., Osuga, D.T. and Fene, R.E. (1984) *Anal. Biochem.*, **139**, 58.

9. Riordan, J.F. and Vallee, B.L. (1967) *Methods Enzymol.*, **11**, 565.

10. Smyth, D.G. (1967) *J. Biol. Chem.*, **242**, 1592.

11. Lomant, A.J. and Fairbanks, G. (1976) *J. Mol. Biol.*, **104**, 243.

12. Anjaneyulu, P.S.R. and Staros, J.V. (1987) *Int. J. Peptide Prot. Res.*, **30**, 117.

13. Dixon, H.B.F. and Perham, R.N. (1968) *Biochem. J.*, **109**, 312.

14. Habeeb, A.F.S.A. and Atassi, M.Z. (1970) *Biochemistry*, **9**, 4939.

15. Goldfarb, A.R. (1966) *Biochemistry*, **5**, 2570.

16. Fields, R. (1971) *Biochem. J.*, **124**, 581.

17. Riordan, J.F. (1973) *Biochemistry*, **12**, 3915.

18. Rogers, T.B., Borresen, T. and Feeney, R.E. (1978) *Biochemistry*, **17**, 1105; Yamasaki, R.B., Vega, A. and Feeney, R.E. (1980) *Analyt. Biochem.*, **109**, 32.

19. Takahashi, K. (1968) *J. Biol. Chem.*, **243**, 6171.

20. Cheung, S.T. and Fonda. M.L. (1979) *Biochem. Biophys. Res. Commun.*, **90**, 940.

21. Melchior, W.B. and Fahrney, D. (1970) *Biochemistry*, **9**, 251.

22. Angeles, T.S., Smanik, P.A., Borders, C.L. and Viola, R.E. (1989) *Biochemistry*, **28**, 8771.

23. Hoare, D.G. and Koshland, D.E. (1967) *J. Biol. Chem.*, **242**, 2447.

24. Huynh, Q.K. (1988) *J. Biol. Chem.*, **263**, 11631.

25. Staros, J.V., Wright, R.W. and Swingle, D.M. (1986) *Anal. Biochem.*, **156**, 220.

26. Jones, J.G., Otieno, S., Barnard, E.A. and Bharagaua, A.K. (1975) *Biochemistry*, **14**, 2396.

27. Crestfield, A.M., Moore, S. and Stein, W.H. (1963) *J. Biol. Chem.*, **238**, 622.

28. Markham, G.D. and Satishchandran, C. (1988) *J. Biol. Chem.*, **263**, 8666.

29. Lewis, C.T., Seyer, J.M. and Carlson, G.M. (1989) *J. Biol. Chem.*, **264**, 27.

30. Ellman, G.L. (1959) *Arch. Biochem. Biophys.*, **82**, 70.

31. Fujioka, M., Takata, Y., Konishi, K. and Ogawa, H. (1987) *Biochemistry*, **26**, 5696.

32. Vanaman, T.C. and Stark, G.R. (1970) *J. Biol. Chem.*, **245**, 3565.

33. Boyer, P.D. (1954) *J. Am. Chem. Soc.*, **76**, 4331.

34. Silverstein, E. and Sulebele, G. (1970) *Biochemistry*, **9**, 274.

35. Stauffer, C. and Etson, D. (1969) *J. Biol. Chem.*, **244**, 5333.

36. Bamezai, S., Banez, M.A.T. and Breslow, E. (1990) *Biochemistry*, **29**, 5389.

37. Edelhoch, H. and Lippoldt, R.E. (1962) *J. Biol. Chem.*, **237**, 2788.

38. Hunter, W.M. and Greenwood, F.C. (1962) *Nature*, **194**, 492.

39. Marchalonis, J.J. (1969) *Biochem. J.*, **113**, 299.

40. Sokolovsky, M., Riordan, J.F. and Vallee, B.L. (1966) *Biochemistry*, **5**, 3582.

41. Riordan, J.F., Wacker, W.E.C. and Vallee, B.L. (1965) *Biochemistry*, **4**, 1758.

42. Viswanatha, T., Lawson, W.B. and Witkop, B. (1960) *Biochem. Biophys. Acta.*, **40**, 216.

43. Riordan, J.F. and Vallee, B.L. (1972) *Methods Enzymol.*, **25**, 494.

44. Benesch, R. and Benesch, R.E. (1956) *J. Am. Chem. Soc.*, **78**, 1597.

45. Klotz, I.M. and Heiney, K.E. (1962) *Arch. Biochem. Biophys.*, **96**, 505.

46. Pendergast, F.G., Meyer, M., Carlson, G.L., Lida, S. and Potter, D.J. (1983) *J. Biol. Chem.*, **258**, 7541.

47. Cavins, J.F. and Friedman, M. (1967) *Biochemistry*, **6**, 3766.

48. Toyo'oka, T. and Imaij, K. (1985) *Anal. Chem.*, **57**, 1931; Kirley, T.L. (1989) *Anal. Biochem.*, **180**, 231.

49. Andrews, J.L., Ghosh, P., Ternai, B. and Whitehouse, M.W. (1982) *Arch. Biochem. Biophys.*, **214**, 386.

50. Lopez, S., Miyashita, Y. and Simmons, S.S. (1990) *J. Biol. Chem.*, **265**, 16039; Knowles, F.C. and Benson, A.A. (1983) *Trends in Biochem. Sci.*, **8**, 178.

51. Fleet, G.W.J., Porter, R.R. and Knowles, J.R. (1969) *Nature*, **224**, 511; Ji, T.H. (1979) *Biochim. Biophys. Acta.*, **559**, 39.

52. Fontana, A. and Scoffone, E. (1972) *Methods Enzymol.*, **143**, 76.

53. Lapis, S.F and Harrison, J.H. (1978) *J. Biol. Chem.*, **253**, 7576.

54. Heilman, H.D. and Holzner, M. (1981) *Biochem. Biophys. Res. Commun.*, **99**, 1146.

55. Kosower, N.S. and Kosower, E.M. (1987) *Methods Enzymol.*, **143**, 76.

56. Cole, R.D. (1967) *Methods Enzymol.*, **11**, 315; Planes, A. and Kirsch, J. (1990) *Prot. Engineer.*, **3**, 625.

57. Erlanger, B.F., Vratsanos, S.M., Wassermann, M. and Cooper, A.G. (1965) *J. Biol. Chem.*, **240**, 3447.

58. Witkop, B. (1961) *Adv. Prot. Chem.*, **16**, 221; Ybarra, J., Prasad, A.R.S and Nishimura, J.S. (1986) *Biochemistry*, **25**, 7174.

59. Riordan, J.F. (1973) *Biochemistry*, **12**, 3915.

60. Buechler, J.A. and Taylor, S.S. (1989) *Biochemistry*, **28**, 2065.

61. Carrawya, K.L. and Koshland, D.E. (1972) *Methods Enzymol.*, **25**, 616.

62. Sela, M. and Arnon, R. (1972) *Methods Enzymol.*, **25**, 616.

63. Schechter, Y., Burstein, Y. and Patchornik, A. (1975) *Biochemistry*, **15**, 5071.

64. Wu, J.C. and Wang, J.H. (1986) *Biochemistry*, **25**, 7991.

65. Schechter, Y., Burstein, Y. and Gertler, A. (1977) *Biochemistry*, **15**, 5071.

66. Atassi, M.Z. and Habeeb, A.F.S.A. (1972) *Methods Enzymol.*, **25**, 546.

67. Stark, G.R. (1965) *Biochemistry*, **4**, 588, 1030, 2363.

68. Gross, E. (1967) *Methods Enzymol.*, **11**, 238.

69. Gorbunoff, M.J. (1972) *Methods Enzymol.*, **25**, 506.

70. Patthy, L. and Smith, E.L. (1975) *J. Biol. Chem.*, **250**, 565.

71. Scouten, W.H., Lubcher, R. and Baugman, W. (1974) *Biochim. Biophys. Acta*, **485**, 236.

72. Kosower, N.S. and Kosower, E.M. (1987) *Methods Enzymol.*, **143**, 264.

73. Riehm. J.P. and Scheraga, H.A. (1965) *Biochemistry*, **4**, 772.

74. Kureki, R., Yamada, H., Moriyama, T. and Imoto, T. (1986) *J. Biol. Chem.*, **261**, 13571.

75. Sokolovsky, M. and Valle, B.L. (1966) *Biochemistry*, **5**, 3574; Pielak, G.J., Urdea, M.S., Igi, K. and Legg, J.I. (1984) *Biochemistry*, **23**, 589, 596.

76. Abuchowski, A., van Es, T., Palezuk, N.C. and Davis, F.F. (1977) *J. Biol. Chem.*, **252**, 3578; Clonis, Y.D. and Lowe, C.R. (1980) *Biochem. J.*, **191**, 247.

77. Oh, K.-J. and Churchich, J.E. (1974) *J. Biol. Chem.*, **249**, 4737.

78. Gray, W.R. (1967) *Methods Enzymol.*, **11**, 139.

79. Horton, H.R. and Tucker, W.P. (1970) *J. Biol. Chem.*, **245**, 3397.

80. Hirs, C.H.W., Halmann, M. and Kycia, J.H. (1965) *Arch. Biochem. Biophys.*, **111**, 209.

81. Delpierre, G.R. and Fruton, J.S. (1965) *Proc. Natl. Acad. Sci. USA*, **54**, 1161.

82. Staros, J.V. (1982) *Biochemistry*, **21**, 3950.

83. Grassetti, D.R. and Murray, J.F. (1967) *Arch. Biochem. Biophys.*, **119**, 41.

84. Cleland, W.W. (1964) *Biochemistry*, **3**, 480; Wan-Kyng, L. and Meienhofer, J. (1968) *Biochem. Biophys. Res. Commun.*, **31**, 467.

85. Raftery, M.A. and Cole, R.D. (1963) *Biochim. Biophys. Acta.*, **10**, 467.

86. Brewer, C.F. and Riehm, J.P. (1967) *Anal. Biochem.*, **18**, 248.

87. Bodlaender, P., Feinstein, G. and Shaw, E. (1969) *Biochemistry*, **8**, 4941; Sinha, D. and Brewer, J.M. (1985) *Anal. Biochem.*, **18**, 248.

88. Goldberger, R.F. (1967) *Methods Enzymol.*, **11**, 317.

89. Levine, R.L. (1983) *J. Biol. Chem.*, **258**, 11823; Stadtman, E.R. and Oliver, C.N. (1991) *J. Biol. Chem.*, **266**, 2005.

90. Reddy, G., Nanduri, V.B., Basu, A. and Modak, M.J. (1991) *Biochemistry*, **30**, 8195.

91. Rinderknecht, H. (1960) *Experientiae* **16**, 430.

92. Martin, C.J. and Marini, M.A. (1967) *J. Biol. Chem.*, **242**, 5736; Tome, D., Kozlowski, A. and Mabon. F. (1985) *J. Agric. Food Chem.*, **33**, 449.

93. Peters, T. and Richards, F.M. (1977) *Ann. Rev. Biochem.*, **46**, 523.

94. Jonas, A. and Weber, G. (1971) *Biochemistry*, **10**, 1335.

95. Habeeb, A. (1960) *Can. J. Biochem. Physiol.*, **38**, 493; Robinson, N.C., Nevrath, H. and Walsh, K.A. (1973) *Biochemistry*, **12**, 414.

96. Gurd, F.R.N. (1967) *Methods Enzymol.*, **11**, 532.

97. Koshland, D.E., Karkhanis, Y.D. and Latham, H.G. (1964) *J. Am. Chem. Soc.*, **86**, 1448.

98. Yamasaki, R.B., Vega, A. and Feeney, R.E. (1980) *Anal. Biochem.*, **109**, 32.

99. Blumberg, S. (1979) *Biochemistry*, **18**, 2815; Anjaneyulu, P.S.R. and Staros, J.V. (1987) *Int. J. Peptide Prot. Res.*, **30**, 117.

100. King, T.P., Li, Y. and Kochoumiano, L. (1978) *Biochemistry*, **17**, 1499.

101. Schwartz, W.E., Smith, P.K. and Royer, G.P. (1980) *Anal. Biochem.*, **106**, 43.

102. Mahoney, W.C., Smith, P.K. and Hermodson, M.A. (1981) *Biochemistry*, **20**, 443.

103. Brown, W.E. and Wold, F. (1973) *Biochemistry*, **12**, 828; Lee, C.K. (1976) *J. Biol. Chem.*, **251**, 6226.

104. Edman, P. (1950) *Acta Chem. Scand.*, **4**, 283; Chang, J.-Y. (1989) *J. Biol. Chem.*, **264**, 3111.

105. Butler, P.J.G., Harris, J.I., Hartley, B.S. and Leberman, R. (1969) *Biochem. J.*, **112**, 679.

106. Boyer, P.D. (1954) *J. Am. Chem. Soc.*, **76**, 4331.

107. Doscher, M.S. and Wilcox, P.E. (1961) *J. Biol. Chem.*, **236**, 1328.

108. Tamaoki, H., Murase, Y., Minato, S. and Nakanishi, K. (1967) *J. Biochem. (Tokyo)*, **62**, 7.

109. Klee, W.A. and Richards, F.M. (1957) *J. Biol. Chem.*, **229**, 489; Shields, G.S., Hill, R.L. and Smith, E.L. (1959) *J. Biol. Chem.*, **234**, 1747.

110. Smith, D.J., Maggio, E.T. and Kenyon, G.L. (1975) *Biochemistry*, **14**, 766.

111. Swenson, R.P., Williams, C.H. and Massey, V. (1984) *J. Biol. Chem.*, **259**, 5585.

112. Witt, J.J. and Roskoski, R. (1980) *Biochemistry*, **19**, 143; Kamisaki, Y., Wada, H.,

Yagura, T. and Matsushima, A. (1984) *J. Pharmacol. Exp. Ther.,* **216**, 410.

113. Scoffone, E., Fontana, A. and Rocchi, R. (1968) *Biochemistry,* **7**, 971.

114. Fontana, A. (1972) *Methods Enzymol.,* **25**, 419.

115. Liao, T.H., Ting, R.S. and Yeung, J.E. (1982) *J. Biol. Chem.,* **257**, 5637.

116. Park, J.W and Means, G.E. (1987) *Biochem. Biophys. Res. Commun.,* **145**, 1048.

117. Thannhauser, T.W., Konishi, Y. and Scheraga, H.A. (1984) *Anal. Biochem.,* **138**, 181; Thannhauser, T.W. and Scheraga, H.A. (1985) *Biochemistry,* **24**, 7681.

118. Degani, Y. and Patchornik, A. (1974) *Biochemistry,* **13**, 1; Kindman, L.A. and Jencks, W.P. (1981) *Biochemistry,* **26**, 6980.

119. Kurosky, A. and Hofmann, T. (1972) *Can. J. Biochem.,* **50**, 1282.

120. Duerksen-Hughes, P.J., Xu, X. and Wilkinson, K.D. (1987) *Biochemistry,* **26**, 6980.

121. Gilbert, H.F. and O'Leary, M.H. (1975) *Biochemistry,* **14**, 5194.

122. Hirs, C.H.W. (1967) *Methods Enzymol.,* **11**, 59.

123. Geoghegan, K.F., Dallas, J.L. and Feeney, R.E. (1980) *J. Biol. Chem.,* **255**, 11429; Yamasaki, R.B. Osuga, D.T. and Feeny, R.E. (1982) *Anal. Biochem.,***124**, 272.

124. Liao, K., Hoffman, R.D. and Lane, M.D. (1991) *J. Biol. Chem.,* **266**, 6544; Webb, J.L. (1966) *Enzyme Metab. Inhibitors,* **3**, 595.

125. Neumann, H., Goldberger, R.F. and Sela, M. (1964) *J. Biol. Chem.,* **239**, 1536.

126. Weil, L., Seibles, T.S. and Herskovits, T.T. (1965) *Arch. Biochem. Biophys.,* **111**, 308; Ray, W.J. and Koshland, D.E. (1962) *J. Biol. Chem.,* **237**, 2493.

127. Blaner, W.S. and Churchich, J. (1979) *J. Biol. Chem.,* **254**, 1794; Puri, R.N., Bhatnagar, D. and Roskoski, R. (1985) *Biochemistry,* **24**, 6499.

128. Williams, J.N. and Jacobs, R.M. (1968) *Biochim. Biophys. Acta,* **154**, 323.

129. Light, A., Hardwick, B.C., Hatfield, L.M. and Sondack, D.L. (1969) *J. Biol. Chem.,* **244**, 6289.

130. Cole, R.D. (1967) *Methods Enzymol.,* **11**, 206.

131. Riordan, J.F. and Vallee, B.L. (1964) *Biochemistry,* **3**, 1768; Meighan, E.A. and Schachman, H.K. (1970) *Biochemistry,* **9**, 1163.

132. Boyd, H. , Leach, S.J. and Milligan, B. (1972) *Int. J. Peptide Protein Res.,* **4**, 117.

133. Bolton, A.E. and Hunter, W.M. (1973) *Biochem. J.,* **133**, 529.

134. Carlsson, J., Drevin, H. and Axen, R. (1978) *Biochem. J.,* **173**, 723.

135. Riordan, J.F. and Vallee, B.L. (1972) *Methods Enzymol.,* **25**, 515.

136. Parker, D.J and Allison, W.S. (1969) *J. Biol. Chem.,* **244**, 180; Liu, T.Y. (1967) *J. Biol. Chem.,* **242**, 4029.

137. Colanduoni, J. and Villafranca, J.J. (1985) *J. Biol. Chem.,* **260**, 15042.

138. Burstein, Y. and Patchornik, A. (1972) *Biochemistry,* **11**, 4641.

139. Ruegg, U.T. and Rudinger, J. (1977) *Methods Enzymol.,* **47**, 111.

140. Parsons, S.M., Jao, L., Dahlquist, F.W., Borders, C.L., Groff, T., Racs, J. and Raftery, M.A. (1969) *Biochemistry,* **8**, 700.

141. Ozols, J. and Strittmatter, P.(1966) *J. Biol. Chem.,* **241**, 4793; Fields, R. (1971) *Biochem. J.,* **124**, 581.

142. Crans, D.C. and Simone, C.M. (1991) *Biochemistry,* **30**, 6734; Muhlrad, A., Peyser, Y.M. and Ringel, I. (1991) *Biochemistry,* **30**, 958.

143. Friedman, M., Krull, L.H. and Cavins, J.F. (1970) *J. Biol. Chem.,* **245**, 3868.

144. Stadtman, E.R. and Chock, P.B. (1977) *Proc. Natl. Acad. Sci. USA,* **74**, 2761.

145. Chock, P.B., Rhee, S.G. and Stadtman, E.R. (1980) *Ann. Rev. Biochem.,* **49**, 813.

146. Uy, R. and Wold, F. (1977) *Science,* **198**, 890.

147. Tsunasawa, S. and Sakiyama, F. (1984) *Methods Enzymol.,* **106**, 165.

148. Alfrey, V.G., DiPaola, E.A. and Sterner, R. (1984) *Methods Enzymol.,* **107**, 224.

149. Moss, J., Yost, D.A. and Stanley, S.J. (1983) *J. Biol. Chem.,* **258**, 6466.

150. Hsia, J.A., Tsai, S.-C., Adamik, R., Yost, D.A., Hewlett, E.L. and Moss, J. (1985) *J. Biol. Chem.,* **260**, 16187.

151. Ogata, N., Ueda, K. and Hayaishi, O. (1980) *J. Biol. Chem.,* **255**, 7616.

152. Tatemoto, K. and Mutl, V. (1978) *Proc. Natl. Acad. Sci. USA,* **75**, 4115.

153. Koch, T.H., Christy, M.R., Barkley, R.M., Sluski, R., Bohemier, D., Van Buskirk, J.J. and Kusch, W.M. (1984) *Methods Enzymol.,* **107**, 563.

154. Price, P.A. (1983) *Methods Enzymol.,* **81**, 13.

155. Edge, A.S.B., Faltynek, C.R., Hof, L., Reichert, Jr., J.E. and Weber, P. (1981) *Anal. Biochem.,* **118**, 131.

156. Kornfeld, R. and Kornfeld, S. (1980) in *The Biochemistry, of Glycoproteins and Proteoglycans* (W.J. Ennarz ed.). Plenum, New York, p. 1.

157. Belz, R., Crabb, J.W., Meyer, H.E., Wittig, R. and Duntze, W. (1987) *J. Biol. Chem.*, **262**, 546.

158. Clarke, S., McFadden, P.N., O'Connor, C.M. and Lou, L.L. (1984) *Meth. Enzymol.*, **106**, 330.

159. Clarke, S., Sparrow, K., Panasenko, S. and Koshland, Jr. D.E. (1980) *J. Supramolec. Struct.*, **13**, 315.

160. Carr, S.A., Biemann, K., Shoji, S., Parmalee, D.C. and Titani, K. (1982) *Proc. Natl. Acad. Sci. USA*, **79**, 6128.

161. Rhee, S.G., Chock, P.B. and Stadtman, E.R. (1985) in *The Enzymology of Post-translational Modification of Proteins* (R.B. Freedman and H.C. Hawkins eds). Academic, New York, p. 273.

162. Kaufman, J.F., Krangel, M.S. and Strominger, J.L. (1984) *J. Biol. Chem.*, **259**, 7230.

163. Martenson, T.M. (1984) *Methods Enzymol.*, **107**, 3.

164. Huttner, W.H. (1984) *Methods Enzymol.*, **107**, 200.

165. Davie, E.W., Fujikawa, K. and Kisiel, W. (1991) *Biochemistry*, **30**, 10363.

166. Shacter, E., Chock, P.B. and Stadtman, E.R. (1984) *J. Biol. Chem.*, **259**, 12252.

167. Moss, J. and Vaughan, M. (1977) *J. Biol. Chem.*, **252**, 2455.

168. Yamauchi, O. and Odani, A. (1985) *J. Am. Chem. Soc.*, **107**, 5938.

169. Ringer, D.P., Etheredge, J.L., Dalrymple, B.L. and Niedbalski, J.S. (1990) *Biochem. Biophys. Res. Commun.*, **168**, 267.

170. Huttner, W.H. (1982) *Nature*, **299**, 273.

171. Kataoka, H. Sakiyama, N. and Makita, M. (1991) *J. Biochem. (Tokyo)*, **109**, 577.

172. Capony, J.-P. and Demaille, J.G. (1983) *Anal. Biochem.*, **128**, 206.

173. Cooper, J.A., Sefton, B.M. and Hunter, T. (1983) *Methods Enzymol.*, **99**, 387.

174. West, Jr., R.E., Moss, J., Vaughan, M. and Liu, T.-Y. (1985) *J. Biol. Chem.*, **260**, 14428.

175. Ogata, N., Ueda, K. and Hayaishi, O. (1980) *J. Biol. Chem.*, **255**, 7610.

176. Black, S.D. and Mould, D.R. (1991) *Anal. Biochem.*, **193**, 72.

177. Leach, B.S., Collawn, Jr., J.F. and Fish, W.W. (1980) *Biochemistry*, **19**, 5734.

178. Walsh, K.A. and Sasagawa, T. (1984) *Methods Enzymol.*, **106**, 22.

179. Carr, S.A., Roberts, G.D. and Hemling, M.E. (1990) in *Mass Spectrometry of Biological Molecules* (C.N. McEwan and B.S. Larsen eds). Marcel Dekker, New York, Vol. 8, p. 87.

180. Farnsworth, C.C., Gelb, M.H. and Glomset, J.A. (1990) *Science*, **247**, 320.

181. Maltese, W.A. (1990) *FASEB J.*, **4**, 3319.

182. Glomset, J.A., Gelb, M.H. and Farnsworth, C.C. (1990) *Trends in Biochem. Sci.*, **15**, 139.

183. Rine, J.A. and Kim, S.-H. (1990) *The New Biologist*, **2**, 139.

184. Reiss, Y., Stradley, S.J., Gerasch, L.M., Brown, M.S. and Goldstein, J.L. (1991) *Proc. Natl. Acad. Sci. USA*, **88**, 732.

185. Moores, S.L., Schabers, M.D., Mosser, S.D., Rands, E., O'Hara, M.B., Garsky, V.M., Marshall, M.S., Pompliano, D.L. and Gibbs, J.B. (1991) *J. Biol. Chem.*, **266**, 14603.

186. Beck, L.A., Hosick, T.J. and Sinensky, M. (1990) *J. Cell. Biol.*, **110**, 1489.

187. Tan, E.W., Perez-Sala, D., Canada, F.J. and Rando, R.R. (1991) *J. Biol. Chem.*, **266**, 14603.

188. Hancock, J.F., Cadwallader, K. and Marshall, C.J. (1991) *EMBO J.*, **10**, 641.

189. Kenelly, P.J. and Krebs, E.G. (1991) *J. Biol. Chem.*, **266**, 15555.

190. Kemp, B.E., Graves, D.J., Benjami, E. and Krebs, E.G. (1977) *J. Biol. Chem.*, **252**, 4888.

191. Graves, D.J. (1983) *Methods Enzymol.*, **99**, 268.

192. Marshak, D.R., Vandenberg, M.T., Bae, Y.S. and Yu, I.J. (1991) *J. Cell. Biochem.*, **45**, 391.

193. Marin, O., Meggio, F., Marchiori, F., Borin, G. and Pinna, L.A. (1986) *Eur. J. Biochem.*, **160**, 239.

194. Casnellie, J.E. and Krebs, E.G. (1984) *Adv. Enzyme Regul.*, **22**, 501.

195. Pike, L.J. (1987) *Methods Enzymol.*, **146**, 353.

196. Kornfeld, R. and Kornfeld, S. (1985) *Ann. Rev. Biochem.*, **54**, 631.

197. Lee, F.-J. S., Lin, L.-W. and Smith, J.A. (1990) *J. Biol. Chem.*, **265**, 11576.

198. Bradbury, A.F., Finnie, M.D.A. and Smyth, D.G. (1982) *Nature*, **298**, 686.

199. Price, P.A., Fraser, J.D. and Metz-Virca, G. (1987) *Proc. Natl. Acad. Sci. USA*, **84**, 8335.

200. Cheung, A., Suttie, J.W. and Bernatowicz, M. (1990) *Biochim. Biophys. Acta*, **1039**, 90.

201. Gavel, Y. and von Heijne, G. (1990) *Protein Eng.*, **3**, 433.

202. Ronin, C., Granier, C., Caseti, C., Bouchilloux, S. and Van Rietschoten, J. (1981) *Eur. J. Biochem.*, **118**, 159.

203. Kivirikko, K.I. and Myllyla, R. (1980) in *The Enzymology of Post-translational Modification of Proteins* (R.B. Freedman and H.C. Hawkins eds). Academic, New York, Vol. 1, p. 64.

204. Paik, W.K. and Kim, S. (1985) in *The Enzymology of Post-translational Modification of Proteins* (R.B. Freedman and H.C. Hawkins eds). Academic, New York, Vol. 2, p. 187.

205. Towler, D.A., Adams, S.P., Eubanks, S.R., Towery, D.S., Jackson-Machelski, E., Glaser, L. and Gordon, J.I. (1987) *Proc. Natl. Acad. Sci. USA*, **84**, 2708.

206. Towler, D.A., Eubanks, S.R., Towery, D.S., Adams, S.P., Glaser, L. and Gordon, J.I. (1987) *Proc. Natl. Acad. Sci. USA*, **84**, 2713.

207. Hortin, G.L., Folz, R., Gordon, J.I. and Strauss, A.W. (1986) *Biochem. Biophys. Res. Commun.*, **141**, 326.

208. Repko, E.M. and Maltese, W.A. (1989) *J. Biol. Chem.*, **264**, 9945.

209. Sepp-Lorenzio, L., Rao, S. and Coleman, P.S. (1991) *Eur. J. Biochem.*, **200**, 579.

210. Kinsella, B.T., Erdman, R.A. and Maltese, W.A. (1991) *Proc. Natl. Acad. Sci. USA*, **88**, 8934.

211. Manne, V., Roberts, D., Tobin, A., O'Rourke, E., De Virgilio, M., Meyers, C., Ahmed, N., Kurz, B., Resh, M., Kung, H.-F. and Barbacid, M. (1990) *Proc. Natl. Acad. Sci. USA*, **87**, 7541.

212. Yokoyama, K., Goodwin, G.W., Ghomaschi, F., Glomset, J.A. and Gelb, M.H. (1991) *Proc. Natl. Acad. Sci. USA.*, **88**, 5302.

213. Yoshida, Y., Kawata, M., Katayama, M., Horiuchi, H., Kita, Y. and Takai, I. (1991) *Biochem. Biophys. Res. Commun.*, **175**, 720.

214. Seabra, M.C., Reiss, Y., Casey, P.J., Brown M.S. and Goldstein, J.L. (1991) *Cell*, **65**, 429.

215. Reiss, Y., Seabra, M.C., Armstrong, S.A., Goldstein, J.L. and Brown, M.S. (1991) *J. Biol. Chem.*, **266**, 10672.

216. Kenyon, G.L. and Garcia, G.A. (1987) *Medic. Res. Rev.*, **7**, 389.

217. Ricouart, A., Gesquiere, J.C., Tartar, A. and Sergheraert, C. (1991) *J. Med. Chem.*, **34**, 73.

218. Sinensky, M., Beck, L.A., Leonard, S. and Evans, R. (1990) *J. Biol. Chem.*, **265**, 19937.

219. Whitlock, Jr. J.P., Galeazzi, D. and Schulman, H. (1983) *J. Biol. Chem.*, **258**, 1299.

220. Rankin, P.W., Jacobson, E.L., Benjamin, R.C., Moss, J. and Jacobson, M.K. (1989) *J. Biol. Chem.*, **264**, 4312.

221. Nimmo, H.G. and Cohen, P.T.W. (1987) *Biochem. J.*, **247**, 1.

222. Kinoshita, N., Ohkura, H. and Yanagida, M. (1990) *Cell*, **63**, 405.

223. Maller, J.L. (1983) *Methods Enzymol.*, **99**, 219.

224. Yoshida, M., Kijima, M., Akita, M. and Beppu, T. (1990) *J. Biol. Chem.*, **265**, 17174.

225. Srivaatava, A.K. (1985) *Biochem. Biophys. Res. Commun.*, **131**, 1.

226. Herbert, J.M., Seban, E. and Maffrand, J.P. (1990) *Biochem. Biophys, Res. Commun.*, **171**, 189.

227. Wender, P.A., Cribbs, C.M., Koehler, K.F., Sharkey, N.A., Herald, C.L., Kamano, Y., Pettit, G.R. and Blumberg, P.M. (1988) *Proc. Natl. Acad. Sci. USA*, **85**, 7197.

228. Hsu, C.-Y.J., Persons, P.E., Spada, A.P., Bednar, R.A., Levutzki, A. and Zilberstein, A. (1991) *J. Biol. Chem.*, **266**, 21105.

229. Cullis, P.M., Wolfenden, R., Cousens, L.S. and Alberts, B.M. (1982) *J. Biol. Chem.*, **265**, 8415.

230. Karvonen, K., Ala-Kokko, L., Pihlajaniemi, T., Helaakoski, T., Henke, S., Gunzler, V., Kirivikko, K.I. and Savolainen, E.-R. (1990) *J. Biol. Chem.*, **265**, 8415.

231. Smets, L.A., Loesberg, C., Janssen, M. and Van Rooij, H. (1990) *Biochim. Biophys. Acta*, **1054**, 49.

232. Bradbury, A.F., Mistry, J., Roos, B.A. and Smyth, D.G. (1982) *Eur. J. Biochem.*, **189**, 363.

233. Ren, P. Laliberte, R.E. and Bell, R.G. (1974) *Molec. Pharmacol.*, **10**, 373.

234. Grossman, C.P. and Suttie, J.W. (1990) *Biochem. Pharmacol.*, **40**, 1351.

235. Fischer, T., Thomas, B., Scheurich, P. and Plizenmauer, K. (1990) *J. Biol. Chem.*, **265**, 1710.

236. Hortin, G.L. and Boime, I. (1980) *J. Biol. Chem.*, **255**, 8007.

237. Hussain, M.Z., Ghani, Q.P. and Hunt, T.K. (1989) *J. Biol. Chem.*, **265**, 5157.

238. Leonard, S., Beck, L. and Sinensky, M. (1990) *J. Biol. Chem.*, **265**, 5157.

239. Enouf, J., Lawrence, F., Tempete, C., Robert-Gero, M. and Lederer, E. (1979) *Cancer Res.*, **39**, 4497.

240. Johnson, D.R., Cox, A.D., Solski, P.A., Devades, B., Adams, S.P., Leingruber, R.M., Heukeroth, R.O., Buss, J.E. and Gordon, J.I. (1990) *Proc. Natl. Acad. Sci. USA*, **87**, 8511.

241. Lamb, N.J.C., Fernandez, A., Watrin, A., Labbe, J.-C. and Cavadore, J.-C. (1990) *Cell*, **60**, 151.

242. Davis, P.D., Hill, C.H., Keech, E., Lawton, G., Nixon, J.S., Sedgwick, A.D., Wadsworth, J. and Westmacot, D. (1989) *FEBS Lett.,* **259**, 61.

243. Dvir, A., Milner, Y., Chomsky, O., Gilon, C., Gazit, A. and Levitzki, A. (1991) *J. Cell Biol.,* **113**, 857.

244. Traxler, P.M., Wacker, H., Bach, L., Geissler, J.F., Kump, W., Meyer, O., Regenass, U., Roessel, J.L. and Lydon, N. (1991) *J. Med. Chem.,* **34**, 2328.

245. Hortin, G.L., Schilling, M. and Graham, J.P. (1988) *Biochem. Biophys. Res. Commun.,* **150**, 342.

246. Klemperer, N.S. and Pickart, C.M. (1989) *J. Biol. Chem.,* **264**, 19245.

247. Romanik, E.A. and O'Connor, C.M. (1989) *J. Biol. Chem.,* **264**, 14050.

248. Lamb, N.J.C., Cavadore, J.-C., Labbe, J.-C., Maurer, R.A. and Fernandez, A. (1991) *EMBO J.,* **10**, 1523.

249. Lamb, N.J.C., Fernandez, A., Watrin, A., Labbe, J.-C. and Cavadore, J.-C. (1990) *Cell,* **60**, 151.

250. Brautigan, D.L., Sunwoo, J., Labbe, J.-C., Fernandez, A. and Lamb, N.J.C. (1990) *Nature,* **344**, 74.

251. Picard, A., Capony, J.-P., Brautigan, D.L. and Doree, M. (1989) *J. Cell Biol.,* **109**, 3347.

252. Fernandez, A., Brautigan, D.L., Mumby, M. and Lamb, N.J.C. (1990) *J. Cell Biol.,* **111**, 103.

253. Radhakrishna, G. and Wold, F. (1989) *J. Biol. Chem.,* **264**, 11076.

254. Chang, Y.-C., Soman, G. and Graves, D.J. (1986) *Biochem. Biophys. Res. Commun.,* **139**, 932.

255. Moss, J., Tsai, S.-C., Adamik, R., Chen, H.C. and Stanley, S.J. (1988) *Biochemistry,* **27**, 5819.

256. Tanuma, S.-I. and Endo, H. (1990) *Eur. J. Biochem.,* **191**, 57.

257. Maruta, H., Inageda, K., Acki, T., Nishima, H. and Tanuma, S.-I. (1991) *Biochemistry,* **30**, 5907.

258. Tanuma, S.-I. (1989) *Biochem. Biophys. Res. Commun.,* **163**, 1047.

259. Oka, J., Ueda, K., Hayaishi, O., Komura, H. and Nakanishi, K. (1984) *J. Biol. Chem.,* **259**, 986.

260. Yan, S.-C.B. (1984) *Trends. Biochem. Sci.,* **9**, 331.

261. Nuck, R., Zimmermann, M., Sauvageot, D., Josic, D. and Reutter, W. (1990) *Glycoconjugate J.,* **7**, 279; Fisher, K.J., Tollersud, O.K. and Aronson, Jr., N.N. (1990) *FEBS Lett.,* **269**, 440.

262. Gagnon, C., Harbour, D. and Camato, R. (1984) *J. Biol. Chem.,* **259**, 10212.

263. Veeraragavan, R. and Gagnon, C. (1989) *Biochem. J.,* **260**, 11.

264. Ingebritsen, T.S. and Cohen, P. (1983) *Science,* **221**, 331.

265. Alexander, D.A. (1990) *New Biologist,* **2**, 1049.

266. Agell, N., Ryan, C. and Schlesinger, M.J. (1991) *Biochem. J.,* **273**, 615.

267. Mann, K.G., Jenny, R.J. and Krishnaswamy, S. (1988) *Ann. Rev. Biochem.,* **57**, 915.

268. Reid, K.B.M. and Porter, R.R. (1981) *Ann. Rev. Biochem.,* **50**, 433.

269. Verner, K. and Schatz, G. (1988) *Science,* **241**, 1307.

270. Harris, R.B. (1989) *Arch. Biochem. Biophys.,* **275**, 315.

271. Pohl, T., Zimmer, M., Mugele, K. and Spoess, J. (1991) *Proc. Natl. Acad. Sci. USA,* **88**, 10059.

272. Shimada, Y., Sugihara, A., Tominga, Y., Izumi, T. and Tsunazawa, S. (1989) *J. Biochem. (Tokyo),* **106**, 383.

CHAPTER 11
NUCLEIC ACIDS AND THEIR COMPONENTS
D. Rickwood and D. Patel

1. NUCLEOSIDES AND NUCLEOTIDES

1.1. Basic structures
Nucleosides are conjugates of a nitrogenous compound (a base) and a sugar. Nucleotides are phosphate esters of nucleosides. Nucleic acids, deoxyribonucleic acid (DNA) and ribonucleic acid (RNA), are polymers with a sugar–phosphate backbone with a base attached to the C-1 of the sugar. In RNA the sugar is ribose and in DNA it is deoxyribose. The bases in DNA are adenine and guanine (purines), and cytosine and thymine (pyrimidines). In RNA thymine is replaced by uracil. The structures of the sugars and the common bases are shown in *Figure 1*. *Figure 2* shows the structures of modified bases that occur especially in rRNA and tRNA. The bases normally pair as A:T and G:C (*Figure 3*) although in RNA G:U base pairing is also found.

1.2. Nucleotides as acids
The phosphate groups of nucleotides make them acidic. Hence all stock solutions should be neutralized by adding a calculated amount of cation (e.g. 3.5 moles of Na^+ per mole of nucleoside triphosphate) or by titrating with a basic buffer such as Tris or Tricine.

1.3. Optical density of bases, nucleosides and nucleotides of DNA and RNA
The bases of DNA and RNA absorb UV light strongly. Accordingly, the values of λ_{max} and ε_{max} are useful for determining the concentrations of nucleotides and to check their identity and purity. *Table 1* lists the values for acid solutions; optical densities and 280/260 ratios are pH dependent and so will be different in neutral and alkaline solutions. Nucleotides should be stored at $-20°C$; the sodium salt of GTP is very unstable and so it is preferable to use the more stable lithium salt if at all possible.

1.4. Nucleotide-derived compounds
A wide range of compounds is derived from nucleotides. In a number of cases these compounds are activated intermediates such as acyl adenylates or nucleoside diphosphate sugars, while others are coenzymes. *Table 2* is a selection of nucleotide-derived compounds taken from published sources (1, 2).

1.5. Selected nucleotide analogs
Given the central position of nucleotides in the replication and transcription of DNA, it is not surprising that nucleotide analogs have been investigated for a range of therapeutic activities, including antimicrobial and antiviral agents as well as chemotherapeutic agents. *Table 3* lists a selection of compounds and their activities taken from published sources (1–3).

Figure 1. Bases and sugars of nucleic acids.

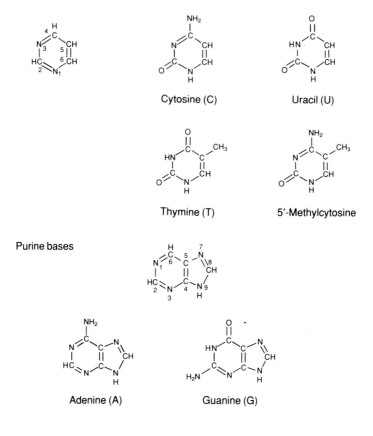

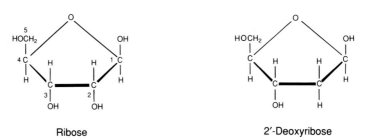

Figure 2. Unusual bases of nucleic acids.

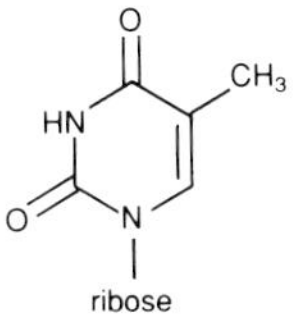

Ribothymidine (T)

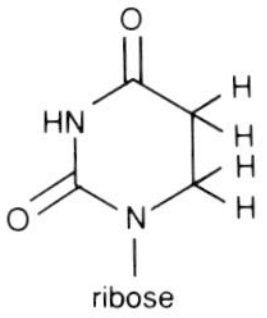

Dihydrouridine (D)

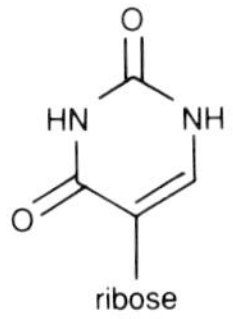

Pseudouridine (ψ)

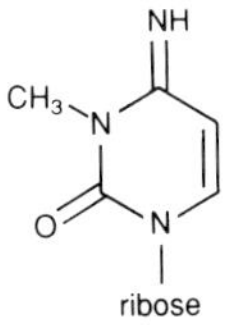

3′-Methylcytidine

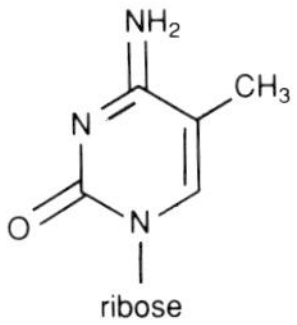

5′-Methylcytidine

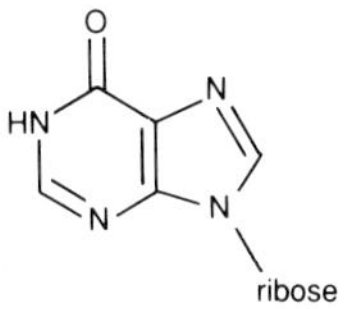

Inosine (I)

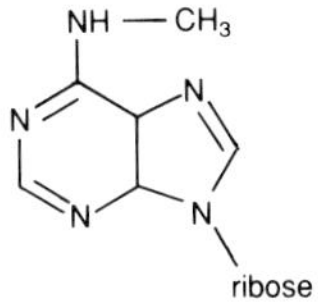

N^6-Methyladenosine

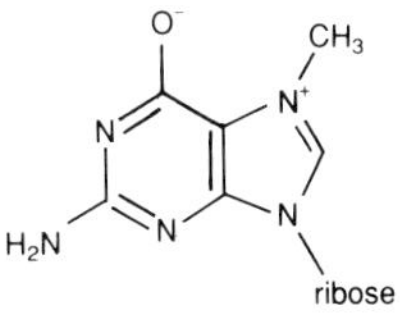

7′-Methylguanosine

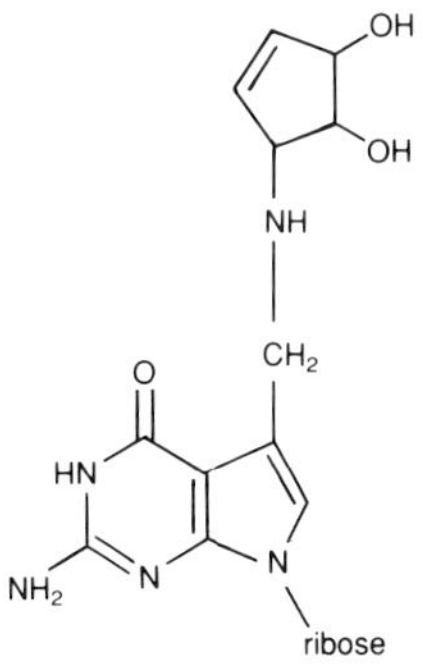

Queuosine (Q)

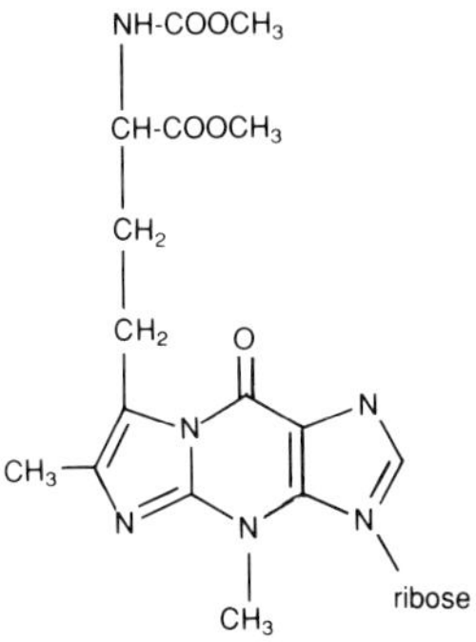

Wyosine (Y)

Figure 3. Base pairing in DNA.

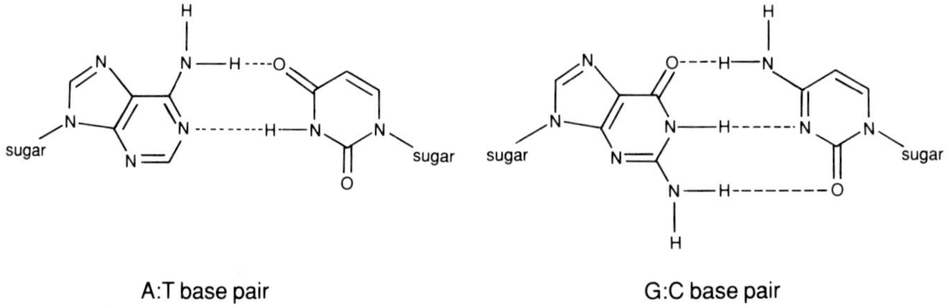

A:T base pair   G:C base pair

Table 1. Physical properties of bases, nucleosides and nucleotides of DNA and RNA

Name	Mol. wt[a]	Solubility (g/100 ml)	λ_{max}	Acid pH ε_{max}^{b} ($\times 10^3$)	ε_{260}^{b} ($\times 10^3$)	Ratio 280/260
Bases						
Adenine	135.1	0.09	263	13.20	13.0	0.37
Cytosine	111.1	0.77	276	10.00	6.0	1.53
Guanine	151.1	0.01	276	7.35	8.0	0.84
Thymine	126.1	0.40	265	7.89	7.4	0.53
Uracil	112.1	0.36	259	8.20	8.2	0.17
Ribonucleosides						
Adenosine	267.2	sol.	257	14.60	14.3	0.22
Cytidine	243.2	v. sol.	280	13.40	6.4	2.10
Guanosine	238.2	0.08	256	12.30	11.8	0.70
Uridine	244.2	sol.	232	10.10	9.9	0.35
Deoxyribonucleosides						
Deoxyadenosine	251.2	sol.	258	14.10	—	0.24
Deoxycytidine	227.2	sol.	280	13.20	6.2	2.16
Deoxyguanosine	267.2	sol.	255	12.30	—	0.70
Deoxythymidine	242.2	sol.	267	9.65	8.8	0.72
Ribonucleotides						
AMP	347.2	v. sol.	257	15.10	14.5	0.22
ADP	427.2	v. sol.	257	15.00	14.5	0.22
ATP	507.2	v. sol.	257	14.70	14.3	0.22
CMP	323.2	v. sol.	281	13.20	6.2	2.09
CDP	403.2	v. sol.	280	12.80	6.2	2.07
CTP	483.2	v. sol.	280	12.80	6.0	2.12
GMP	363.2	v. sol.	256	12.20	11.8	0.67
GDP	443.2	v. sol.	256	12.30	11.8	0.67
GTP	523.2	v. sol.	256	12.30	11.8	0.67

Table 1. Continued

Name	Mol. wt[a]	Solubility (g/100 ml)	Acid pH			Ratio 280/260
			λ_{max}	ε_{max}[b] ($\times 10^3$)	ε_{260}[b] ($\times 10^3$)	
UMP	324.2	v. sol.	262	10.00	9.9	0.39
UDP	404.2	v. sol.	262	10.00	9.9	0.39
UTP	484.2	v. sol.	262	10.00	9.9	0.38
Deoxyribonucleotides						
dAMP	331.2	v. sol.	258	14.30	14.2	0.23
dADP	411.2	v. sol.	258	14.30	14.2	0.23
dATP	491.2	v. sol.	258	14.30	14.2	0.23
dCMP	307.2	v. sol.	280	13.20	6.3	2.12
dCDP	387.2	v. sol.	280	12.90	6.2	2.07
dCTP	467.2	v. sol.	280	12.90	6.2	2.14
dGMP	347.2	v. sol.	255	11.80	10.6	0.70
dGDP	427.2	v. sol.	255	11.80	10.6	0.70
dGTP	507.2	v. sol.	255	11.80	10.6	0.70
dTMP	322.2	v. sol.	267	9.64	8.4	0.73
dTDP	402.2	v. sol.	267	9.64	8.9	0.71
dTTP	482.2	v. sol.	267	9.64	8.9	0.72

[a] Molecular weight does not include the presence of cations or water of crystallization.
[b] Molar extinction coefficient.
Abbreviations: sol., soluble; v. sol., very soluble.
Modified from Sambrook, J., Fritsch, E.F. and Maniatis, T. (1989) *Molecular Cloning: A Laboratory Manual (2nd edn)*, with permission from Cold Spring Harbor Laboratory Press.

2. DNA OF SELECTED ORGANISMS

Tables 4–8 list the characteristics of the DNA of selected organisms. These data have been derived from previous publications (4, 5).

3. THE GENETIC CODE

The genetic code is essentially the same in all organisms, from bacteria to baboons. However, an accumulating number of variations has been reported, particularly in the mitochondrial genomes of a wide range of eucaryotes, although, in addition, occasional exceptions have been found in the nuclear genes of protozoa. The assignment of codons in the 'universal' genetic code is shown in *Table 9*.

Table 2. Nucleotide-derived compounds

Derived compounds	Structure	Mol. wt	Comments
Acyl adenylates	R = amino acid side chain		Activated amino acids intermediate in the charging of tRNA
Adenosine 3'-phosphate 5'-sulfatophosphate (PAPS)		507.3	Sulfate donor; APS similar but without 3' phosphate
S-adenosylmethionine (SAM)		Free cation, 398.4	Methyl-group donor

Table 2. Continued

Derived compounds	Structure	Mol. wt	Comments
Coenzyme A		767.6	Acyl-group donor
FAD		785.6	Redox coenzyme
FMN		456.4	Redox coenzyme

Table 2. Continued

Derived compounds	Structure	Mol. wt	Comments
NAD$^+$/NADP$^+$	Reduced form; R (at 2' of adenosine) = −H for NAD$^+$; R (at 2' of adenosine) = phosphate for NADP$^+$	NAD$^+$ 663.4 NADP$^+$ 743.4	Redox coenzyme
Nucleoside diphosphate (NDP) sugars, e.g. UDP-glucose; UDPG			Activated sugar NDP linked at C-1

Abbreviations: APS, adenosine phosphosulfate; FAD, flavin adenine dinucleotide; FMN, flavin mononucleotide; NAD, nicotinamide adenine dinucleotide; NADP, nicotinamide adenine dinucleotide phosphate; tRNA, transfer RNA.

Table 3. A selection of nucleotide analogs

Analog	Structure	Mol. wt	Comments
Azacytidine		244.2	Chemotherapeutic and antibacterial action
Cordycepin		251.2	Chemotherapeutic and antibacterial action
Formycin		267.2	Chemotherapeutic, antiviral and antibacterial action
Puromycin		471.52	Antibacterial action; inhibits protein synthesis
Showdomycin		229.2	Antibacterial and chemotherapeutic action

Table 3. Continued

Analog	Structure	Mol. wt	Comments
Toyocamycin		291.3	Antibacterial and antiviral action
Tubercidin		266.25	Antibacterial action

Table 4. DNA of cells: bacteria

Organism	DNA/cell (pg)	Mole % $G+C$[b]	Refs
Bacillus			
cereus (spores)	0.0108	36(D)	12[a], 13
(veg. cells)	0.0129	36(D)	14, 13
licheniformis		46(T)	15
megaterium	0.025	45.5(C)	16, 17
subtilis		44.5(T)	18
(spores)	0.00542		12[a]
ws (veg. cells)	0.0048		19
Clostridium			
acidiurici		25.3(T)	20
pasteurianum		30.5(T)	20
perfringens		26.5(T)	21
Desulfovibrio			
aestuarii		55(D)	21
orientis		41.7(D)	22
Escherichia coli	0.009	51.5(T)	23, 24
B (log phase)	0.0137		25
(stat. phase)	0.0078		25
Salmonella			
typhimurium	0.011		26
typhi (typhosa)		50(T)	27

[a] Content of DNA estimated from the phosphorus analysis cited in the reference and a value of 9.23% phosphorus in DNA.
[b] Mole % guanosine and cytosine (G + C) was determined by one of the following methods, which are given in their abbreviated form in the table: T_m, T; chemical analysis, C; buoyant density, D. Values are given to the nearest 0.1%.

Table 5. DNA of cells: protozoa, algae, fungi, echinoderms, arthropods and Insecta

Organism	DNA/cell (pg)	Mole % G + C[b]	Unique % DNA	Refs
Protozoa				
Euglena gracilis	2.9			28, 29
var. bacillaris		50.2(T)[c]		30
Paramecium				
aurelia[d]		29(D)	85	31, 32
caudatum	1.53–1.67	28.2(C)		33, 34
Tetrahymena	13.6	25(C)	80	32, 35, 36
pyriformis				
Trichomonas		29(D)		37
vaginalis				
Trypanosoma		47.9(D)		38, 39
equiperdum				
Algae				
Anabaena	0.036			40
variabilis				
Anacystis nidulans	0.030			40
Chlamydomonas		64(D)[c]	70[e]	41, 42
reinhardi				
Fungi				
Aspergillus nidulans		47(C)		43, 44
Green haploid	0.0438			
(per conidium)				45[a]
Candida albicans		32.5(T)		46
Dictyostelium		22(C)		21
discoideum				
whole cell DNA			40[e]	47
nuclear DNA			60[e]	47
Neurospora crassa		55(T)	80	48, 49, 50
Physarum		42.5(D)[c]	58[e]	51, 52, 53
polycephalum				
(slime mold)				
Saccharomyces	0.046	39.5(T)		54, 55, 56
cerevisiae				
haploid	0.0245			57[a]
Echinodermata				
Sea urchin				
Lytechinus, sperm	0.90			58
Paracentrotus,	0.70	36.1(C)		59, 60
sperm				
Stronglyocentrotus			38	61, 62
purpuratus				
Arthropoda				
Aedes caspius,	0.988			63
spermatid				
Artemia salina				64
haploid nucleus	3.0			65
Insecta				
Drosophila		39.8(C)	78[e,f]	66, 67, 68
melanogaster				
sperm	0.18			69

[a] See footnote [a] for Table 4.
[b] See footnote [b] for Table 4.
[c] Satellite DNA present.
[d] Incubation at 50°C.
[e] Reassociation monitored optically by the decrease in hyperchromicity.
[f] Incubation at 66–67°C.

Table 6. DNA of cells: Chordata

Organism	DNA/nucleus (pg)	Mole % G + C[b]	Unique % DNA	Refs
Fish (Teleostei)				
Salmo gairdneri irideus				70
erythrocytes	4.9			23
sperm	2.45	43.1(C)		71, 72
Amphibians and reptiles				
Rana clamitans			22	73
(green frog)				
Rana pipiens		47.3(C)		74, 75
erythrocytes	15.0			76
sperm	6.48			77
Xenopus laevis		42.2(C)	54	61, 75, 78, 79
(African clawed toad)				
diploid	8.4			80
liver	7.5			81
erythrocytes	6.3			76
Toad				
erythrocytes	7.33			76
sperm	3.70			76
Bufo bufo				
liver	14.2			81
erythrocytes	14.6			82
Uredele, *Axolotl*				
(*Siredon mexicanum*)				
erythrocyte	6.88			83
sperm	3.36, 4.8			83
Newt				
Triturus cristatus				78
liver	52.5			81
T. vulgaris, liver	71.0			81
Ambystoma mexicanum	77.0			81
Ambystoma tigrinum			≤ 24	73, 84
(tiger salamander)				
Alligator	4.98			23, 85
Birds				
Canary (*Serinus canaria*)				86
liver	4.1			87
Chicken			70	88, 89
(*Gallus domesticus*)				
erythrocytes	2.34	41(D)		13, 76
sperm	1.26	41(D)		13, 76
Duck				90
erythrocytes	2.65			23
liver	2.1			59
Gallus gallus, sperm	0.73			91

Table 6. Continued

Organism	DNA/nucleus (pg)	Mole % G+C[b]	Unique % DNA	Refs
Sparrow				92
diploid	1.9			93
Sturnis vulgaris (Starling)				
liver	3.6			87
Mammals				
Boar, sperm	3.44			109
Dog				105
liver	5.5			59
testes		41.9(C)		110
Guinea-pig				105
adrenal		42.1(C)		106
liver	5.9			59
Homo sapiens (human)			64	117, 118
lymphocyte	6.50	40(D)		13, 119[a]
sperm	2.44	40(D)		13, 120
Horse				115
liver	5.8			59
spleen		43(C)		116
Mus musculus (mouse)			60	97, 98, 99
Ehrlich ascites cell	14.0	40(D)		13, 96
fibroblast embryo	5.1	40(D)		13, 94
fibroblast, L-P3		40(D)		13
leukocytes		43.5(C)		100
liver, diploid	6.0			59
lymphocyte	5.31			95[a]
Ox				
liver	6.4	39(D)		13, 107
sperm	3.3	39(D)		13, 107
Pig				
liver	5.0	41(C)		107, 108
spleen		41.2(C)		108
Rabbit				
liver	5.3	39(D)		13, 59
sperm	3.25	39(D)		13, 106
Rattus (rat)			65	101, 102
lymphocyte	6.05			103
sperm	3.11			103
spleen		43(C)		104
Sheep				111
leukocyte	6.83			112
sperm	2.9	44(C)		113, 114

[a] See footnote [a] for *Table 4*.
[b] See footnote [b] for *Table 4*.

Table 7. DNA of cells: animal viruses

Organism	DNA/particle ($\mu g \times 10^{12}$)	Mole % $G+C^b$	Refs
Adenovirus, 5	120	57.5	121[a], 122
Bovine papilloma (BPV)	8.22	45.5	123[a], 124
Cow pox	27.5	36	125, 126
Equine abortion	417	56.5	119, 127
Herpes simplex	113.6		119
type 1		68	128
Paravirus, H-1 of mice	2.8		129[a]
Parvovirus, HVM of mice	2.8		129[a]
Polyoma virus	5.67	48	130, 131
Pseudorabies	113.6	72	119, 128

[a] See footnote[a] for *Table 4*.
[b] See footnote[b] for *Table 4*.

Table 8. DNA of cells: plants[a]

Organism	DNA/genome (pg)	Mole % $G+C^b$	Unique % DNA	Refs
Allium cepa (onion)	16.8	36.6(C)		132, 133, 134
Arabidopsis thaliana (thale cress)	0.2			132, 133
Avena sativa (oat)	13.7			132, 133
Beta vulgaris (beet)	1.2			132, 133
Brassica napus (turnip)	1.6			132, 133
Cucurbita melo (melon)	1.0			132, 133
Glycine max (soybean)	0.9			132, 133
Gossypium hirsutum (cotton)	3.0	37.2(C)		132, 133, 135
Hordeum vulgare (barley)	5.6	43.1(C)		132, 133, 136
Lilium davidii	43.2			132, 133
Lolium perenne (ryegrass)	4.9			132, 133
Lycopersicon esculentum (tomato)	0.75			132, 133
Nicotiana tabacum (tobacco)	3.9	40.5(C)		132, 133, 136
Oryza sativa (rice)	1.0			132, 133
Pisum sativum (pea)	4.9	41.9(C)		132, 133, 136
Raphanus sativus (radish)	0.4			132, 133
Secale cereale (rye)	9.5			132, 133
Solanum tuberosum (potato)	2.1			132, 133
Spinacia oleracea (spinach)	1.0	37.4(C)		132, 133, 137
Triticum aestivum (hexaploid wheat)	17.3		20	132, 133, 138
Vicia faba (broad bean)	13.3		≤15	132, 133, 139
Zea mays (maize)	3.9	46(C)		132, 133, 140

[a] The data in this table have been kept to a minimum, further details can be obtained from *Plant Molecular Biology Labfax*.
[b] See footnote[b] for *Table 4*.

Table 9. The genetic code

First position (5' end)	Second position				Third position (3' end)
	U	C	A	G	
U	Phe	Ser	Tyr	Cys	U
	Phe	Ser	Tyr	Cys	C
	Leu	Ser	Stop (Ochre)	Stop (Umber)	A
	Leu	Ser	Stop (Amber)	Trp	G
C	Leu	Pro	His	Arg	U
	Leu	Pro	His	Arg	C
	Leu	Pro	Gln	Arg	A
	Leu	Pro	Gln	Arg	G
A	Ile	Thr	Asn	Ser	U
	Ile	Thr	Asn	Ser	C
	Ile	Thr	Lys	Arg	A
	Met	Thr	Lys	Arg	G
G	Val	Ala	Asp	Gly	U
	Val	Ala	Asp	Gly	C
	Val	Ala	Glu	Gly	A
	Val	Ala	Glu	Gly	G

4. ABBREVIATIONS OF AMINO ACIDS

Table 10. Amino acid abbreviations

Amino acid	Three-letter abbreviation	One-letter abbreviation
Alanine	Ala	A
Arginine	Arg	R
Asparagine	Asn	N
Aspartic acid	Asp	D
Cysteine	Cys	C
Glutamic acid	Glu	E
Glutamine	Gln	Q
Glycine	Gly	G
Histidine	His	H
Isoleucine	Ile	I
Leucine	Leu	L
Lysine	Lys	K
Methionine	Met	M
Phenylalanine	Phe	F
Proline	Pro	P
Serine	Ser	S
Threonine	Thr	T
Tryptophan	Trp	W
Tyrosine	Tyr	Y
Valine	Val	V

5. ASSAYS FOR NUCLEIC ACIDS

The accurate measurement of nucleic acid concentration is most important. It can be done simply by UV measurement at 260 nm if there are no other UV absorbing substances present; a native DNA solution of 1 mg ml^{-1} has an absorbance at 260 nm of 22.5 while the same strength solution of RNA has an absorbance of 25.0. Chemical assays for nucleic acids can be divided up into fluorimetric and spectrophotometric assays. As a general rule, the former are more sensitive and tend to suffer less from interference by other substances. The methodologies are given in a number of manuals, (e.g. ref. 6). *Table 11* summarizes the different types of methods that are available.

Table 11. Assays for DNA and RNA

Assay	Type S/F[a]	Minimum detectable	Interfering compounds	Refs
DNA				
UV absorption (260 nm)	S	1 μg	UV-absorbing compounds	
Diphenylamine	S	50 μg	Sugars, protein	7
DAPI	F	20 ng	Fluorescence depends on DNA base composition	8
Ethidium bromide	F	50 ng		9, 10
Methyl green	S	2 μg	Impurities in methyl green, detergents	11
RNA				
UV absorption (260 nm)	S	1 μg	UV-absorbing compounds	
Orcinol	S	20 μg	Sugars, nucleotides, protein	7
Ethidium bromide	F	100 ng		9, 10

[a] S, spectrophotometric assay; F, fluorometric assay.
Abbreviation: DAPI, 4,6, diamidino-2-phenylindole.

6. REFERENCES

1. Dawson, R.M.C., Eliott, D.C., Elliott, W.H. and Jones, K.M. (1969) in *Data for Biochemical Research (2nd edn).* Oxford University Press, Oxford.

2. Fasman, G.D. (1975) *Handbook of Biochemistry and Molecular Biology (3rd edn) Nucleic Acids.* CRC Press, Cleveland, Vol. 1.

3. Greenwood, D. (1989) *Antimicrobial Chemotherapy (2nd edn).* Oxford University Press, Oxford.

4. Brown, T.A. (1990) *Molecular Biology Labfax.* BIOS Scientific Publishers, Oxford.

5. Dealtry, G.D. and Rickwood, D. (1992) *Cell Biology Labfax.* BIOS Scientific Publishers, Oxford.

6. Rickwood, D. (1984) *Centrifugation: A Practical Approach.* IRL Press at OUP, Oxford.

7. Schneider, W.C. (1957) *Methods Enzymol.,* **3**, 680.

8. Brunk, C.F., Jones, K.C. and James, T.W. (1979) *Anal. Biochem.,* **92**, 497.

9. Karsten, U. and Wollenberger, A. (1972) *Anal. Biochem.,* **46**, 135.

10. Karsten, U. and Wollenberger, A. (1977) *Anal. Biochem.,* **77**, 464.

11. Peters, D.L. and Dahmus, M.E. (1979) *Anal. Biochem.,* **93**, 306.

12. Fitz-James, P.C. and Young, I.E. (1959) *J. Bacteriol.,* **78**, 743.

13. Szybalski, W. (1968) *Methods Enzymol.,* **12B**, 330.

14. Hodson, P.H. and Beck, J.V. (1960) *J. Bacteriol.,* **79**, 661.

15. Dubnau, D., Smith, I., Morrell, P. and Marmur, J. (1965) *Proc. Natl. Acad. Sci. USA,* **54**, 491.

16. Spiegelman, S., Aronson, A. and Fitz-James, P.C. (1958) *J. Bacteriol.,* **75**, 102.

17. Ikeda, Y., Saito, H., Miura, K.I., Takagi, J. and Aoki, H. (1965) *J. Gen. Appl. Microbiol.*, **11**, 181.

18. Welker, N.E. and Campbell, L.L. (1967) *J. Bacteriol.*, **94**, 1124.

19. Aubert, J.P., Ryter, A. and Schaeffer, P. (1968) *Ann. Inst. Pasteur*, **115**, 989.

20. Tonomura, B., Malkin, R. and Rabinowitz, J.C. (1965) *J. Bacteriol.*, **89**, 1438.

21. Schildkraut, C.L., Marmur, J. and Doty, P. (1962) *J. Mol. Biol.*, **4**, 430.

22. Saunders, G.F., Campbell, L.L. and Postgate, J.R. (1964) *J. Bacteriol.*, **87**, 1073.

23. Vendrely, R. (1958) *Ann. Inst. Pasteur*, **94**, 142.

24. De Ley, J. (1970) *J. Bacteriol.*, **101**, 738.

25. Gillies, N.E. and Alper, T. (1960) *Biochim. Biophys. Acta*, **43**, 182.

26. Lark, K.G. and Maaløe, O. (1956) *Biochim. Biophys. Acta*, **21**, 448.

27. Baptist, J.N., Shaw, C.R. and Mandel, M. (1969) *J. Bacteriol.*, **99**, 180.

28. Leedale, G.F. (1958). *Nature*, **181**, 502.

29. Brawerman, G., Rebman, C.A. and Chargaff, E. (1960) *Nature*, **187**, 1037.

30. Edelman, M., Schiff, J.A. and Epstein, H.T. (1965) *J. Mol. Biol.*, **11**, 769.

31. Suyama, Y. and Preer, J.R. (1965) *Genetics*, **52**, 1051.

32. Allen, S. and Gibson, I. (1972) *Biochem. Genet.*, **6**, 293.

33. Gintsburg, G.I. (1961) *Zh. Obshch. Biol.*, **22**, 452.

34. Gintsburg, G.I. (1963), cited in Antonov (1965) *Usp. Sovrem. Biol.*, **60**, 161.

35. Scherbaum (1957) *Exp. Cell Res.*, **13**, 24.

36. Swartz, M.N., Trautner, T.A. and Kornberg, A. (1962) *J. Biol. Chem.*, **237**, 1961.

37. Mandel, M. and Honigberg, B.M. (1964) *J. Protozool.*, **11**, 114.

38. Roskin, G. and Schischliaiewa, S. (1928) *Arch. Protistenk.*, **60**, 460.

39. Riou, G., Pautrizel, R. and Paoletti, C. (1966) *C.R. Acad. Sci. (Paris)*, **262**, 2367.

40. Craig, Leach and Carr (1969) *Arch. Microbiol.*, **65**, 218.

41. Chun, E.H.L., Vaughan M.H. and Rich, A. (1963) *J. Mol. Biol.*, **7**, 130.

42. Wells, R. and Sager, R. (1971) *J. Mol. Biol.*, **58**, 611.

43. Elliott, C.G. (1960) *Genet. Res.*, **1**, 462.

44. Dutta, S.K., Richman, N., Woodward, V.W. and Mandel, M. (1967) *Genetics*, **57**, 719.

45. Heagy, F.C. and Roper, J.A. (1952) *Nature*, **170**, 713.

46. Nakase and Komataga (1968) *J. Gen. Appl. Microbiol.*, **14**, 345.

47. Firtel, R.A. and Bonner, J. (1972) *J. Mol. Biol.*, **66**, 339.

48. Fincham, J.R.S. and Day, P.R. (1963) *Fungal Genetics*. F.A. Davis, Philadelphia.

49. Luck, D.J.L. and Reich, E. (1964) *Proc. Natl. Acad. Sci. USA.*, **52**, 931.

50. Brooks and Huang (1972) *Biochem. Genet.*, **6**, 41.

51. Guttes, E. (1972) Unpublished. University of Texas, Dallas.

52. Mandel, M. (1970) in *A Handbook of Biochemistry (2nd edn)* (Sober, ed.). Chemical Rubber Co., Cleveland, H-75.

53. Fouquet, H., Bierweiler, B. and Sauer, H.W. (1974) *Eur. J. Biochem.*, **44**, 407.

54. Hawthorne, D.C. and Mortimer, R.K. (1968) *Genetics*, **60**, 735.

55. Williamson, D. and Scopes, A.W. (1961) *Exp. Cell Res.*, **24**, 151.

56. Storck, R. (1966) *J. Bacteriol.*, **91**, 227.

57. Ogur, Minckler, Lindegren and Lindegren (1952) *Arch. Biochem. Biophys.*, **40**, 175.

58. Mirsky, A.E. and Ris, H. (1951) *J. Gen. Physiol.*, **31**, 451.

59. Vendrely, R. and Vendrely, C. (1949) *Experientia*, **5**, 327.

60. Chargaff, Lipshitz and Green (1952) *J. Biol. Chem.*, **195**, 155.

61. Davidson, E.H. and Britten, R.J. (1973) *Q. Rev. Biol.*, **48**, 565.

62. Britten, R.J., Graham and Henrey (1972) *Carnegie Inst. Wash. Year Book*, **71**, 270.

63. Jost, E. and Mameli, M. (1972) *Chromosoma*, **37**, 201.

64. Artom, C. (1928) *C.R. Sol. Biol.*, **99**, 29.

65. Rheinsmith, E.L., Hinegardner, R. and Bachmann, K. (1974) *Comp. Biochem. Physiol.*, **48B**, 343.

66. Guyénot, E. and Naville, A. (1929) *Cellule*, **39**, 25.

67. Argyrakis, M.P. and Bessman, M.J. (1963) *Biochim. Biophys. Acta*, **72**, 122.

68. Wu, J., Hurn, J. and Bonner, J. (1972) *J. Mol. Biol.*, **64**, 211.

69. Rasch, Barr and Rasch (1971) *Chromosoma*, **33**, 1.

70. Ohno, S., Stenius, C., Faisst, E. and Zenzes, M.T. (1965) *Cytogenetics*, **4**, 117.

71. Vendrely, R. and Vandrely, C. (1953) *Nature*, **172**, 30.

72. Felix, K., Jilke, I. and Zahn, R.K. (1956) *Hoppe-Seyler's Z. Physiol. Chem.*, **303**, 140.

73. Straus, N.A. (1971) *Proc. Natl. Acad. Sci. USA*, **68**, 799.

74. Porter, K.R. (1941) *Biol. Bull.*, **80**, 238.

75. Dawid, I.B. (1965) *J. Mol. Biol.*, **12**, 581.

76. Mirsky, A.E. and Ris, H. (1949) *Nature*, **163**, 666.

77. England, M.C. and Mayer, D.T. (1957) *Exp. Cell Res.*, **12**, 249.

78. Mikamo, K. and Witschi, E. (1966) *Cytogenetics*, **5**, 1.

79. Davidson, E.H., Hough, B.R., Amenson, C.S. and Britten, R.J. (1973) *J. Mol. Biol.*, **77**, 1.

80. Birstow and Deuchar (1964) *Exp. Cell Res.*, **35**, 580.

81. Conger, A.D. and Clinton, J.H. (1973) *Radiat. Res.*, **54**, 69.

82. Bachmann, K. (1970) *Chromosoma*, **29**, 365.

83. Edström, J.E. (1964) *Biochim. Biophys. Acta*, **80**, 399.

84. Carrick, R. (1934) *Trans. Roy. Soc. Edinburgh*, **58**, 63.

85. Cohen, M.M. and Gans, C. (1970) *Cytogenetics*, **9**, 81.

86. Ohno, S., Stenius, C., Christia, L.C., Becak, W. and Becak, M.L. (1964) *Chromosoma*, **15**, 280.

87. Bachmann, K., Harrington, B.A. and Craig, J.P. (1972) *Chromosoma*, **37**, 405.

88. Owen, J.T.T. (1965) *Chromosoma*, **16**, 601.

89. Sanchez de Jimenéz, E., González, J.L., Domínguez, J.L. and Saloma, E.S. (1974). *Eur. J. Biochem.*, **45**, 25.

90. Hammer, B. (1966) *Hereditas*, **55**, 367.

91. Eapen and Raza Nasir (1963) *Indian Vet. J.*, **40**, 803.

92. Brink, J.M. van (1959) *Chromosoma*, **10**, 1.

93. Vendrely, R. and Vendrely, C. (1950) *C.R. Acad. Sci. Ser. D,* **230**, 788.

94. Bassleer, R. (1964) *C.R. Soc. Biol.*, **158**, 384.

95. Menton, Willms and Wright (1953) *Cancer Res.*, **13**, 729.

96. Leuchtenberger, Klein and Klein (1952) *Cancer Res.*, **12**, 480.

97. Makino, S. (1941) *J. Fac. Sci. Hokkaido Imp. Univ.*, **VI**, **7**, 305.

98. Straus, N.A. (1976) in *Handbook of Biochemistry & Molecular Biology (3rd edn)* (G.D. Fasman ed.). CRC Press, Ohio, Vol. II, p. 319.

99. Straus, N.A. and Birboim, H.C. (1974) *Proc. Natl. Acad. Sci. USA*, **71**, 2992.

100. Penn, N.W., Suwalski, R., O'Riley, C., Bojanowski, K. and Yura, R. (1972) *Biochem. J.*, **126**, 781.

101. Makino, S. and Asana, J.J. (1948) *Chromosoma*, **3**, 208.

102. Holmes, D.S. and Bonner, J. (1974) *Biochemistry*, **13**, 841.

103. Sandritter, Müller and Gensecke (1960) *Acta Histochemica*, **10**, 139.

104. Kleinschmidt and Manthey (1958) *Arch. Biochem. Biophys.*, **73**, 52.

105. Awa, A. *et al.* (1959) *Jap. J. Zool.*, **12**, 257.

106. Bransome, Jr., E.D. and Chargaff, E. (1964) *Biochim. Biophys. Acta*, **91**, 180.

107. Vendrely, R. and Vendrely, C. (1948) *Experienta*, **4**, 434.

108. Chargaff, E. and Lipshitz (1953) *J. Am. Chem. Soc.*, **75**, 3658.

109. Ivanov, Korban and Sharobaiko (1969) *Bull. Eksper. Biol. Med.*, **67**, 46.

110. Busch, E.W., von Borcke, I.M., Greve, H. and Thorn, W. (1968) *Hoppe-Seyler's Z. Physiol. Chem.*, **349**, 801.

111. Melander, Y. (1959) *Hereditas*, **45**, 649.

112. Mandel, M., Metais and Cuny (1950) *C.R. Acad. Sci.*, **231**, 1172.

113. Aberg, B. and Gillner, M. (1966) *Acta Physiol. Scand.*, **66**, 106.

114. Wyatt, G.R. (1951) *Biochem. J.*, **48**, 584.

115. Sasaki, M.S. and Makino, S. (1962) *J. Hered.*, **53**, 157.

116. Daly, Allfrey and Mirsky (1950) *J. Gen. Physiol.*, **33**, 497.

117. Tijio, J.H. and Levan, A. (1956) *Hereditas*, **42**, 1.

118. Saunders, G.F., Shirakawa, S., Saunders, P.P., Arrighi, F.E. and Hsu, T. (1972) *J. Mol. Biol.*, **63**, 323.

119. Darlington, R.W. and Randall, C.C. (1963) *Virology*, **19**, 322.

120. Leuchtenberger, Leuchtenberger and Davis (1954) *Am. Pathol.*, **30**, 65.

121. Allison, A.C. and Burke, D.C. (1962) *J. Gen. Microbiol.*, **27**, 181.

122. Piña, M. and Green, M. (1965) *Proc. Natl. Acad. Sci. USA*, **54**, 547.

123. Lang, D., Bujard, H., Wolff, B. and Russell, D. (1967) *J. Mol. Biol.*, **23**, 163.

124. Thomas, C.A. and MacHattie, L.A. (1967) *Ann. Rev. Biochem.*, **36**, 485.

125. Joklik, W.K., cited in Shapiro, H.S. (1976) in *Handbook of Biochemistry & Molecular Biology (3rd edn)* (G.D. Fasman ed.). CRC Press, Ohio, Vol. II, p. 310.

126. Joklik, W.K. (1962) *J. Mol. Biol.*, **5**, 265.

127. Soehner, R.L., Gentry, G.A. and Randall, C.C. (1965) *Virology*, **26**, 394.

128. Plummer, G., Goodheart, C.R., Henson, D. and Bowling, C.P. (1969) *Virology*, **39**, 134.

129. McGoech, D.J., Crawford, L.V. and Follett, E.A.C. (1970) *J. Gen. Virol.*, **6**, 33.

130. Crawford, L.V. (1974) *Virology*, **22**, 149.

131. Yamagishi, H., Yoshizako, F. and Sato, K. (1966) *Virology*, **30**, 29.

132. Bennett *et al.* (1982) *Proc. R. Soc. London, Series B*, **216**, 179.

133. Bennett and Smith (1976) *Phil. Trans. R. Soc. London, Series B*, **274**, 227.

134. Uryson, S.O. and Belozerskii, A.N. (1959) *Dokl. Akad. Nauk SSSR*, **125**, 1144.

135. Sulimova, G.E., Mazin, A.L., Vanyushin, B.F. and Belozerskii, A.N. (1970) *Dokl. Akad. Nauk SSSR*, **193**, 1422.

136. Vanyushin, B.F., Kadyrova, D.Kh., Karimov, Kh.Kh. and Belozerskii, A.N. (1971) *Biokhimya*, **36**, 1251.

137. Bard, S.A. and Gordon, M.P. (1969) *Plant Physiol.*, **44**, 377.

138. Bendich, A.J. and McCarthy, B.J. (1970) *Genetics*, **65**, 545.

139. Straus (1972) *Carnegie Inst. Wash. Year Book*, **71**, 257.

140. Ergle, D.R. and Katterman, F.R. (1961) *Plant Physiol.*, **36**, 811.

CHAPTER 12
LIPIDS
T. J. Walton

1. INTRODUCTION

Lipids are those naturally occurring substances that are readily soluble in organic solvents of low polarity, such as chloroform and hydrocarbons, but insoluble in water. In biological systems they have major functions in membrane structure, storage and metabolism, including roles as physiological and pathophysiological mediators. Individual lipids frequently have more than one of these functions. The diverse group of substances encompassed by this definition can be divided into two broad structural classes: acyl lipids, which contain hydrocarbon chains derived from fatty acids and terpenoids, or polyisoprenoids, which contain one or more characteristic five-carbon, branched-chain isoprene units.

2. STRUCTURE AND CHARACTERISTICS OF ACYL LIPIDS

2.1. Fatty acids

Fatty acids are long-chain saturated and unsaturated monocarboxylic aliphatic acids. They are not normally found in tissue in their nonesterified fatty acid (NEFA) forms. These have strong detergent properties and also exhibit considerable affinity for proteins. As components of acyl lipids, however, it is the hydrocarbon chains of the fatty acids that confer their characteristic solubility properties on this group of lipids. The fatty acids, present as their *O*-acyl esters or *N*-acyl amides in the acyl lipid, typically contain an even number of carbon atoms with chain lengths of C16 and C18 predominating (*Table 1*). Frequently one (monoenoic fatty acids) or more (2-6) (polyunsaturated fatty acids; PUFAs) *cis*-(Z) olefinic double bonds are present. Monoenoic fatty acids characteristically contain a *cis* double bond, frequently though not invariably between carbon atoms 9 and 10(9). *Trans*-(E) isomers are rare components of natural lipids, with the exception of *trans*-3-hexadecenoic acid, a characteristic fatty acid of chloroplasts (*Table 1*). In the commonly occurring PUFAs, the double bonds are invariably separated from each other by a methylene grouping — the so-called methylene interrupted, non-conjugated double bond (*cis*-(Z), *cis*-(Z)-1,4-diene) system (*Table 2*).

Fatty acids are commonly referred to by their trivial names rather than by the systematic ones. In addition, two systems of shorthand nomenclature are in wide use. In the first, chain length, the number of double bonds and their position in the molecule relative to the carboxyl carbon atom (C-1) are specified numerically; for example oleic acid, a C18 acid containing one *cis* double bond between C-9 and C-10 (*Table 1*), is represented as 18:1(9). Double bonds are assumed to be *cis*; the presence of a *trans* double bond is specifically indicated; thus *trans*-3-hexadecenoic acid is represented as 16:1(3*t*). The second shorthand convention is of particular value in considering structural and metabolic relationships of polyunsaturated fatty acids. Under this system the position of the first unsaturation from the methyl end of the molecule is designated. Thus α-linolenic acid (see *Table 2*) is described as 18:3(*n*-3) where *n* (ω in older literature) indicates numbering from the methyl terminus.

Table 1. Selected saturated and monoenoic fatty acids

Structure	Shorthand nomenclature	Common name	m.p. °C	Comments
Saturated				
$CH_3(CH_2)_{10}COOH$	12:0	Lauric	44	Dominant in seed fat of Lauracea, e.g. palm kernel oil
$CH_3(CH_2)_{12}COOH$	14:0	Myristic	54	Widespread; dominant in seed fat of Myristicacea; characteristic acyl moiety in protein acylation
$CH_3(CH_2)_{14}COOH$	16:0	Palmitic	63	Major component; most widespread fatty acid; important in protein acylation
$CH_3(CH_2)_{16}COOH$	18:0	Stearic	70	Widespread, common major component
$CH_3(CH_2)_{18}COOH$	20:0	Arachidic	77	Widespread minor component, major in some seed oils
$CH_3(CH_2)_{20}COOH$	22:0	Behenic	82	Minor and widespread; occasionally major in some seed oils
$CH_3(CH_2)_{22}COOH$	24:0	Lignoceric	83	Widespread as a minor component in seed fat and plant waxes
Monoenoic				
$CH_3(CH_2)_5\overset{c}{CH{=}CH}{-}(CH)_7COOH$	16:1(9)	Palmitoleic	1	Widespread minor component, major in some seed oils
$CH_3(CH_2)_{11}\overset{t}{CH{=}CH}{-}CH_2COOH$	16:1(3t)	trans-3-Hexadecenoic	53	Major fatty acid in phosphatidyl glycerol of chloroplasts
$CH_3(CH_2)_7\overset{c/t}{CH{=}CH}{-}(CH_2)_7COOH$	18:1(9)	Oleic	13	Most widespread and abundant unsaturated fatty acid of animals and plants
	18:1(9t)	Elaidic	44	Rare natural component of fats
$CH_3(CH_2)_5\overset{c}{CH{=}CH}{-}(CH_2)_9COOH$	18:1(11)	Vaccenic	44	Major bacterial fatty acid
$CH_3(CH_2)_7{-}\overset{c}{CH{=}CH}{-}(CH_2)_{11}COOH$	22:1(13)	Erucic	35	Present in high levels of seed oils of Cruciferae; toxic in mammals, causes heart muscle lesions

Abbreviation: m.p., melting point.
Modified from data originally published in ref. 1, with permission from Chapman and Hall.

Table 2. Selected naturally occurring polyunsaturated fatty acids (PUFAs)

Fatty acid family	Shorthand nomenclature	Common name	m.p. °C	Comments
n-3	16:3 (7, 10, 13)			Present in green algae, e.g. *Euglena*
	16:4 (4, 7, 10, 13)			
	18:3 (9, 12, 15)	α-Linolenic	−16	Major fatty acid of chloroplasts and some seed oils
	20:5 (5, 8, 11, 14, 17)	Eicosapentaenoic (EPA)	−54	Major fatty acid of marine algae and fish oil; eicosanoid precursor
	22:5 (7, 10, 13, 16, 19)	Clupanodonic		Abundant in fish
	22:6 (4, 7, 10, 12, 16, 19)	Docosahexaenoic (DHA)	−44	Major component of marine algae and fish oil
n-6	18:2 (9, 12)	Linoleic acid	−5	Ubiquitous in vegetable fats, major component of many seed oils; essential fatty acid (EFA) for animals
	18:3 (6, 9, 12)	γ-Linolenic acid (GLA)		Traces in animal fats and fish oil; significant in some plants, e.g. evening primrose
	20:3 (8, 11, 14)	Homo-γ-linolenic		Important intermediate in arachidonate formation
	20:4 (5, 8, 11, 14)	Arachidonic	−49	Widespread in animals and fish oils, rare in plants; major eicosanoid precursor in mammals
n-9	18:2 (6, 9)			Minor fatty acid in animals
	20:3 (5, 8, 11)		−11	Accumulates in animal lipids in EFA deficiency

Abbreviation: m.p., melting point.

The fatty acid composition of acyl lipid components is a major factor in determining membrane fluidity, a central tenet of the fluid-mosaic model of membane structure (2). The fluidity of a lipid membrane is primarily a consequence of lipid dynamics — lateral and rotational diffusion of entire molecules as well as rotation about single carbon–carbon bonds — and acyl chain order (lipid packing), which refers to the average orientation of each carbon atom along the chain. Saturated fatty acids assume extended, straight-chain conformations, and the melting point increases with increased chain length (*Table 1*). Double-bond systems restrict rotation of the acyl chain. A single *cis* configuration unsaturation introduces a fixed 120° kink into the average molecular shape and substantially reduces the effective length of the acyl chain compared with the saturated analog. A fatty acid with a *trans* double bond has properties more similar to those of the equivalent saturated acid. *Cis* forms are less stable thermodynamically than the corresponding *trans* isomers, and thus have lower melting points than both the latter and their saturated analogs (cf. 18:0, 18:1(9) and 18:1(9*t*)) (*Table 1*). Comprehensive details of the physical and chemical properties of individual fatty acids have been compiled (3). Acyl chain order therefore increases with acyl chain length but decreases with *cis*-unsaturation, and in more disordered membranes the acyl chains do not pack as tightly, facilitating the diffusion of molecules. At physiological temperatures cholesterol increases acyl chain order, as does sphingomyelin, while the nature of the phospholipid head group also has a significant effect on fluidity (4).

A new structural role for fatty acids has recently been established in the post-translational modification of proteins. It is now apparent (reviewed in ref. 5) that acylated proteins containing myristic acid and palmitic acid covalently bonded via *O*-acyl and *N*-acyl linkages are widespread in eucaryotes and viruses, whereas the prenylation of proteins with retinoic acid (6), farnesol and geranylgeraniol (7, 8) has also been described recently. Roles proposed for protein acylation and prenylation include the localization and retention of hydrophilic proteins in plasma membrane systems and signal transduction roles analogous to post-translational phosphorylation of proteins.

'Unusual' fatty acids and fatty acid derivatives
In addition to the straight-chain, even carbon-number fatty acid classes described above, many other classes of fatty acid occur in tissue, either in trace amounts of wide distribution, or of major proportion but in limited distribution. Such classes include allenic and acetylenic fatty acids, *iso-*, *anteiso-* and internally methyl-substituted fatty acids, carbocyclic fatty acids and hydroxy- and epoxy-fatty acids. Details of the structures and distribution of these compounds are outside the scope of this article, but may be found in ref. 9. Fatty alcohols and fatty aldehydes — even-chain-length primary alcohols and aldehydes — derived from the corresponding fatty acid are important metabolic intermediates and widespread components of the surface waxes of mammals (10), insects (11) and higher plants (12).

2.2. Neutral acyl lipids — wax esters, acylglycerols and glycerol ethers
Wax esters are fatty acyl esters of long-chain primary and secondary alcohols (see *Table 3*), and are common, though often not major, components of surface (epicuticular) waxes, although they are major components of uropygial gland wax in birds (13) and of the internal waxes of marine animals (14), bacteria (15) and fungi (16).

Acylglycerols, also known as glycerides, are esters of glycerol (propane-1,2,3-triol) with fatty acids. In triacylglycerols (TAGs), still commonly referred to as triglycerides, all three glycerol hydroxyl groups are esterified with a fatty acid group, whereas the partial acylglycerols mono- and diacylglycerols (MAGs and DAGs, respectively) contain one and two fatty acyl substituents, respectively (see *Table 3* for structures).

Class	General structure	Comments
Wax esters		
Alkan-1-ol acyl esters	$CH_3(CH_2)_{18}COOCH_2(CH_2)_{24}CH_3$	Chain lengths C_{32-72} typically alcohols C_{26-28} fatty acid C_{16-24}. Widespread occurrence. Alternative energy reserve to triacylglycerols
Acylglycerols		
Monoacylglycerols (monoglycerides; MAG)	$CH_2O\cdot CO\cdot R$ | HOCH | CH_2OH 1-monoacyl-*sn*-glycerol CH_2OH | $R\cdot CO\cdot OCH$ | CH_2OH 2-monoacyl-*sn*-glycerol	Important intermediates in metabolism, particularly in retailoring of dietary triacylglycerol in animal adipose tissue and enterocytes
Diacylglycerols (diglycerides; DAG)	$CH_2O\cdot CO\cdot R_1$ | $R_2\cdot CO\cdot OCH$ | CH_2OH 1,2-diacyl-*sn*-glycerol $CH_2O\cdot CO\cdot R_1$ | HOCH | $CH_2O\cdot CO\cdot R_2$ 1,3-diacyl-*sn*-glycerol	Important metabolic intermediates in triacylglycerol, PC and PE formation. 1,2-*sn*-DAG released by phospholipase C hydrolysis of glycerophospholipids is an intracellular second messenger (*q.v.*) activating protein kinase C
Triacylglycerol (triglyceride; TAG)	$CH_2O\cdot CO\cdot R_1$ | $R_2\cdot CO\cdot OCH$ | $CH_2O\cdot CO\cdot R_3$ 1,2,3-triacyl-*sn*-glycerol	Chief constituents of natural fats (solids) and oils (liquids), the major storage lipids in animals and plants. Most abundant fatty acids are 16:0, 18:0, 18:1(9) and 18:2(9, 12). Plant acylglycerols usually have a higher proportion of unsaturated fatty acids

In the IUPAC convention (17) for designation of the structure of glycerol-containing lipids, the carbon atoms of glycerol are numbered stereospecifically. Glycerol is drawn in the Fischer projection with the secondary hydroxyl group to the left of the central, prochiral carbon atom, then the carbons are numbered 1, 2 and 3 from the top to the bottom (see *Table 3*). Use of this convention is described by the prefix '*sn*' immediately prior to the term 'glycerol' in the name of the glycerolipid. This system avoids the confusion of earlier nomenclature systems in which glycerolipid structures were related to the stereoisomer of glyceraldehyde yielded by oxidation of the glycerolipid.

Glycerol ethers contain hydrocarbon chains linked to hydroxyl groups of glycerol by ether rather than ester linkages. They are of restricted distribution but are abundant in fish liver oils, where typically *sn*-1-alkenyl-ethers predominate over *sn*-1-alkyl- forms.

2.3. Glycerophospholipids

Glycerophospholipids (phosphoglycerides or phospholipids) are the major lipid components of most biological membranes, the only general exceptions being the photosynthetic membranes of plants, algae and cyanobacteria. The simplest glycerophospholipid is the phosphoric acid mono-ester of diacylglycerol, 3-*sn*-phosphatidic acid (*Table 4*). When additional substituents are esterified to the phosphate group, the resulting compounds are named as derivatives of phosphatidic acid (*Table 4*). These compounds are amphipathic molecules because they have a hydrophobic acyl chain domain and a polar, hydrophilic head group. They fall into two broad classes: those with no net charge, and the anionic phospholipids with net negative charge (*Table 4*). Glycerophospholipid nomenclature is also based on the unambiguous stereochemical numbering (*sn*) system described above. As an example, under this system, phosphatidylcholine (X = choline, *Table 4*) is described as 1,2-diacyl-*sn*-glycerol-3-phosphocholine or 3-*sn*-phosphatidylcholine.

There are several significant additional classes of glycerophospholipids which have variations on the phosphatidic acid-based structures described above. These include lyso- or monoacyl forms, plasmalogens and alkyl ethers and the phosphonolipids (see *Table 5* for structures).

Table 4. Structure and distribution of important membrane glycerophospholipids
General formula:

$$\begin{array}{c} \text{H}_2\text{CO}-\overset{\text{O}}{\overset{\|}{\text{C}}}-\text{R}' \\ \text{R}''-\overset{\text{O}}{\overset{\|}{\text{C}}}-\text{OCH} \\ \text{H}_2\text{CO}-\text{O}-\overset{\text{O}}{\underset{\text{O}^-}{\overset{\|}{\text{P}}}}-\text{X} \end{array}$$

Substituent (X)	Phospholipid	Comments
OH	Phosphatidic acid (PA)	Negatively charged lipid; important metabolic intermediate, only occurring in trace amounts
$OCH_2CH_2\overset{+}{N}H_3$	Phosphatidylethanolamine (PE)	Widespread and major lipid; major component of fraction formerly termed cephalin; frequent major component of bacterial membranes; N-acyl derivative found in certain seeds
$OCH_2CH(\overset{+}{N}H_3)COO^-$	Phosphatidylserine (PS)	Negatively charged lipid; serine is the L-isomer; widespread but minor lipid; minor component of old 'cephalin fraction'
$OCH_2CH_2\overset{+}{N}((CH_3)_3$	Phosphatidylcholine (PC)	Has a net neutral charge; the major animal phospholipid; formerly lecithin; found in large amounts in plants, rare in bacteria
$OCH_2CH(OH)CH_2OH$	Phosphatidylglycerol (PG)	Negatively charged lipid; head-group glycerol has sn-1 configuration; the major phospholipid in higher plant photosynthetic membranes and many bacteria; some bacteria contain O-aminoacyl groups (lysine, ornithine, arginine or alanine) attached to position 3 of the base glycerol; bisphosphatidic acid, the fully acylated analog of PG, has been found in some plant tissues
(inositol structure)	Phosphatidylinositol (PI, PtdIns)	Negatively charged lipid; inositol is the *myo*-isomer; widespread and usually minor lipid

Table 4. Continued

Substituent (X)	Phospholipid	Comments
(inositol-4-phosphate structure)	Phosphatidyl-*myo*-inositol 4-phosphate (PtdIns(4)P)	Strongly negatively charged lipid; minor component in animals and eucaryotic plants; PtdIns(3)P also occurs in much smaller amounts
(inositol-4,5-bisphosphate structure)	Phosphatidyl-*myo*-inositol 4,5-bisphosphate (PtdIns(4,5)P$_2$)	Strongly negatively charged lipid; minor component (<10% PI levels) in eucaryotic cells; key role in agonist-stimulated Ca^{2+} mobilization and protein kinase C activation; PtdIns(3,4)P$_2$ and PtdIns(3,4,5)P$_3$ present in some transformed cells
$-OH_2C-CH(OH)-CH_2O-\overset{\overset{O}{\|}}{P}-O-$ $R.COOH_2C-HC-CH_2O$ $\quad\quad\quad\quad OOC.R'$	Diphosphatidylglycerol (cardiolipin, DPG)	Negatively charged lipid; common and major in bacteria; localized in the inner mitochondrial membrane of eucaryotes

Abbreviations: R, R' and R'' are long-chain alkyl groups.
Modified with permission from data originally published in ref. 9, with permission from Chapman and Hall.

Table 5. Glycerophospholipids containing structural variations of the phosphatidyl moiety

Class and general/representative structure	Remarks
(i) Lysophospholipids sn-1-acyl isomer / sn-2-acyl isomer	Lyso(monoacyl) forms of phosphatidyl-choline, -ethanolamine and -serine occur as intermediates in phospholipid metabolism; possess potent detergent properties
(ii) Ether glycerophospholipids 1-O-alkenyl-2-acyl ethers (plasmalogens) 1-O-alkyl-2-acyl ethers 1,2-diphytanyl ethers	Ether lipid analogs of PE and PC are usually minor components of animal tissue, but are substantial in spermatozoa and protozoan cilia, and occur at elevated concentration in many tumors; alkenyl ethers (plasmalogens) more widespread than alkyl ethers; C16:0, C18:0 and C18:1 alkyl chains common; the alkyl ether platelet activating factor (PAF; 1-O-alkyl-2-acetyl-sn-glycerol-3-phosphorylcholine) is a bioactive phosphoglyceride (q.v.) Polar lipids of Archebacteria are composed almost exclusively of phytanyl ether lipids: halophiles contain phytanyl diethers; thermoacidophiles, tetraethers; and methanogens, both di- and tetraethers
(iii) Phophonolipids 1,2-diacyl-sn-glycero-3-phosphono-2-ethylamine	Significant in lower animals, e.g. molluscs, coelenterates and protozoa; very minor in mammals and bacteria; phosphonolipids are particularly resistant to chemical and enzymic attack. 1-O-alkyl analogs also occur

Abbreviations: PC, phosphatidylcholine; PE, phosphatidylethanolamine.

2.4. Glyceroglycolipids

Glycolipids containing 1–4 sugar residues linked glycosidically to 1,2-diacyl-sn-glycerol are widely distributed. They are minor constituents of animal membranes (or lipids) but are the main lipid components of the photosynthetic membranes of eucaryotic algae, cyanobacteria and higher plants and so are the most abundant membrane lipids in nature. They comprise two major, galactose containing lipids — monogalactosyl-diacylglycerol (I, *Figure 1*) and digalactosyldiacylglycerol (II, *Figure 1*), that together typically represent approximately 40% of the dry weight of photosynthetic membranes of higher plants. In the disaccharide moiety of digalactosylglycerol the galactose residues are linked α-1,6. In addition to the galactosyldiacylglycerols, sulfoquinovosyldiacylglycerol (III, *Figure 1*), the plant sulfolipid, is also a substantial characteristic component of chloroplasts and cyanobacteria. The plant sulfolipid is negatively charged because of the highly acidic sulfonic acid group of the sulfoquinose moiety, but the galactosyldiacylglycerols are electroneutral. Gram-positive bacteria also contain glycosyldiacylglycerols, of which mono-, di- and triglycosyldiacylglycerols are the most widespread. Monoglucosyldiacylglycerol (IV, *Figure 1*) is the characteristic glycosylglyceride of *Pneumococcus* and *Mycoplasma*. In the diglycosyldiacylglycerol group, typical saccharide moieties are two glucose, two galactose or two mannose units linked α-1,2, α-1,3 or β-1,6 (e.g. V, VI, *Figure 1*). Diglycosyldiacylglycerols are either absent or very minor components in Gram-negative bacteria.

Figure 1. Structures of major glycosylglycerolipids of higher plants and bacteria.

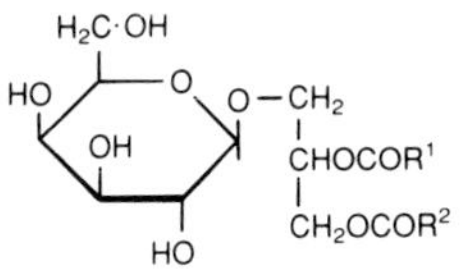

I. Monogalactosyldiacylglycerol (MGDG)
(1,2-diacyl-[β-D-galactopyranosyl-(1→3)]-*sn*-glycerol

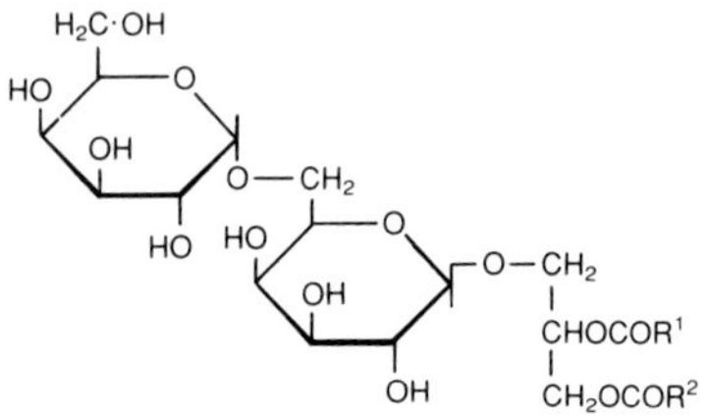

II. Digalactosyldiacylglycerol (DGDG)
(1,2-diacyl-[α-galactopyranosyl-(1'→6')-β-D-galactopyranosyl-(1'→3)]-*sn*-glycerol

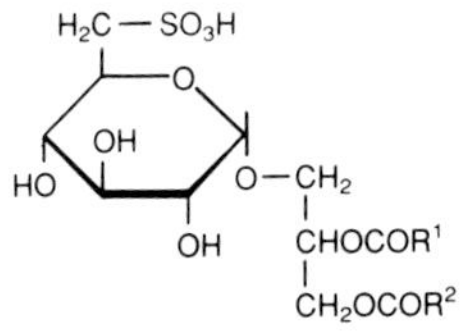

III. Sulfoquinovosyldiacylglycerol (SQDG; plant sulfolipid) (1,2-diacyl-[6-sulfo-α-D-quinovosopyranosyl-(1'→3)]-*sn*-glycerol)

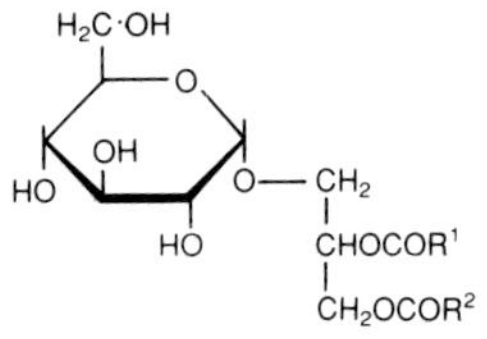

IV. Monoglucosyldiacylglycerol
(*Pneumococcus, Mycoplasma*) (1,2-diacyl-[α-D-glucopyranosyl-(1'→3)]-*sn*-glycerol

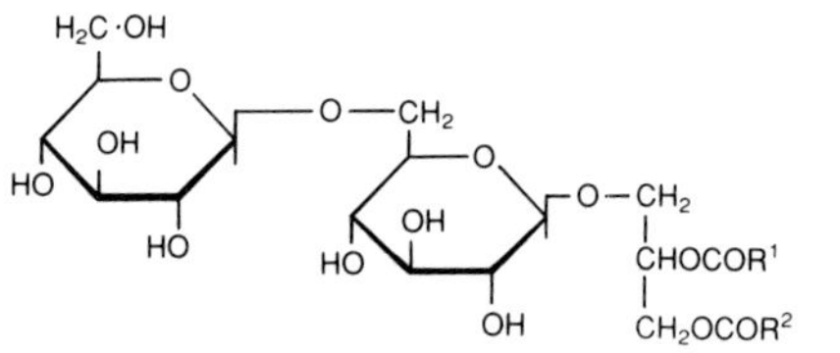

V. Diglycosyldiacylglycerol *(Staphylococcus)*
(1,2-diacyl-[β-D-glucopyranosyl-(1'-6')-β-D-glucopyranosyl (1'→3)]-*sn*-glycerol

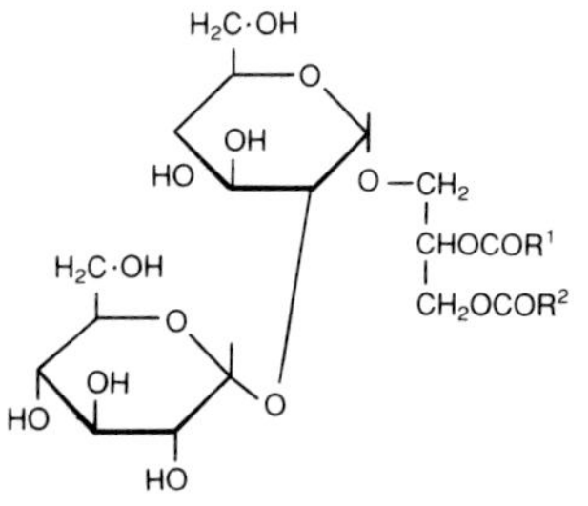

VI. Diglucosyldiacylglycerol *(Mycoplasma, Streptococcus)* (1,2-diacyl-[α-D-glucopyranosyl-(1'→2')-α-D-glucopyranosyl-(1'→3)]-*sn*-diacylglycerol

2.5. Sphingolipids

Sphingolipids are based on the long-chain amino alcohol sphingosine (D-*erythro*-2-amino-*trans*-4-octadecane-1,3-diol (4D-sphingenine)); several other amino alcohols with structures closely related to sphingosine are also found in naturally occurring sphingolipids (*Figure 2a*). Attachment of an *N*-acyl group to sphingosine yields ceramide (Cer), the structure of which is shown together with the general structures of other sphingolipid structural classes in *Figure 2b*. This class of lipids appears to be of major importance in animals but of minor importance in plants and micro-organisms.

Sphingomyelin (SM), the major sphingophospholipid, has an acyl group attached via an amide linkage to sphingosine and a phosphocholine moiety esterified at the primary hydroxyl of the base (*Figure 2c*). Under the general classification of sphingolipid structure (*Figure 2b*), SM can also be named ceramide-1-phosphocholine. SM from most tissues contains not only sphingosine but also the dihydrosphingosine (*Figure 2a*) analog, while SM from kidney contains the phytosphingosine (*Figure 2a*) analog.

In animals, glycosphingolipids are the major groups of glycolipid constituents and can be divided broadly into two classes. These are the glycosyl ceramides, which contain one or

Figure 2. Sphingolipid structures: (a) structures of sphingosine and related sphingosyl alcohols; (b) sphingolipid structural classes; and (c) sphingomyelin.

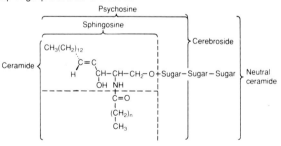

more hexoside units glycosidically linked to ceramide, the *N*-acyl amide of sphingosine (*Figure 2b*), and the gangliosides, which contain one or more molecules of sialic acid (*N*-acetylneuraminic acid, NANA) in addition to hexosides linked to ceramide. The simplest glycosyl ceramides are the monoglycosyl ceramides or cerebrosides, of which monogalactosylcerebroside (Cer-gal) is widely distributed and particularly abundant in brain and myelin sheath. Glucosyl cerebroside (Cer-glu) predominates in serum. Cerebroside sulfate or sulfatide (Cer-gal-3-sulfate) is also widely distributed in animal tissue. The cerebrosides characteristically are rich in fatty acids of longer chain length than those typically found in glycerolipids, including behenic (22:0), lignoceric (24:0) and nervonic (24:1) acids. They also contain high concentrations of α-hydroxyfatty acids (2-hydroxyfatty acids) (e.g. 2-OH-24:0 (cerebronic acid)). The higher neutral glycosyl ceramides contain di-, tri- or tetrasaccharide residues. Lactosyl ceramide (Cer-glc-(4←1)gal) is widely distributed, and the triglycosylceramide (Cer-glc-(4←1)gal-(4←1)gal) is a significant component of kidney, lung, spleen and liver tissue. Tetraglycosylceramides include cer-glc-(4←1)gal-(4←1)gal-(3←1)-β-*N*-acetylgalactosamine (aminoglycolipid globoside) and the 'Fossman antigen', cer-glc-(4←1)gal-(4←1)gal-(3←1)α-*N*-acetylgalactosamine.

Gangliosides, which contain one or more molecules of sialic acid, which is *N*-acetylneuraminic acid (NANA or NeuAc), appear to be confined to the animal kingdom. They occur mainly in the gray matter of brain, but also in spleen, erythrocyte, liver and kidney. In brain, gangliosides containing one (I, *Figure 3*), two (III and IV, *Figure 3*) and three (V, *Figure 3*) molecules of NANA occur, whereas in other tissues the monosialoganglioside (II, *Figure 3*) is the major ganglioside. The predominant (86–95%) fatty acid in these sphingolipids is stearic acid (18:0). The complexities of these structures has led to the use of shorthand nomenclatures to describe them. The most widely used system, introduced by Svennerholm (18) relates ganglioside structure to monosialoganglioside structure I (*Figure 3*) as the parent compound, whereas the convention proposed by Wiegandt (19) relates ganglioside structure to its respective sialic acid-free oligosaccharide.

Figure 3. Structures of some major gangliosides.

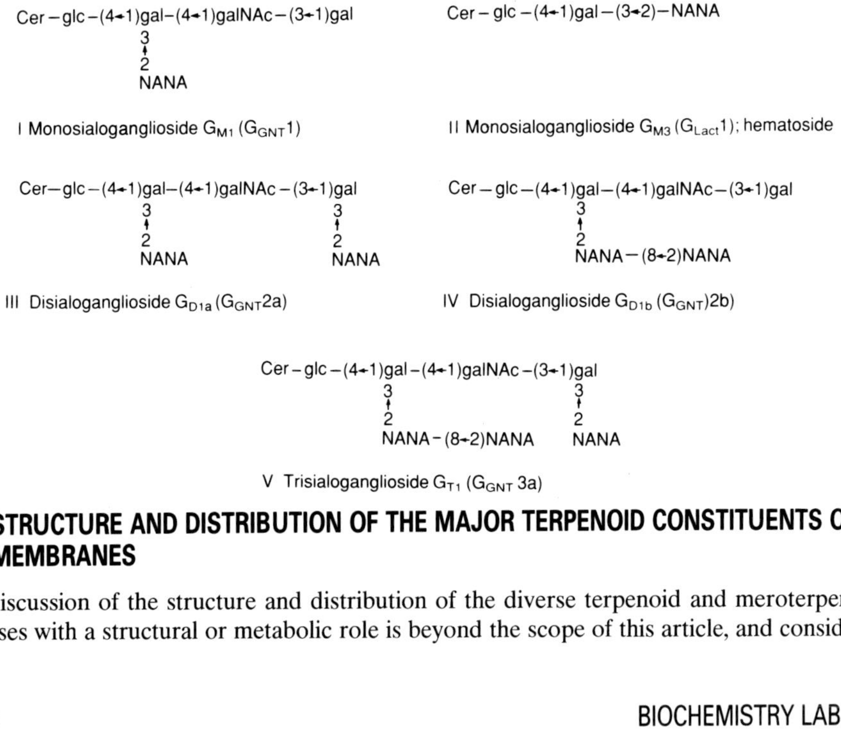

```
Cer–glc–(4←1)gal–(4←1)galNAc–(3←1)gal           Cer–glc–(4←1)gal–(3←2)–NANA
              3
              ↑
              2
            NANA

I Monosialoganglioside G_M1 (G_GNT1)              II Monosialoganglioside G_M3 (G_Lact1); hematoside


Cer–glc–(4←1)gal–(4←1)galNAc–(3←1)gal           Cer–glc–(4←1)gal–(4←1)galNAc–(3←1)gal
              3              3                                                3
              ↑              ↑                                                ↑
              2              2                                                2
            NANA           NANA                                       NANA–(8←2)NANA

III Disialoganglioside G_D1a (G_GNT2a)            IV Disialoganglioside G_D1b (G_GNT)2b)


              Cer–glc–(4←1)gal–(4←1)galNAc–(3←1)gal
                            3                    3
                            ↑                    ↑
                            2                    2
                   NANA–(8←2)NANA              NANA

                   V Trisialoganglioside G_T1 (G_GNT 3a)
```

3. STRUCTURE AND DISTRIBUTION OF THE MAJOR TERPENOID CONSTITUENTS OF MEMBRANES

A discussion of the structure and distribution of the diverse terpenoid and meroterpenoid classes with a structural or metabolic role is beyond the scope of this article, and consideration will be limited to a brief review of the structure and distribution of the sterols which are major structural components of animal and fungal membranes, and to the chlorophylls and carotenoid pigments as major lipid components of photosynthetic membranes.

3.1. Sterols

Sterols are significant components of many natural membranes but are rare in bacteria. In animals, cholesterol (I, *Figure 4*) is found almost entirely in the plasma membrane, where it typically constitutes about 45% of the total lipid. The 24-methyl sterol, ergosterol (II, *Figure 4*), plays a similar role in the plasma membrane of fungi. In plants the major sterols are typically the 24-ethyl sterols sitosterol (~ 70%) and stigmasterol (~ 20%) (III and IV, respectively, *Figure 4*), with much smaller amounts of cholesterol and its 24-methyl analog, campasterol (V, *Figure 4*), and these are much more evenly distributed within subcellular membrane fractions. Sterol derivatives, including fatty acyl esters of sterols, 3-hydroxysterylglycosides and acylated 3-hydroxy glycosides, are widespread and significant components of the total sterol fraction of higher plants (20) and fungi (21, 22). An excellent monograph describing procedures for analysis of diverse sterol classes has been published recently (23).

3.2. Chlorophylls and carotenoids

The structures of the major chlorophylls and carotenoids of photosynthetic membranes of higher plants, algae and photosynthetic bacteria are presented in *Figures 5* and *6*, respectively. Cogdell has reviewed the role of these pigments in chloroplasts (24), and the distribution of carotenoids in non-photosynthetic tissue and their biological roles other than in photosynthesis have also been reviewed (25, 26, 27). Comprehensive accounts of the

▶ p. 282

Figure 4. Structures of the major membrane sterols.

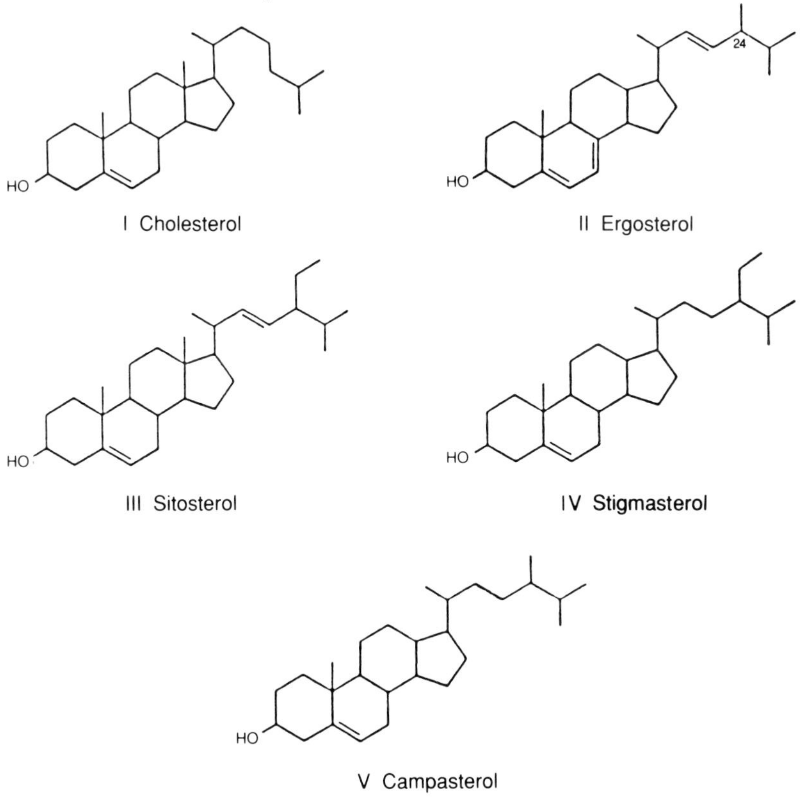

I Cholesterol

II Ergosterol

III Sitosterol

IV Stigmasterol

V Campasterol

Figure 5. Structures of plant and bacterial chlorophylls.

Chlorophylls of higher plants, algae and cyanobacteria

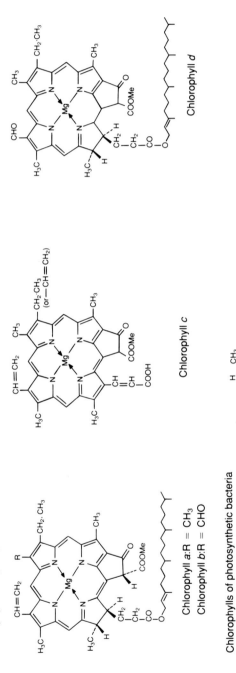

Chlorophylls of photosynthetic bacteria

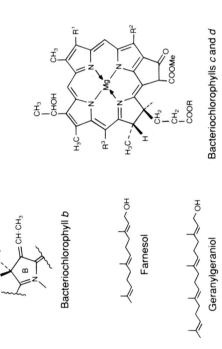

Figure 6. Structures of the characteristic carotenoids of plants, algae and bacteria.

Major carotenoids of higher plants and green algae

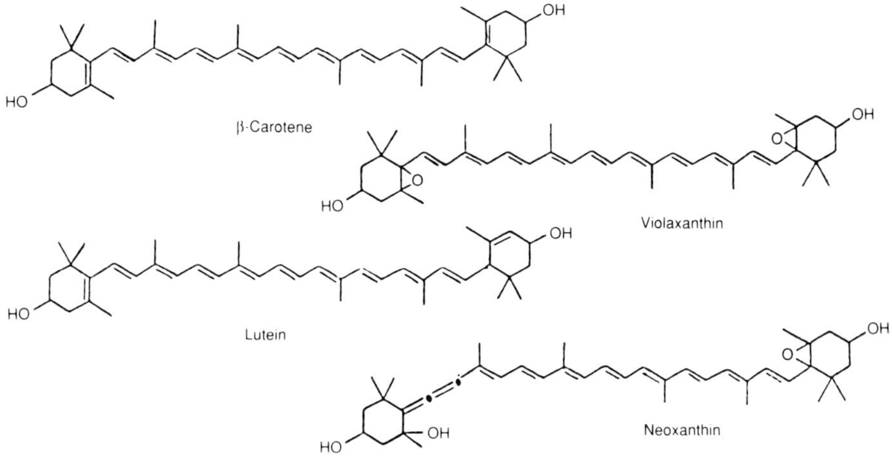

β-Carotene

Violaxanthin

Lutein

Neoxanthin

Characteristic algal carotenoids

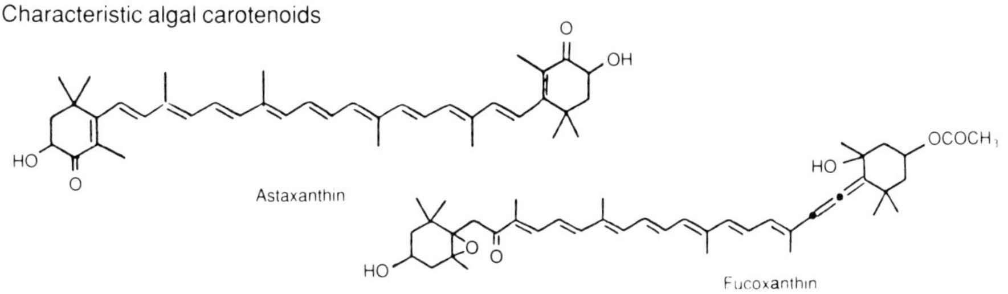

Astaxanthin

Fucoxanthin

Characteristic bacterial carotenoids
(i) Cyanobacteria and non-photosynthetic bacteria

Echinenone

Myxoxanthophyll

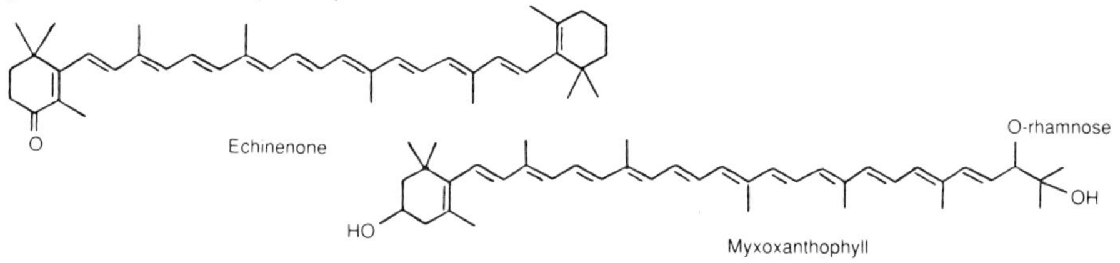

(ii) Purple, non-sulfur photosynthetic bacteria

Spirilloxanthin

Hydroxyspheroidene

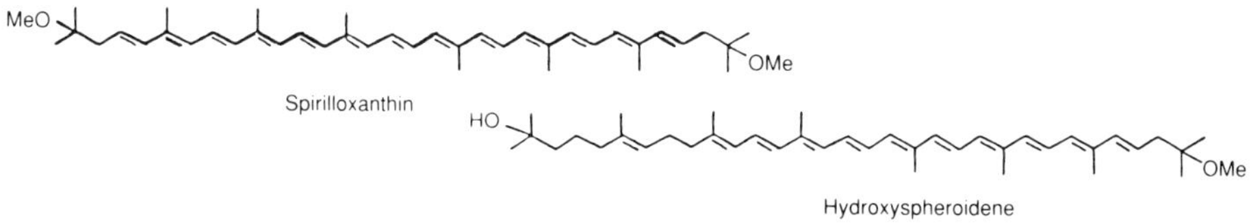

(iii) Purple and green sulfur bacteria

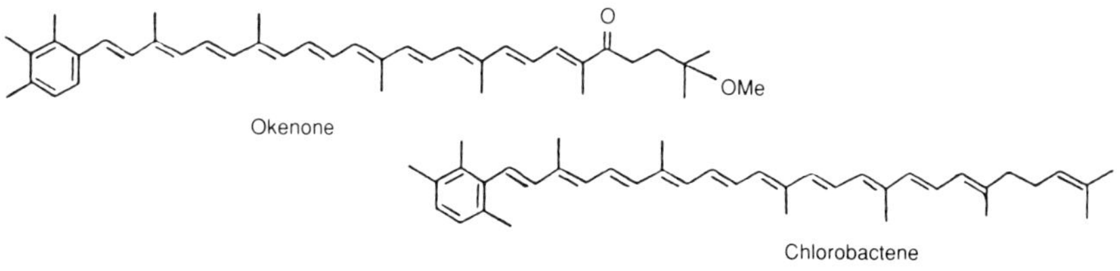

Okenone

Chlorobactene

methodology for the extraction and analysis of carotenoids have been presented (28, 29) and developments in the analysis of chlorophylls recently reviewed (30).

4. COMPOSITION AND DISTRIBUTION OF LIPIDS IN MEMBRANES

4.1. Composition of membrane lipids

Extensive collections of data on the overall lipid composition have been compiled covering mammalian (31), plant (32, 33), fungal (21) and bacterial (34) membrane systems as well as on the asymmetric distribution of both acyl lipids (35) and cholesterol (36) between the lipid bilayers observed in the majority of eucaryotic and procaryotic membrane systems, in which choline-containing lipids are generally enriched in the outer leaflet and aminolipids and cholesterol enriched in the inner monolayer. There are a number of comprehensive monographs on the general methodology for the extraction and analysis of acyl lipids (37–40), together with accounts of the specialist procedures for the structural identification of fatty acids (41) and the analysis of ether lipids (42), gangliosides (43) and inositol phospholipids (44).

4.2. Fatty acid composition of membrane lipids

The most complete analysis of fatty acid composition is obtained directly on the total lipid extracted from a tissue or subcellular preparation, but more valuable information is usually gained by chromatographic separation of such extracts into their individual neutral and polar lipid classes prior to fatty acid analysis. The fatty acid composition of the individual lipid classes isolated from whole tissue and subcellular membrane fractions from selected mammalian (*Tables 6* and *7*), plant (*Tables 8* and *9*), fungal (*Table 10*) and bacterial (*Table 11*) species and tissues are presented, and more comprehensive data are contained in the references from which these tables were compiled.

Individual lipid classes do not normally contain single pure compounds, but are complex mixtures of species in which the composition of the fatty acyl or alkyl moieties varies from molecule to molecule. A complete structural analysis of a lipid class therefore ideally requires that it is separated into individual molecular species that have identical acyl and/or alkyl moieties in all relevant positions of the molecule. Methodology for the separation and analysis of molecular species of triacylglycerols and for diacylglycerols obtained by phospholipase C hydrolysis of phosphoglycerides is well established (37, 50) and high-pressure liquid chromatography (HPLC) methods have recently been developed for separation of molecular species for some glycerophospholipids and glyceroglycolipid classes (51, 52).

The composition of the fatty acyl substituents at each carbon atom within the glycerolipid can be determined by procedures involving analysis of the fatty acids released by enzymic hydrolysis with lipase, specific for primary ester bonds (*sn*-1 and *sn*-3) and phospholipase A_2, specific for secondary ester bonds (*sn*-2). The principles and practice of such stereospecific analysis procedures are discussed in ref. 37.

Bacteria, plants and poikilothermic animals such as fish have a battery of mechanisms by which the fatty acid composition of their membrane lipids adapts to the ambient temperature in a process termed homeoviscous adaptation, to maintain membrane fluidity. Amongst the molecular mechanisms involved are phospholipid retailoring, in which fatty acids bound to a population of phospholipids are redistributed to generate molecular species with fatty acid pairings different from those originally present (53), and the introduction of double bonds into pre-existing fatty acid chains (54). As a result, the fatty acid composition of microorganisms shows marked dependence on growth temperature (34, 35).

Table 6. Fatty acid composition of the whole tissue and subcellular preparations from rat liver

	14:0	16:0	16:1 (n-7)	18:0	18:1 (n-9)	18:2 (n-6)	20:3 (n-9)	20:4 (ω-6)	22:6 (n-3)
Phosphatidylcholine									
whole tissue	0.5	29.7	1.0	16.8	10.4	16.8	1.5	18.3	3.4
plasma membrane	—	32.8	2.9	34.9	10.2	8.1	1.1	8.4	1.6
endoplasmic reticulum	0.7	24.8	3.3	21.0	12.3	17.7	1.2	15.8	2.9
Golgi apparatus	0.9	34.7	—	22.5	8.7	18.1	—	14.5	—
mitochondria	0.4	27.0	3.9	21.6	13.0	12.4	—	17.7	2.9
Phosphatidylethanolamine									
whole tissue	—	25.0	2.0	14.0	6.0	6.0	—	10.0	20.5
plasma membrane	—	30.6	1.2	31.3	10.4	6.5	0.9	16.5	2.9
endoplasmic reticulum	0.3	22.6	2.3	23.4	9.8	10.3	—	23.1	7.2
Golgi apparatus	0.7	33.5	0.4	31.8	5.1	10.0	—	18.3	—
mitochondria	0.3	26.6	3.2	27.3	12.0	5.4	—	22.0	3.2
Phosphatidylinositol									
whole tissue	tr	11.0	2.0	59.0	6.0	1.0	—	27.0	—
plasma membrane	—	30.7	8.4	36.6	13.2	2.9	—	8.0	—
endoplasmic reticulum	2.5	19.3	1.8	45.0	7.2	3.2	—	21.4	—
Golgi apparatus	3.9	36.3	—	19.9	21.9	1.6	—	10.2	—
mitochondria	—	26.3	5.8	38.4	14.0	4.2	—	7.6	3.2
Phosphatidylserine									
whole tissue	0.9	22.8	2.2	15.3	15.0	30.9	0.8	7.6	2.0
plasma membrane	1.5	38.7	—	46.1	8.4	1.0	—	4.2	—
endoplasmic reticulum	9.0	11.1	1.0	4.7	21.8	52.3	—	—	—
Golgi apparatus	7.1	29.6	11.8	8.2	40.3	2.9	—	—	—
Diphosphatidylglycerol (cardiolipin)									
whole tissue	—	4.4	2.6	1.6	14.9	77.5	—	—	—
mitochondria	0.2	7.0	7.6	3.6	19.9	58.8	—	1.2	1.8

Modified from data originally published in ref. 31 with permission from Elsevier Science Publishers.

Table 7. Fatty acid composition of sphingolipids of rat liver cells

Lipid class	Fatty acid (weight %)									
	16:0	18:0	18:1	20:0	22:0	22:1	23:0	23:1	24:0	24:1	24:2
Rat liver											
sphingomyelin	15	13	1	2	10	1	9	2	25	21	—
Rat liver plasma membranes											
ceramides	14	9	9	5	6	6	10	5	21	14	—
sphingomyelin	18	13	5	2	8	1	10	3	22	16	—

Modified from data originally published in ref. 45 with permission from Plenum Publishing Corp.

Table 8. Fatty acid composition of representative higher plants, algae and cyanobacteria

Plant	Fatty acids (% of total acids)								
	16:0	16:1(9c)	16:1(3t)	16:3	18:0	18:1	18:2	18:3	Other
Higher plant leaves									
maize	8	1	3	0	2	7	8	66	
barley	13	1	3	0	2	6	6	64	
broad bean	12	1	4	0	2	7	14	56	
pea	12	1	2	0	1	2	25	53	
spinach	13	—	3	5	tr.	7	16	56	
Non-photosynthetic tissue									
turnip root	13	1	—	—	1	9	19	57	
potato tuber	20	—	—	—	3	4	60	15	
apple fruit	25	—	—	—	4	7	54	11	
Green algae									
Scenedesmus obliquus	35	2	—	—	—	9	6	30	15(16:4)
Euglena gracilis (light grown)	13	2	tr.	5	tr.	4	3	17	11(16:4) 8(20:4) 7(20:5)
Chlorella vulgaris	26	8	tr.	2	—	2	34	20	
Marine phytoplankton									
Monodus subterraneus	24	24	—	—	—	9	9	tr.	29(20:5) 5(20:4)
Cryptomonas maculata	15	7	—	—	—	3	—	6	17(20:5) 16(18:4) 1(20:4)

Table 8. Continued

Plant	Fatty acids (% of total acids)									
	16:0	16:1(9c)	16:1(3t)	16:3	18:0	18:1	18:2	18:3	Other	
Marine macroalgae										
Fucus vesiculosus	21	2	—	—	—	26	10	7	15(20:4) 8(20:5) 4(18:4)	
Polysiphona lanosa	32	8	—	—	—	15	—	—		
Cyanobacteria										
Anabaena variabilis	32	32	—	—	1	11	17	16	1(16:2)	
Anacystis nidulans	46	46	—	—	1	3	—	—	3(14:1) 1(14:0)	

Data compiled from refs 32 and 33 (higher plants), 46 (green algae), 47 (marine algae) and 48 (cyanobacteria).

Table 9. The fatty acid composition of chloroplast lipids from spinach (a 16:3-plant)[a] and barley (an 18:3-plant)[a]

Plant	Lipid (%)	Fatty acid composition (% of total fatty acids)							
		16:0	16:1(9c)	16:1(3t)	16:3	18:0	18:1	18:2	18:3
Spinach	MGDG (39)	tr.	—	—	25	—	1	2	72
	DGDG (26)	3	—	—	5	—	2	2	87
	SQDG (11)	29	1	—	—	1	7	26	36
	PG (14)	11	—	32	—	—	2	4	47
Barley	MGDG (42)	3	1	—	—	1	1	3	91
	DGDG (28)	9	2	—	—	1	3	7	78
	SQDG (12)	32	3	—	—	1	2	5	55
	PG (10)	18	—	27	—	3	2	11	38

[a] Reproduced from data originally published in ref. 49 with permission from The Biochemical Society, which also contains an explanation of '16:3-' and '18:3-' terminology.
Abbreviations: DGDG, digalactosyldiacylglycerol; MGDG, monogalactosyldiacylglycerol; PG, phosphatidylglycerol; SQDG, sulfoquinovosyldiacylglycerol; tr., trace.

Table 10. Fatty acid composition of selected fungi

Species	Fatty acid composition (% of total fatty acids)								
	12:0	14:0	16:0	16:1	18:0	18:1	18:2	18:3	20:4
Phycomycetes									
Blastocladiella emersonii	—	1	13	3	3	39	17	11	16
Entomophthora muscae	—	6	14	17	3	38	5	2	14
Mucor strictus	—	6	22	2	9	35	12	14	—
Phycomyces blakesleeanus	—	tr.	11	2	21	30	35	2	—
Rhizophlyctis rosea (aquatic)	—	tr.	14	—	6	74	3	3	—
Ascomycetes									
Candida albicans	—	1	12	8	7	36	25	10	—
Saccharomyces cerevisiae	—	3	18	52	2	20	—	—	—
Torulopsis candida	—	tr.	28	4	9	43	12	3	—
Fungi imperfecti									
Aspergillus niger	—	tr.	16	1	7	21	38	16	—
Penicillium chrysogenum	—	3	13	—	12	19	43	6	—
Homobasidiomyces									
Calvatia gigantea	12	3	9	1	tr.	9	62	tr.	1
Lactarius vellerns	—	tr.	7	tr.	54	22	14	2	—
Polyporus hirsutus	1	1	20	7	2	22	50	2	tr.
Heterobasidiomyces									
Ustilago scitaminca	1	2	20	3	7	30	32	2	—

Reproduced from data originally published in ref. 9, with permission from Chapman and Hall.
Abbreviation: tr., trace.

Table 11. Fatty acid composition of selected bacterial membrane systems

Phospholipid source	12:0	14:0	16:0	16:1	17:0 cyc.	18:0	18:1	19:0 cyc.	D-3-OH 14:0
Escherichia coli									
total	—	3	35	33	2	1	25	1	—
inner membrane	—	3	34	33	2	1	26	1	77
outer membrane	—	4	37	31	3	1	23	1	—
lipid A[a]	9	10	2	1	—	—	1	—	77
Rhodopseudomonas sphaeroides									
non-photosynthetic membrane	—	—	7	—	—	15	76	—	—
chromatophores	—	—	7	—	—	12	80	—	—

Compiled from data in ref. 33.
[a] Lipid A is a complex glycolipid found in the cell envelope of Gram-negative bacteria, which is not extracted by standard procedures used for phospholipid isolation.

5. STRUCTURE AND PROPERTIES OF BIOACTIVE LIPIDS

5.1. Eicosanoids

Eicosanoid is the generic term for a variety of C20 oxygenated fatty acids with potent autocrine and paracrine effects that are derived from C20-polyunsaturated fatty acids (eicosapolyenoic acids), in particular arachidonic acid. The group may be divided into two classes on the basis both of structure and biogenesis. The first is the prostanoids which include the prostaglandins, thromboxanes and prostacyclins which are formed via the cyclo-oxygenase pathway, and the second is the non-prostanoids, comprised of leukotrienes (slow reacting substances of anaphylaxis, SRS-A) and hydroxyeicosatetraenoic acids, formed via the lipoxygenase pathway. Detailed descriptions of established techniques for the study of arachidonate metabolites are provided in ref. 56.

Prostanoid structure
For their systematic nomenclature prostanoids are regarded as derivatives of the notional C20 fatty acid, prostanoic acid (I, *Figure 7*). However, this leads to rather cumbersome names, and the terminology system in wide use classifies prostaglandins (PGs) firstly into groups PGA through PGI according to the pattern of oxygen substituents on the cyclopentane ring (*Figure 7*), and then into series 1, 2 or 3, denoted by subscript, indicating the total number of double bonds in side-chains of the cyclopentane ring (*Figure 7*). The number of double bonds in the prostanoid also relates to the number of double bonds in the precursor eicosapolyenoic acid; thus series 2 PGs, which are generally the major series, are formed from arachidonic acid (20:4 (5, 8, 11, 14)) while series 1 and 3 PGs arise from the eicosatrienoic acid 20:3 (5, 8, 11) and eicosapentaenoic acid 20:5 (5, 8, 11, 14, 17), respectively. Partly for historical reasons and partly because of their biogenetic relationship (*Figure 8*) to the cyclic endoperoxides PGG and PGH (VIII, *Figure 7*), PGDs, PGEs and PGFs (V, VI, VII, respectively, *Figure 7*) are commonly referred to as primary prostaglandins. The bioactive thromboxane TXA_2, containing an oxane and oxetane ring (X, *Figure 7*) undergoes spontaneous degradation to TXB_2 (XI, *Figure 7*) which is stable and does not affect platelet aggregation. Prostacyclin (PGI_2) is also unstable ($t_{1/2} \simeq 10$ min) because of the 5,6-enol ether linkage (IX, *Figure 7*), degrading to 6-oxo-$PGF_{1\alpha}$ which is stable and relatively inert.

Figure 7. Structures of the prostanoids — prostaglandins, thromboxanes and prostacyclins.

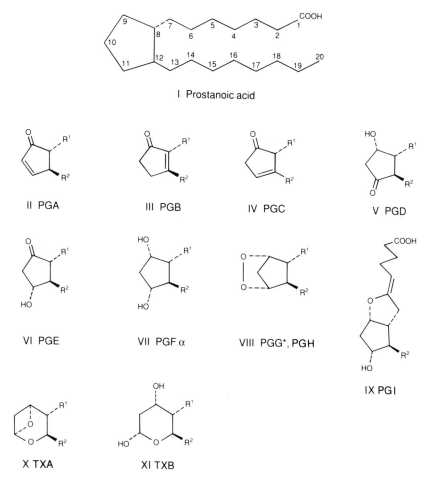

	R¹	R²
Series 1 from 20 : 3 8c11c14c	$[CH_2]_6COOH$	$CH\overset{t}{=}CHCH(OH)[CH_2]_4CH_3$
Series 2 from 20 : 4 5c8c11c14c	$CH_2CH\overset{c}{=}CH[CH_2]_3COOH$	$CH\overset{t}{=}CHCH(OH)[CH_2]_4CH_3$
Series 3 from 20 : 5 5c8c11c14c17c	$CH_2CH\overset{c}{=}CH[CH_2]_3COOH$	$CH\overset{t}{=}CHCH(OH)CH_2CH\overset{c}{=}CHCH_2CH_3$

*PGG is 15-OOH (hydroperoxy) rather than 15-OH

Biological activity of prostanoids
At concentrations as low as pg g^{-1} tissue, prostanoids produce a wide range of biological responses, including effects on smooth muscle contraction, notably in the vascular system, respiratory tract and reproductive system, and on renal function, platelet aggregation and the central nervous system (57). The molecular mechanism of prostanoid action is currently best understood in platelet/blood vessel wall interaction (58, 59) and in kidney function (60), where prostanoid effects are mediated by second messengers, signal transduction involving prostanoid receptors linked to G-proteins.

Prostaglandins are not stored in cells but are generated from arachidonate (see *Figure 8*) produced either directly by the action of phospholipase A_2, or indirectly by the action of phospholipase C and diacylglycerol lipase (49), released immediately on synthesis and then rapidly catabolized to the corresponding 15-oxo derivative, which is typically 10–100-fold less active than the corresponding parent prostaglandin. For this reason inhibitors of eicosanoid formation, particularly the non-steroidal anti-inflammatories aspirin and indomethacin, that selectively inhibit cyclooxygenase activity (*Figure 8*) have been used extensively to investigate the biological effects of prostanoids (61) and, by comparison of their effects with those of general inhibitors of eicosanoid formation such as eicosatetraynoic acid 20:4 (5a, 8a, 11a, 14a) that inhibit (62) the cyclo-oxygenase and lipoxygenase activities (*Figure 8*) and with those of selective inhibitors of lipoxygenase activity (63–67), to distinguish effects of prostanoids from those of leukotrienes and hydroxyeicosapentaenoic acids.

Structure of leukotrienes and hydroxyeicosatetraenoic acids
The biological activity of the eicosanoids now termed leukotrienes (68) was originally recognized (69) and described as 'slow reacting substance of anaphylaxis (SRS-A)' some 40 years before chemical structures were first established (70, 71). The term leukotriene reflects both the origin of these compounds in leukocytes or cells related to leukocytes, and the presence of three conjugated double bonds in these fatty acid derivatives (*Figure 9*). The nomenclature convention distinguishes structurally different leukotrienes into groups A through E, and within groups into series, designated by subscript, containing a total of three, four or five double bonds, which in an analogous manner to prostanoids is indicative of their biogenetic origin, via the 5-lipoxygenase pathway (72) from 20:3 (n-9), 20:4 (n-6) and 20:5 (n-3), respectively (*Figure 8*). Leukotrienes A and B (I and II, *Figure 9*) contain only oxygen substituents, while leukotrienes of the C, D and E groups contain, respectively, glutathione (LTC, IIIa *Figure 9*), glycylcysteine (LTD, IIIb *Figure 9*) and cysteine (LTE, IIIc *Figure 9*) ligated to the eicosaenoic acid at C-6 via a C–S linkage.

The most recent group of eicosanoids to receive recognition as potentially important biological effectors are the hydroxyeicosatetraenoic acids (HETEs) which are also products of arachidonate metabolism via lipoxygenase pathways (*Figure 8*). In this group three isomers, 5-, 12- and 15-HETE, of which 12-HETE (V, *Figure 9*) is the most widely distributed, constitute the main forms, although a number of other isomers are known (72). In addition to these monohydroxyeicosatetraenoic acids, a family of bioactive trihydroxyeicosatetraenoic acids containing a conjugated tetraene system, termed lipoxins, have also been described (73, 74), of which LXA (VII, *Figure 9*; 5,6,15-L-trihydroxy 7,9,11,13-eicosatetraenoic acid) is the major compound.

Biological activity of leukotrienes and hydroxyeicosatetraenoic acids
Unlike the eicosanoids formed via the cyclo-oxygenase pathway, which are generally biological regulators and mediators of tissue homeostasis, the biological effects of the lipoxygenase products derived from arachidonate appear to date to be largely pathophysiological in nature and involved in response to tissue injury, particularly inflammatory and allergic responses, as summarized in *Table 12*. A model for the molecular events in receptor-mediated LTD_4 signal transduction has been proposed recently (76).

Figure 8. Outline of the pathways of eicosanoid formation from arachidonic acid.

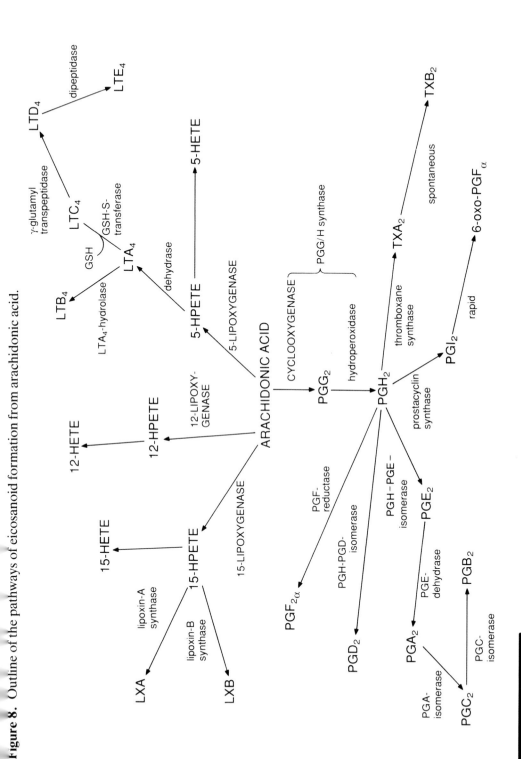

Lipids

Figure 9. Structures of the leukotrienes and hydroxyeicosatetraenoic acids.

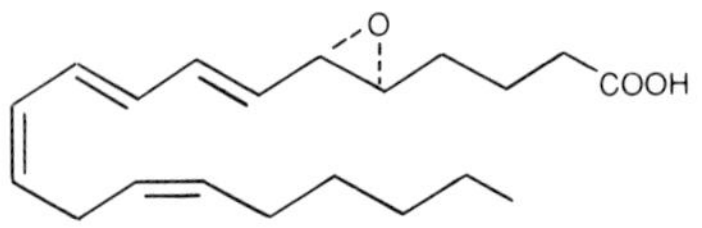

I LTA$_4$

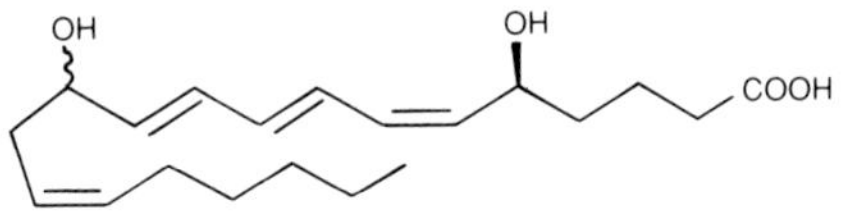

II LTB$_4$

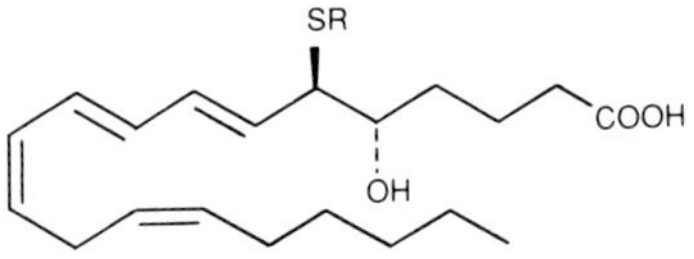

III LTC$_4$, LTD$_4$, LTE$_4$

a : LTC$_4$, R = CH$_2$CH(NHCOCH$_2$CH$_2$CH(NH$_2$)COOH)CONHCH$_2$COOH

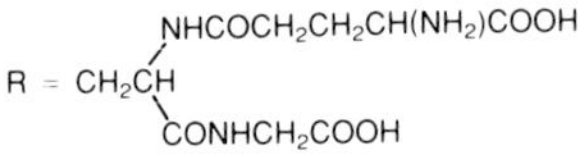

b : LTD$_4$, R = CH$_2$CH(NH$_2$)CONHCH$_2$COOH

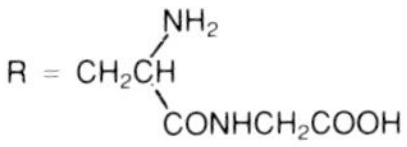

c : LTE$_4$, R = CH$_2$CH(NH$_2$)COOH

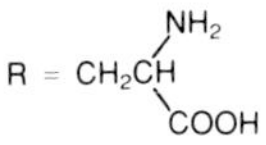

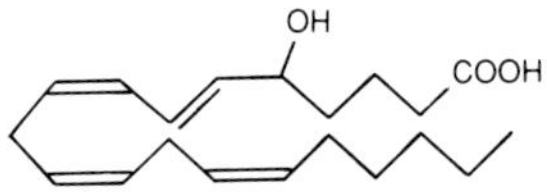

IV 5-HETE

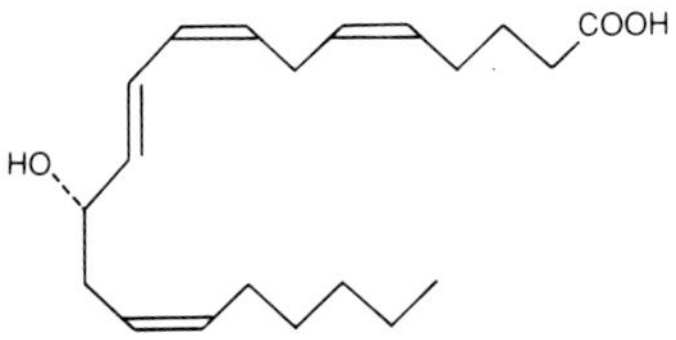

V 12-HETE

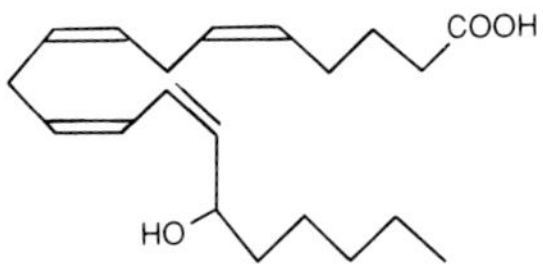

VI 15-HETE

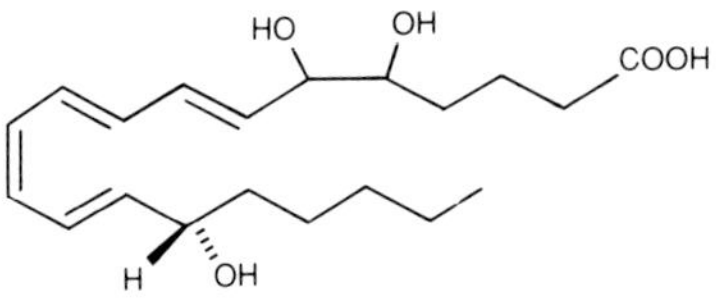

VII LXA*

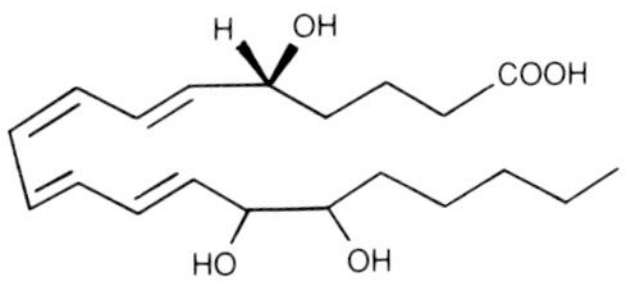

VIII LXB*

*Assignment of double bond configuration in lipoxins is tentative

Table 12. Some biological effects of leukotrienes and hydroxyeicosatetraenoic acids

Compound	Cells in which produced or active	Biological effects and comments
LTA_4	All LT-producing cells	Precursor of biologically active LTs; chemically very labile
LTB_4	Rat and guinea-pig PMNLs; rabbit and human neutrophils	Major mediator of leukocyte activation; chemotactic at pM–nM concn; stimulates cell aggregation and lysosomal enzyme release; increases vascular permeability in presence of PGE_2
LTC_4, LTD_4, LTE_4 (6-sulfido LTs — the SRS-As)	Macrophages and mast cells	Formation and release stimulated by IgE or IgG antibodies with intestinal, bone marrow and lung mast cells; prolonged LTs stimulation of smooth muscle contraction (activity $LTD:LTC:LTE$ 1:0.15:0.1) in guinea-pig ileum; LTC_4 and LTD_4 equally active in guinea-pig tracheal spirals; human bronchus sensitive to 10^{-10} M LTC_4, LTD_4; pulmonary parenchyma of man 10 × sensitivity of bronchus; LTC and LTD increase vascular permeability, cause transient rapid constriction of microvasculature, and at 4×10^{-10} M stimulate mucus secretion in human bronchial strips
5-, 12- and 15-HETE	Neutrophils and macrophages	Induce neutrophil degranulation at μM concn and potentiate PAF degranulating action; mediate hypersecretion of mucus and glycoproteins in respiratory tract in response to irritants or allergens; stimulate leukocyte migration and attachment in tissue injury
LXA	Leukocytes	Stimulates superoxide anion generation and degranulation in human leukocytes at sub-μM concn without promoting aggregation

Compiled from refs 72, 74 and 75 and references therein.
Abbreviations: HETE, hydroxyeicosatetraenoic acid; LT, leukotriene; LXA, 5,6,15-L-trihydroxy 7,9,11,13-eicosatetraenoic acid; PAF, platelet activating factor; PGE, prostaglandin E; PMNLs, polymorphonuclear leukocytes; SRS-As, slow reacting substances of anaphylaxis.

5.2. Platelet activating factor (PAF)

The structure of a bioactive compound released from leukocytes which causes platelet degranulation, termed platelet activating factor (PAF), was shown to be 1-O-alkyl-2-acetyl-sn-glycero-3-phosphocholine (*Table 6*), representing the first example of a biologically active phosphoglyceride, the properties of which are summarized in *Table 13*. Although not yet structurally characterized, a bioactive phospholipid with chromatographic properties similar to those of PAF has recently been detected in higher plant tissue (80).

Table 13. Some biological effects of platelet activating factor

Cell/tissue	Biological effect	Comments
Platelet	Very potent aggregation and degranulation agent (active at 5×10^{-11} M)	sn-1-Fatty acyl analogs have only 1/300th PAF activity; sn-2-butyryl lowers activity 1000-fold
Neutrophil	Chemotaxis, aggregation, superoxide generation	Platelets have 250 PAF-specific receptors of a single class, neutrophils have high-affinity and low-affinity receptors
Alveolar macrophages	Respiratory bursts, superoxide generation	Phagocytic cells are better sources of PAF than platelets; formation in platelets is stimulated by thrombin or Ca^{2+} ionophore A-23187, in neutrophils by A-23187 or zymosan, and in endothelial cells by interleukin-2
Liver	Inositide turnover, stimulation of glycogenolysis	

Compiled from refs 77–79 and references therein.

5.3. Diacylglycerol

sn-1,2-Diacylglycerol (DAG; *Table 3*) generated in the plasma membrane of eucaryotic cells by receptor-stimulated hydrolysis of membrane phospholipids activates protein kinase C (PKC), the multifunctional serine/threonine-specific kinase that phosphorylates a range of cellular proteins (81–83). Thus DAG is a second messenger, transmitting the external signal of agonist–receptor interaction to the interior of the cell. DAG activation of PKC is mimicked by phorbol ester tumor promoters (e.g. phorbol myristate acetate, PMA) that interact competitively at the same site as DAG (84). sn-1,2-DAG molecular species with sn-1-unsaturated acyl groups are most active, while the 'membrane permeable' DAGs, sn-1,2-dioctanoyl- and sn-1-oleoyl-2-acetyl-DAG, are widely used for *in vitro* studies; sn-1,3- and sn-2,3-DAGs are inactive. The best-characterized signal transduction system is phosphatidylinositide-specific phospholipase C hydrolysis of phosphatidylinositol 4,5-bisphosphate (*Table 4*) to yield DAG and inositol 1,4,5-trisphosphate. The latter molecule is also a second messenger, involved in calcium mobilization (85). More recently it has been shown that many agonists also induce rapid hydrolysis of phosphatidylcholine (*Table 4*), either by phospholipase C, to generate DAG directly, or by phospholipase D hydrolysis, to yield phosphatidic acid (*Table 4*), which is further metabolized to DAG by phosphatidate phosphohydrolase (86).

5.4. Steroid hormones

Five classes of steroid hormone are derived from cholesterol by well-established biosynthetic pathways involving initial removal of the sterol side-chain (87): the glucocorticoids, mineralocorticoids, and progestins — all of which are C21-steroids — and the androgens and estrogens, which are C19- and C18-steroids, respectively. The structures of the principal members of each class are shown in *Figure 10* and the biological properties of the steroid classes are summarized in *Table 14*. In the light of the similarity of its structure (*Figure 10*) and analogy of its mechanisms of action at the subcellular level (*Table 14*), the seco-steroid 1,25-dihydroxycholecalciferol (1,25-dihydroxyvitamin D_3) — the active form of vitamin D_3 — is now generally considered as a steroid hormone (88). The insect molting hormones or ecdysteroids may be classified similarly. These are C27-sterols (*Figure 10*), formed from cholesterol by extensive modification of the nucleus followed by sequential hydroxylation, and are widely distributed in arthropods (zooecdysteroids) (89) as well as occurring in many plant species (phytoecdysteroids) (90); their biochemical modes of action in insects appear to be analogous to those of the steroid hormones of vertebrates (*Table 14*). Reviews of the structure and structure–function relationships of the steroid hormones appear in several texts (e.g. 91, 92), and details of the systematic nomenclature of steroids and vitamin D are provided in the appropriate IUPAC-IUB recommendations (93). Descriptions and discussion of the methodology for the identification and quantification of steroid hormones and their metabolites by chromatographic and physical methods (91, 94–98) and by competitive binding procedures (91, 94, 99), for the study of vitamin D metabolism (100) and for analysis of ecdysteroids (101) are available. Structural analogs which are antagonists or agonists of the natural steroid hormones (for representative examples see *Figure 10*) are not only of value in establishing the physiological effects and molecular mechanisms of steroid activity, but are also in widespread clinical use as, respectively, antisteroidal and steroidal compounds (91, 102).

Steroid hormones control a wide variety of physiological responses in their target tissues, by regulating the rates of gene transcription at precise stages of tissue differentiation and development and by their more recently recognized effects on cell surface events (*Table 14*). In the general model for their action upon transcription (103), the steroid binds with a soluble, intracellular receptor-protein, and the steroid receptor complex interacts with a specific DNA sequence termed hormone response element which, upon activation, functions as a transcriptional enhancer. These *trans*-acting receptors are typically phosphoproteins (104) and are frequently associated with 70 kDa and 90 kDa members of the heat-shock family of proteins (105). Properties of the steroid–receptor systems for the principal classes of steroid hormone have recently been reviewed (88, 106–108), and methods for their isolation and purification described (94, 109). The rapid effects of steroids on cell-surface events and properties, which are particularly prominent in excitable tissue and neurosecretory processes (reviewed in ref. 110) involve membrane-associated binding of steriods, although it is not established whether high-affinity membrane-associated steroid receptors are generally involved in these phenomena.

Figure 10. Structures of the principal steroid hormones and analogs with steroidal and antisteroidal activities

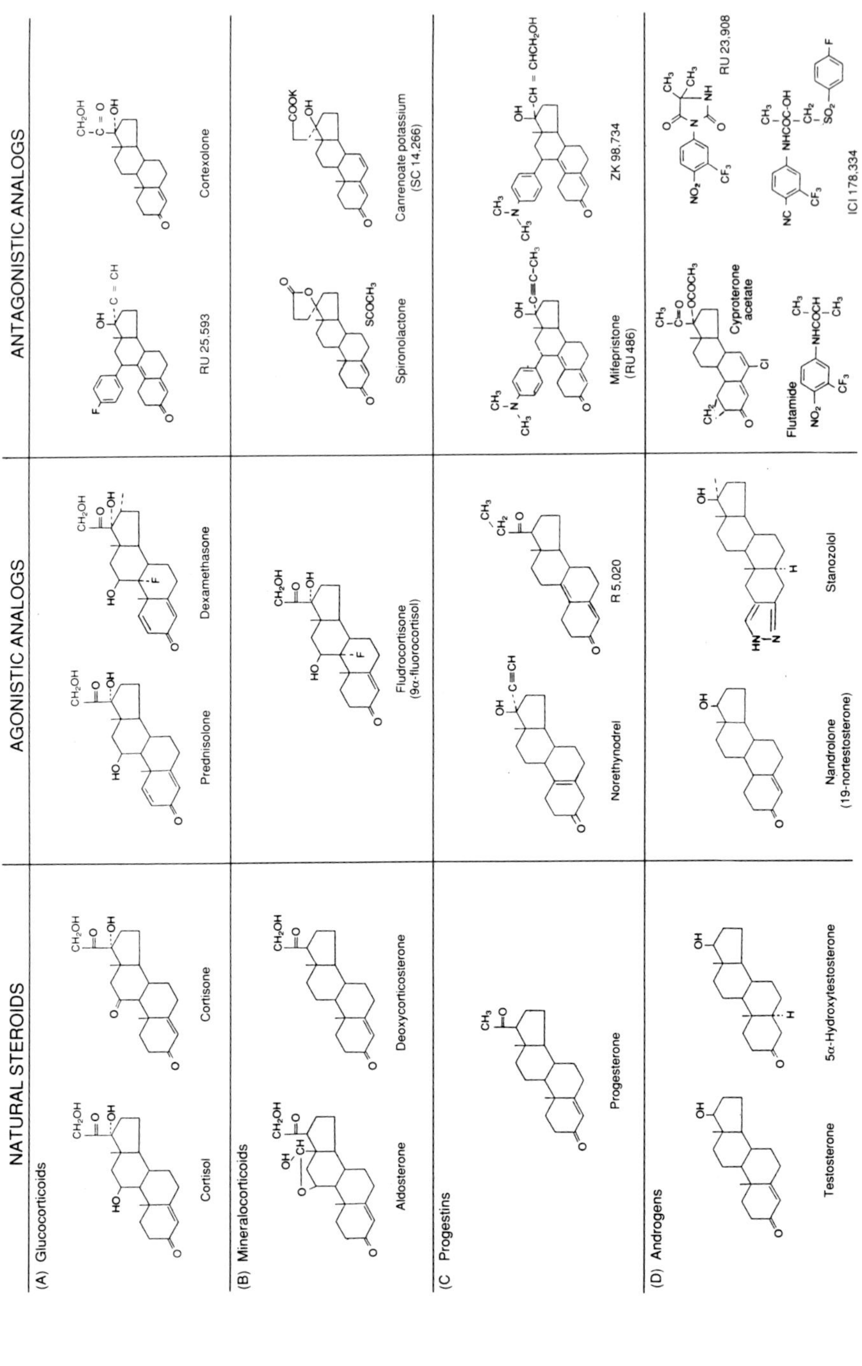

Figure 10. Continued.

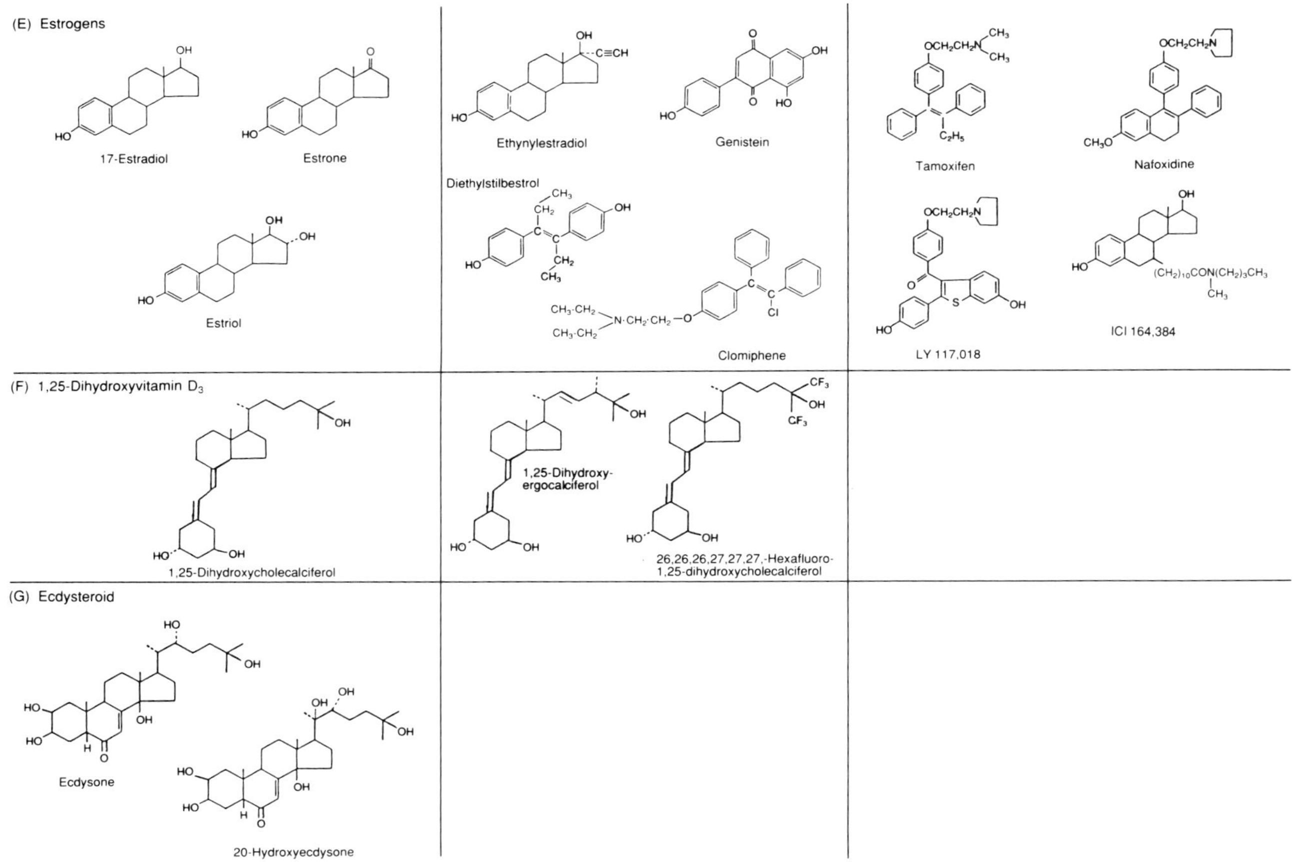

Table 14. Some properties of steroid hormone systems

Steroid hormone class	Steroidogenic tissue (principal sources)	Target organ/tissue	Genomic effects: protein induced (or *repressed)	Non-genomic biochemical effects	Physiological effects
A. Glucocorticoids	Adrenal cortex	Liver Kidney Pituitary	Tyrosine aminotransferase, tryptophan oxygenase Phosphoenolpyruvate carboxykinase Corticotropin*	Bind to neural membranes	Stimulate gluconeogenesis Reduce inflammatory reactions
B. Mineralocorticoids	Adrenal cortex	Kidney Bladder	Citrate synthase Na^+/K^+ ATPase Na^+ channel protein Na^+/K^+ ATPase	Bind to neural membranes; modulate level of β-adreno-receptors in smooth muscle cells	Promote unidirectional transepithelial Na^+ transport Regulation of electrolyte metabolism
C. Progestins	Corpus luteum, adrenal cortex, placenta	Oviduct Uterus	Ovalbumin, avidin, ovomucoid Uteroglobin	+ve modulation of GABA-induced Cl^- flux Release dopamine Effects on Ca^{2+} entry/mobilization	Mediate menstrual cycle and maintain pregnancy
D. Androgens	Testes, adrenal cortex, ovaries, placenta	Prostate Liver Kidney	Aldolase Urinary protein, α-2 globulin β-glucuronidase	Bind to neural membranes	Maturation and function of secondary sex organs, esp. in males; male sexual differentiation
E. Estrogens	Ovary, placenta, adrenal cortex, testes and other tissues	Oviduct Liver Pituitary	Ovalbumin, conalbumin, ovomucoid, lysozyme Vitellogenin, transferrin Prolactin	Modulate activity of medial amygdala neurons Effects on Ca^{2+} entry/mobilization	Maturation and function of secondary sex organs, esp. in females

Table 14. Continued

Steroid hormone class	Steroidogenic tissue (principal sources)	Target organ/tissue	Genomic effects; protein induced (or * repressed)	Non-genomic biochemical effects	Physiological effects
F. 1,25-dihydroxycholecalciferol	Kidney, activated monocytes, placenta	Intestine	Ca^{2+}-binding protein (calbindin)	Internalization of Ca^{2+} at brush border membrane (transcaltachia)	Stimulates intestinal Ca^{2+} absorption, renal Ca^{2+} reabsorption and Ca^{2+} mobilization from bone
G. Ecdysterone	Prothoracic glands, ovaries, eggs	Epidermis Fat body	DOPA-decarboxylase Vitellogenin	May be involved in control of ion flow at plasma membrane	Controls phases of insect molting and embryogenesis

Compiled from refs 88, 89, 91, 106, 107 and 110 and references therein.
Abbreviations: ATPase, adenosine triphosphatase; DOPA, 3,4-dihydroxyphenylalanine; esp., especially; GABA, γ-aminobutyric acid.

6. STRUCTURE AND COMPOSITION OF BILE ACIDS AND BILE SALTS

The formation of bile acids is the most important pathway in mammals for the metabolism and excretion of cholesterol. The major primary bile acids are the C24-compounds, cholic and chenodeoxycholic acid (*Figure 11*), which are formed in the liver by well-defined biosynthetic pathways (111). These are then conjugated with glycine or taurine to form the corresponding bile salts (*Figure 11*) that are secreted with bile into the intestine. Here they may be transformed by micro-organisms (112) into secondary bile acids, such as deoxycholic acid and lithocholic acid (*Figure 11*), which, together with primary bile salts, are largely returned to the liver in the enterohepatic circulation and further metabolized, re-conjugated and re-excreted in the bile. The physicochemical properties of bile salts and their role in intestinal lipid digestion and absorption have been reviewed (113, 114). The bile of vertebrates more primitive than mammals characteristically contains sulfuric acid esters of C27-bile alcohols and taurine-conjugated or unconjugated C27-bile acids, which only occur as minor components in humans (115). An overview of chromatographic analysis of bile acids in biological fluids is available (116); more recently, procedures have been described for HPLC analysis of bile salts either directly (117) or following derivatization (118), and for differentiation of bile acid-isomers by tandem mass spectrometry (119, 120).

Figure 11. Structures of the primary and secondary bile acids and bile salts.

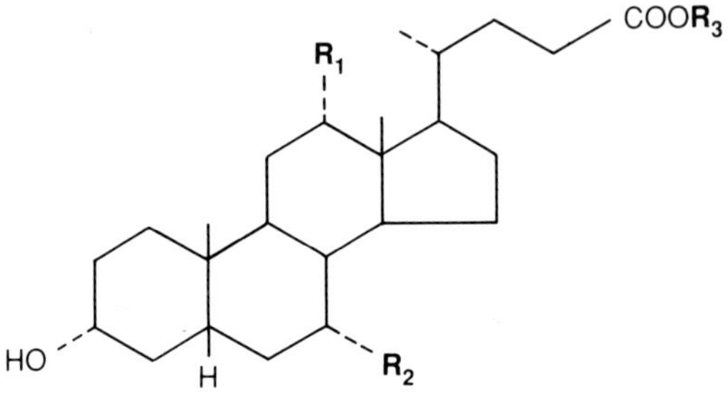

Primary bile acids and bile salts

	R_1 = OH	R_1 = H
	R_2 = OH	R_2 = OH
R_3 = H	cholic acid	chenodeoxycholic acid
R_3 = NH – CH$_2$ –COOH	glycocholic acid	glycochenodeoxycholic acid
R_3 = NH – CH$_2$ –CH$_2$ –SO$_3$ H	taurocholic acid	taurochenodeoxycholic acid

Secondary bile acids and bile salts

	R_1 = OH	R_1 = H
	R_2 = H	R_3 = H
R_3 = H	deoxycholic acid	lithocholic acid
R_3 = NH – CH$_2$ –COOH	glycodeoxycholic acid	glycolithocholic acid
R_3 = NH – CH$_2$ –CH$_2$ –SO$_3$ H	taurodeoxycholic acid	taurolithocholic acid

7. CHEMICAL ABSTRACTS REGISTRY NUMBERS OF LIPIDS CITED

(a) Saturated fatty acids

Lauric	143-07-7	Arachidic	506-30-9
Myristic	544-63-8	Behenic	112-85-6
Palmitic	57-10-3	Lignoceric	557-59-5
Stearic	57-11-4		

(b) Monoenoic fatty acids

Palmitoleic	373-49-9	Elaidic	112-79-8
trans-3-Hexadecenoic	1686-10-8	Vaccenic	693-72-1
Oleic	12-80-1	Erucic	112-86-7

(c) Polyunsaturated fatty acids

α-Linolenic	463-40-1	Linoleic	60-33-3
Eicosapentaenoic	10417-94-4	γ-Lineolenic	506-26-3
Clupanodonic	2548-85-8	Homo-γ-linolenic	506-23-1
Docosahexaenoic	6217-54-5	Arachidonic	506-32-1

(d) Leukotrienes

LTA_4	72059-45-1	12-HETE	71030-37-0
LTB_4	71160-24-2	15-HETE	71030-36-9 or
LTC_4	72025-60-6		54845-95-3
LTD_4	73836-78-9	Lipoxin A	89663-86-5
LTE_4	75715-89-8	Lipoxin B	92950-25-9
5-HETE	70608-72-9	Prostanoic acid	25151-81-9

(e) Carotenoids

β-Carotene	7235-40-7	Echinenone	432-68-8
Violaxanthin	126-29-4	Myxoxanthophyll	11004-68-5
Lutein	127-40-2	Spirilloxanthin	34255-08-8
Neoxanthin	14660-91-4	Okenone	16840-70-3
Astaxanthin	472-61-7	Chlorobactene	2932-09-4
Fucoxanthin	3351-86-8		

(f) Chlorophylls

Chlorophyll a	479-61-8	Bacteriochlorophyll a	17499-98-8
Chlorophyll b	519-62-0	Bacteriochlorophyll b	53199-29-4
Chlorophyll c	11003-45-5	Bacteriochlorophyll c	53986-51-4
Chlorophyll d	519-63-1	Bacteriochlorophyll d	8067-29-6

(g) Steroids

Cholesterol	57-88-5	Sitosterol	12002-39-0
Ergosterol	57-87-4	Stigmasterol	83-48-7

8. REFERENCES

1. Gurr, M.I. and Harwood, J.L. (1991) *Lipid Biochemistry: An Introduction (4th edn)*. Chapman and Hall, London.

2. Singer, S.J. and Nicholson, G.L. (1972) *Science*, **175**, 720.

3. Sober, H.A. (ed.) (1970) in *Handbook of Biochemistry — Selected Data for Molecular Biology Research (2nd edn)*. CRC Press, Cleveland Ohio, p. E2.

4. Houslay, M.D. and Stanley, K.K. (1982) *Dynamics of Biological Membranes*. John Wiley and Sons, Chichester.

5. Grand, R.J.A. (1989) *Biochem. J.*, **258**, 625.

6. Takahashi, N. and Breitman, T.R. (1989) *J. Biol. Chem.*, **264**, 5159.

7. Schafer, W.R., Kim, R., Sterne, R., Thorner, J., Kim, S.-H. and Rine, J. (1989) *Science*, **245**, 379.

8. Goldstein, J.L. and Brown, M.S. (1990) *Nature*, **343**, 425.

9. Gunstone, F.D., Harwood, J.L. and Padley, F.B. (1986) *The Lipid Handbook*. Chapman and Hall, London.

10. Downing, D.T. (1976) in *Chemistry and Biochemistry of Natural Waxes* (P.E. Kolattukudy ed.). Elsevier, Amsterdam, p. 17.

11. Jackson, L.L. (1976) in *Chemistry and Biochemistry of Natural Waxes* (P.E. Kolattukudy ed.). Elsevier, Amsterdam, p. 201.

12. Walton, T.J. (1990) in *Methods in Plant Biochemistry, Vol. 4: Lipids, Membranes and Aspects of Photobiology* (J.L. Harwood and J.R. Bowyer eds). Academic Press, London, p. 105.

13. Jacob, J. (1976) in *Chemistry and Biochemistry of Natural Waxes* (P.E. Kolattukudy ed.). Elsevier, Amsterdam, p. 93.

14. Sargent, J.R., Lee, R.F. and Nevenzel, J.C. (1976) in *Chemistry and Biochemistry of Natural Waxes* (P.E. Kolattukudy ed.). Elsevier, Amsterdam, p. 49.

15. Albro, P.W. (1976) in *Chemistry and Biochemistry of Natural Waxes* (P.E. Kolattukudy ed.). Elsevier, Amsterdam, p. 419.

16. Weete, J.D. (1976) in *Chemistry and Biochemistry of Natural Waxes* (P.E. Kolattukudy ed.). Elsevier, Amsterdam, p. 349.

17. The Nomenclature of Lipids (Recommendations 1976) IUPAC-IUB Commission of Biochemical Nomenclature (1978) *Biochem. J.*, **171**, 21.

18. Svennerholm, L. (1970) in *Comprehensive Biochemistry, Vol. 18: Lipid Metabolism* (M. Florkin and E.H. Stotz eds). Elsevier, Amsterdam, p. 201.

19. Wiegandt, H. (1968) *Angew. Chem.*, **7**, 87.

20. Mudd, J.B. (1980) in *Biochemistry of Plants, Vol. 14: Lipids: Structure and Function* (P.K. Stumpf ed.). Academic Press, New York, p. 509.

21. Weete, J.D. (1980) *Lipid Biochemistry of Fungi and Other Organisms*. Plenum Press, New York.

22. Weete, J.D. (1989) *Adv. Lipid Res.*, **23**, 115.

23. Nes, W.D. and Parish, E.J. (1989) *Analysis of Sterols and Other Biologically Significant Steroids*. Academic Press, London.

24. Cogdell, R. (1988) in *Plant Pigments* (T.W. Goodwin ed.). Academic Press, London, p. 183.

25. Goodwin, T.W. (1980) *Biochemistry of the Carotenoids (2nd edn)*. Chapman and Hall, London, Vol. 1.

26. Goodwin, T.W. (1984) *Biochemistry of the Carotenoids (2nd edn)*. Chapman and Hall, London, Vol. 2.

27. Rau, W. (1988) in *Plant Pigments* (T.W. Goodwin ed.). Academic Press, London, p. 231.

28. Davies, B.H. (1976) in *Chemistry and Biochemistry of Plant Pigments (2nd edn)*. (T.W. Goodwin ed.). Academic Press, London, Vol. 2, p. 38.

29. Goodwin, T.W. and Britton, G. (1988) in *Plant Pigments* (T.W. Goodwin ed.). Academic Press, London, p. 62.

30. Rudiger, W. and Schoch, S. (1988) in *Plant Pigments* (T.W. Goodwin ed.). Academic Press, London, p. 1.

31. White, D.A. (1973) in *Form and Function of Phospholipids* (G.B. Ansell, J.N. Hawthorne and R.M.C. Dawson eds). Elsevier, Amsterdam, p. 441.

32. Galliard, T. (1973) in *Form and Function of Phospholipids* (G.B. Ansell, J.N. Hawthorne and R.M.C. Dawson eds). Elsevier, Amsterdam, p. 253.

33. Harwood, J.L. (1980) in *The Biochemistry of Plants, Vol. 4: Lipids: Structure and Function* (P.K. Stumpf ed.). Academic Press, New York, p. 1.

34. Harwood, J.L. and Russel, N.J. (1984) *Lipids in Plants and Microbes*. George Allen and Unwin, London.

35. Op den Kamp, J.A.F. (1979) *Ann. Rev. Biochem.*, **48**, 47.

36. Schroeder, F. and Nemecz, G. (1990) in *Advances in Cholesterol Research* (M. Esfahani and J.B. Swaney eds). The Telford Press, New Jersey, p. 47.

37. Christie, W.W. (1982) *Lipid Analysis (2nd edn)*. Pergamon Press, Oxford.

38. Kates, M. (1986) *Techniques in Lipidology*. Elsevier, Amsterdam; Hamilton, R.J. and Hamilton, S. (eds) (1992) *Lipid Analysis: A Practical Approach*. IRL Press, Oxford.

39. Christie, W.W. (1987) *HPLC and Lipids: A Practical Guide*. Pergamon Press, Oxford.

40. Christie, W.W. (1989) *Gas Chromatography and Lipids: A Practical Guide*. The Oily Press, Ayr, Scotland.

41. Gunstone, F.D. (1990) in *Methods in Plant Analysis, Vol. 4: Lipids, Membranes and Aspects of Photobiology* (J.L. Harwood and J.R. Bowyer eds). Academic Press, London, p. 1.

42. Mangold, H.K. and Totani, N. (1983) in *Ether Lipids — Biochemistry and Biomedical Aspects* (H.K. Mangold and F. Paltau eds). Academic Press, New York, p. 377.

43. Ledeen, R.W. and Yu, R.K. (1982) *Methods Enzymol.*, **83**, 139.

44. Irvine, R.F. (ed.) (1990) *Methods in Inositide Research*. Raven Press, New York.

45. Kuksis, A. (1978) in *Handbook of Lipid Research, Vol. 1: Fatty Acids and Glycerides* (A. Kuksis ed.). Plenum Press, New York, p. 381.

46. Hitchcock, C. and Nichols, B.W. (1971) *Plant Lipid Biochemistry*. Academic Press, London.

47. Pohl, P. and Zurheide, F. (1979) in *Marine Algae in Pharmaceutical Science* (H.A. Hoppe, T. Levring and Y. Tanaka eds). Walter de Gruyter, Berlin, p. 473.

48. Murata, N. and Nishida, I. (1987) in *The Biochemistry of Plants, Vol. 9: Lipids* (P.K. Stumpf and E.E. Conn eds). Academic Press, New York, p. 315.

49. Harwood, J.L. (1988) in *Plant Membranes — Structure, Assembly and Function* (J.L. Harwood and T.J. Walton eds). The Biochemical Society, London, p. 113.

50. Breckenridge, W.C. (1978) in *Handbook of Lipid Research, Vol. 1: Fatty Acids and Glycerides* (A. Kuksis ed.). Plenum Press, New York, p. 197.

51. Demandre, C., Tremolieres, A., Justin, A.-M. and Mazliak, P. (1985) *Phytochem.*, **24**, 481.

52. Kesselmeir, J. and Heinz, E. (1987) *Methods Enzymol.*, **148**, 650.

53. Thompson, Jr., G.A. (1989) *Biochem. Soc. Trans.*, **17**, 286.

54. Thompson, Jr., G.A. and Martin, C.E. (1984) in *Physiology of Membrane Fluidity* (M. Shinitzky ed.). CRC Press, Boca Raton, Florida, Vol. 1, p. 99.

55. Sato, N. and Murata, N. (1980) *Biochim. Biophys. Acta*, **619**, 353.

56. Lands, W.E.M. and Smith, W.L. (eds) (1982) *Methods Enzymol.*, **86**.

57. Pace-Asciak, C. and Gryglewski, R. (1983) in *Prostaglandins and Related Substances* (C. Pace-Asciak and E. Granstrom eds). Elsevier, Amsterdam, p. 95.

58. Gorman, R.R., Fitzpatrick, F.A. and Miller, O.U. (1978) *Adv. Cyclic Nucleotide Res.*, **9**, 597.

59. Arita, H., Nakano, T. and Hanasaki, K. (1989) *Prog. Lipid Res.*, **28**, 273.

60. Smith, W.L. (1989) *Biochem. J.*, **259**, 315.

61. Samuelsson, B., Goldyne, M., Granstrom, E., Hamberg, M., Hammarstrom, S. and Malmsten, C. (1978) *Ann. Rev. Biochem.*, **47**, 997.

62. Hamberg, M. and Samuelsson, B. (1974) *Proc. Natl. Acad. Sci. USA*, **71**, 3400.

63. Walker, J.R. and Dawson, W. (1979) *J. Pharm. Pharmacol.*, **31**, 778.

64. Walker, J.R., Boot, J.R., Beverley, C. and Dawson, W. (1980) *J. Pharm. Pharmacol.*, **32**, 866.

65. Sun, F.F., McGuire, J.C., Morton, D.R., Pike, J.E., Sprecher, H. and Kunan, W.H. (1981) *Prostaglandins*, **21**, 333.

66. Vanderhoek, J.Y., Bryant, R.W. and Bailey, J.M. (1980) *J. Biol. Chem.*, **255**, 5996.

67. Vanderhoek, J.Y., Bryant, R.W. and Bailey, J.M. (1980) *J. Biol. Chem.*, **255**, 10064.

68. Samuelsson, B. and Hammarstrom, S. (1980) *Prostaglandins*, **19**, 645.

69. Feldberg, W. and Kellaway, C.H. (1938) *J. Physiol. (London)*, **94**, 187.

70. Hammarstrom, S., Murphy, R.C., Samuelsson, B., Clark, D.A., Mioskowski, C. and Corey, E.J. (1979) *Biochem. Biophys. Res. Commun.*, **91**, 1266.

71. Murphy, R.C., Hammarstrom, S. and Samuelsson, B. (1979) *Proc. Natl. Acad. Sci. USA*, **76**, 4275.

72. Spector, A.A., Gordon, J.A. and Moore, S.A. (1988) *Progr. Lipid Res.*, **27**, 271.

73. Serhan, C.N., Hamberg, M. and Samuelsson, B. (1984) *Biochem. Biophys. Res. Commun.*, **118**, 943.

74. Serhan, C.N., Hamberg, M. and Samuelsson, B. (1984) *Proc. Natl. Acad. Sci. USA*, **81**, 5335.

75. Piper, P.J. (ed.) (1983) *Leukotrienes and Other Lipoxygenase Products*. Research Studies Press, Letchworth, England.

76. Crooke, S.T., Saran, H., Saussy, D., Winkler, J. and Foley, J. (1990) in *Advances in Prostaglandin, Thromboxane and Leukotriene Research* (B. Samuelsson, S.-E. Dahlen, J. Fritsch and P. Hedqvist, eds). Raven Press, New York, Vol. 20, p. 127.

77. Benveniste, J. and Vargaftig, B.B. (1983) in *Ether Lipids — Biochemical and Biomedical Aspects* (H.K. Mangold and F. Paltauf eds). Academic Press, New York, p. 355.

78. Hanahan, D.J. (1986) *Ann. Rev. Biochem.*, **55**, 483.

79. Sturk, A., Cate, J.W.T., Hosford, D., Mencia-Huerta, J.-M. and Braquet, P. (1989) *Adv. Lipid Res.*, **23**, 219.

80. Scherer, G.F.E., Martiny-Baron, G. and Stoffel, B. (1988) *Planta*, **175**, 241.

81. Nishizuka, Y. (1988) *Nature*, **334**, 661.

82. Parker, P.J., Bajaj, M., Marais, R., Mitchell, F., Pears, C. and Stabel, S. (1989) *Biochem. Soc. Trans.*, **17**, 279.

83. Jaken, S. (1989) in *Inositol Lipids in Cell Signalling* (R.H. Mitchell, A.H. Drummond and C.P. Downes eds). Academic Press, London, p. 163.

84. Leach, K.L. and Blumberg, P.M. (1989) in *Inositol Lipids in Cell Signalling* (R.H. Michell,

A.H. Drummond and C.P. Downes eds). Academic Press, London, p. 179.

85. Berridge, M.J. and Irvine, R.F. (1989) *Nature*, **341**, 197.

86. Billah, M.M. and Anthes, J.C. (1990) *Biochem. J.*, **269**, 281.

87. Gower, D.B. (1988) in *Hormones and their Actions, Part I* (B.A. Cooke, R.J.B. King and H.J. van der Molen eds). Elsevier, Amsterdam, p. 3.

88. Cancela, L., Theofan, G. and Norman, A.W. (1988) in *Hormones and their Actions, Part I* (B.A. Cooke, R.J.B. King and H.J. van der Molen eds). Elsevier, Amsterdam, p. 269.

89. Rees, H.H. (1989) in *Ecdysone: from Chemistry to Mode of Action* (J. Koolman ed.). Georg Thième, Stuttgart, p. 28.

90. Lafont, R. and Horn, D.H.S. (1989) in *Ecdysone: From Chemistry to Mode of Action* (J. Koolman ed.). Georg Thieme, Stuttgart, p. 39.

91. Makin, H.L.J. (eds) (1984) *Biochemistry of Steroid Hormones (2nd edn)*. Blackwell Scientific Publications, Oxford.

92. Heftmann, E. (1970) *Steroid Biochemistry*. Academic Press, New York.

93. IUPAC-IUB 1971 Definitive Rules for Steroid Nomenclature (1972) *Pure Appl. Chem.*, **31**, 285; IUPAC-IUB 1981 Recommendations for Nomenclature of Vitamin D (1982) *Eur. J. Biochem.*, **124**, 223.

94. O'Malley, B.W. and Hardman, J.G. (eds) (1975) *Methods Enzymol.*, **36**.

95. Görög, S. (1978) in *Analysis of Steroid Hormone Drugs* (S. Gorog and Gy. Szász eds). Elsevier, Amsterdam, p. 15.

96. Hirai, T. (1991) *J. Chromatog. Libr.*, **50**, 255.

97. Serio, M., Guarna, A., Salerno, R., Orlando, C., Pazzagli, M., Calabresi, E. and Moneti, G. (1990) *Ann. NY Acad. Sci.*, **595**, 275.

98. Proc. 2nd Int. Symp. on Analysis of Anabolizing and Doping Agents in Biosamples, Ghent (Belgium) (1991) *J. Chromatog.*, **564**, 361.

99. Van Ginkel, L.A. (1991) *J. Chromatog.*, **564**, 363.

100. Lawson, D.E.M. (1980) in *Vitamin D: Molecular Biology and Clinical Nutrition* (A.W. Norman ed.). Marcel Dekker, New York, p. 93.

101. Thompson, N.J., Svoboda, J.A. and Feldlaufer, M.F. (1989) in *Analysis of Sterols and Other Biologically Significant Steroids* (W.D. Nes and E.J. Parish eds). Academic Press, New York, p. 81.

102. Wakeling, A.E. (1988) in *Hormones and their Actions, Part I* (B.A. Cooke, R.J.B. King and H.J. van der Molen eds). Elsevier, Amsterdam, p. 151.

103. Parker, M.G. (1988) in *Hormones and their Actions, Part I* (B.A. Cooke, R.J.B. King and H.J. van der Molen eds). Academic Press, New York, p. 39.

104. Moudgil, V.K. (1990) *Biochim. Biophys. Acta*, **1055**, 243.

105. Baulieu, E.E., Binart, N., Cadepond, F., Catelli, M.G., Chambraud, B., Garnier, J., Gasc, J.M., Groyer-Schweizer, G., Rafestin-Oblin, M.E., et al. (1990) *Symp. Soc. Exp. Biol.*, **44**, 3.

106. Spelsberg, T.C. and Kumar, R. (eds) (1987) *Steroid and Sterol Hormone Action*. Martin Nijhoff, Boston.

107. Cooke, B.A., King, R.J.B. and van der Molen (eds) (1988) *Hormones and their Actions, Part I*. Elsevier, Amsterdam, p. 293.

108. Duax, W.L. and Griffin, J.F. (1989) *Adv. Drug Res.*, **18**, 115.

109. Blankenstein, M.A. and Muelder, E. (1988) in *Hormones and their Actions, Part I*. (B.A. Cooke, R.J.B. King and H.J. van du Molin eds). Elsevier, Amsterdam.

110. McEwen, B.S. (1991) *Trends Pharm. Sci.*, **12**, 141.

111. Björkhem, I. (1985) in *Sterols and Bile Acids* (H. Danielsson and J. Sjövall eds). Elsevier, Amsterdam, p. 231.

112. Hyelmon, P.B. (1985) in *Sterols and Bile Acids* (H. Danielsson and J. Sjövall eds). Elsevier, Amsterdam, p. 331.

113. Carey, M. (1985) in *Sterols and Bile Acids* (H. Danielsson and J. Sjövall eds). Elsevier, Amsterdam, p. 345.

114. Borgström, B., Barrowman, J.A. and Lindström, M. (1985) in *Sterols and Bile Acids* (H. Danielsson and J. Sjövall eds). Elsevier, Amsterdam, p. 405.

115. Hoshita, T. (1985) in *Sterols and Bile Acids* (H. Danielsson and J. Sjövall eds). Elsevier, Amsterdam, p. 279.

116. Goto, J. (1990) *Yakugaku Zasshi*, **110**, 807.

117. Legrand-Defretin, V., Juste, C., Robert, H. and Corring, T. (1991) *Lipids*, **26**, 578.

118. Ferreira, H.E.C. and Elliot, W.H. (1991) *J. Chromatog.*, **562**, 697.

119. Eckers, C., New, A.P., East, P.B. and Haskins, N.J. (1990) *Rapid Commun. Mass Spectrom.*, **4**, 449.

120. Wood, K.V., Sun, Y. and Elkin, R.G. (1991) *Anal. Chem.*, **63**, 247.

CHAPTER 13
CARBOHYDRATES AND SUGARS
A. S. Ball

1. INTRODUCTION

Carbohydrates are among the most abundant constituents of plants, animals and microorganisms. They serve as sources of energy (sugars) and as stores of energy (starch and glycogen); they also form major structural constituents in plants (hemicellulose, cellulose), fungi (chitin) and bacteria (peptidoglycan). Photosynthetic organisms assimilate carbohydrates from CO_2 and H_2O, while heterotrophs largely rely on photosynthetic organisms to supply carbohydrates.

Carbohydrates may be defined as polyhydroxy aldehydes or ketones, or as substances that yield any one of these compounds on hydrolysis. The term 'carbohydrates' applies to a large number of organic compounds, monomeric, oligomeric and polymeric in nature, all of which can be either synthesized from or hydrolyzed to monomers. Many carbohydrates have the empirical formula $(CH_2O)_n$ where n is 3 or greater. However, other carbohydrates also contain sulfur and nitrogen.

Carbohydrates may be classified according to their degree of polymerization into two main groups:

(i) monomeric carbohydrates — including monosaccharides and their derivatives;
(ii) polymeric carbohydrates — which comprise oligosaccharides, polysaccharides, DNA and RNA.

Polymeric carbohydrates differ in the type of bridge linking their monosaccharide units. While oligosaccharides and polysaccharides are polyacetals, linked by acetal oxygen bridges, DNA and RNA have polyphosphate esters linked by phosphate bridges. In addition to these well-defined groups of carbohydrates, derivatives, such as antibiotics, exist which need to be addressed separately because some of their members may be monomeric while others are oligomeric.

2. STRUCTURES AND CHARACTERISTICS OF MONOMERIC CARBOHYDRATES

2.1. Classification of monosaccharides

Monosaccharides are divided into two groups depending on whether their acyclic form possesses an aldehyde (aldoses) or keto group (ketoses) (*Table 1*). Further classification of both groups is based on the number of carbon atoms in the monosaccharide chain (usually 3–10) and finally according to the size of their rings; five-membered furanoses or six-membered pyranoses. In order to form a furanose ring, four carbon atoms and one oxygen atom are required and therefore only aldotetroses and pentuloses and higher derivatives can form furanoses. Similarly, in order to form a pyranose ring, five carbon atoms and one oxygen atom are required and so only aldopentoses and hexuloses and higher analogs can form pyranoses. Using this classification technique a combination name may be used which defines without ambiguity the group to which the monosaccharide belongs (*Table 1*).

Table 1. Classification of monosaccharides

Number of carbon atoms	Aldofuranose/ketofuranose	Aldopyranose/ketopyranose
Aldoses[a]		
3 — Aldotriose	—	—
4 — Aldotetrose	aldotetrofuranose	—
5 — Aldopentose	aldopentofuranose	aldopentapyranose
6 — Aldohexose	aldohexafuranose	aldohexopyranose
7 — Aldoheptose	aldoheptofuranose	aldoheptopyranose
8 — Aldo-octose	aldo-octofuranose	aldo-octopyranose
9 — Aldononose	aldononofuranose	aldononopyranose
10 — Aldodecose	aldodecofuranose	aldodecopyranose
Ketoses[b]		
4 — Tetrulose	—	—
5 — Pentulose	pentulofuranose	—
6 — Hexulose	hexulofuranose	hexulopyranose
7 — Heptulose	heptulofuranose	heptulopyranose
8 — Octulose	octulofuranose	octulopyranose
9 — Nonulose	nonulofuranose	nonulopyranose
10 — Deculose	deculofuranose	deculopyranose

[a] Aldobiose is not considered a saccharide because, by definition, a saccharide must contain an asymmetric carbon atom.
[b] Triulose is not considered to be a saccharide because it lacks an asymmetric carbon atom.

2.2. Distribution and properties of some important monosaccharides

Monosaccharides are monomeric in nature and therefore cannot be hydrolyzed to simple sugars. Monosaccharides (and oligosaccharides) are soluble in water and are often sweet tasting (hence the term sugars). D-Glucose, an aldohexose, is the most abundant monosaccharide and therefore the most extensively studied. It occurs free in the juices of fruits and in honey. Another common monosaccharide is fructose, a ketohexose which is often found in association with glucose. Both fructose and glucose are commercially available along with other monosaccharides (*Table 2*). Although other saccharides not listed here are available, their high price usually restricts their use to analytical work. The pentose, D-ribose, is a component of ribonucleic acid, where it exists as a furanose ring. Four other monosaccharides play important roles in the metabolism of carbohydrates during photosynthesis. They are the aldotetrose D-erythrose; the ketopentoses, D-xyulose and D-ribulose; and the ketoheptose, D-sedoheptulose. Two other deoxysugars are found in nature as components of cell walls. These are L-rhamnose (β-deoxy-L-mannose) and L-fucose (6-deoxy-L-galactose).

Two common amino sugars are D-glucosamine and D-galactosamine, in which the hydroxyl group at C-2 is replaced by an amino group. D-Glucosamine is a major component of chitin while D-galactosamine is a major component of cartilage.

Table 2. Origin and properties of some naturally occurring monosaccharides

Class	Monosaccharide	Source	m.p. (°C)	Structure
Aldopentose	D-Ribose	Nucleic acids	87	D-Xylose / D-Arabinose / D-Ribose (see below)
Aldopentose	D-Xylose	Xylan	145	
Aldopentose	L-Arabinose	Gums	160	
Aldohexose	D-Glucose	Starch	146	D-Mannose / D-Glucose / D-Galactose (see below)
Aldohexose	D-Mannose	Mannan	133	
Aldohexose	D-Galactose	Gums	167	

Structures (Fischer projections):

D-Xylose:
$$\begin{array}{c} HC=O \\ H-C-OH \\ HO-C-H \\ H-C-OH \\ CH_2OH \end{array}$$

D-Arabinose:
$$\begin{array}{c} HC=O \\ HO-C-H \\ H-C-OH \\ H-C-OH \\ CH_2OH \end{array}$$

D-Ribose:
$$\begin{array}{c} HC=O \\ H-C-OH \\ H-C-OH \\ H-C-OH \\ CH_2OH \end{array}$$

D-Mannose:
$$\begin{array}{c} HC=O \\ HO-C-H \\ HO-C-H \\ H-C-OH \\ H-C-OH \\ CH_2OH \end{array}$$

D-Glucose:
$$\begin{array}{c} HC=O \\ H-C-OH \\ HO-C-H \\ H-C-OH \\ H-C-OH \\ CH_2OH \end{array}$$

D-Galactose:
$$\begin{array}{c} HC=O \\ H-C-OH \\ HO-C-H \\ HO-C-H \\ H-C-OH \\ CH_2OH \end{array}$$

Table 2. Continued

Class	Monosaccharide	Source	m.p. (°C)	Structure
6-Deoxysugar	L-Rhamnose	Glycosides	124	L-Fucose
6-Deoxysugar	L-Fucose	Seaweeds	145	L-Rhamnose
Ketohexose	D-Fructose	Insulin	102	L-Sorbose
Ketohexose	L-Sorbose	Rowans	159	D-Fructose

L-Fucose:
$$\begin{array}{c} HC=O \\ HO-C-H \\ H-C-OH \\ H-C-OH \\ HO-C-H \\ CH_3 \end{array}$$

L-Rhamnose:
$$\begin{array}{c} HC=O \\ H-C-OH \\ H-C-OH \\ HO-C-H \\ HO-C-H \\ CH_3 \end{array}$$

L-Sorbose:
$$\begin{array}{c} CH_2OH \\ C=O \\ H-C-OH \\ HO-C-H \\ HO-C-H \\ CH_2OH \end{array}$$

D-Fructose:
$$\begin{array}{c} CH_2OH \\ C=O \\ HO-C-H \\ H-C-OH \\ H-C-OH \\ CH_2OH \end{array}$$

Abbreviation: m.p., melting point.

2.3. Stereoisomerism

The study of carbohydrates introduces the topic of isomerism. The subject may be divided into structural isomerism and stereoisomerism. Structural isomers have the same molecular formula but differ from each other by having different structures; stereoisomers have the same molecular formula and the same structure, but differ in configuration, that is the arrangement of their atoms in space. The subject of stereoisomerism can be divided into optical isomerism and *cis–trans* isomerism. Optical isomerism is common amongst carbohydrates, usually occurring when a molecule contains one or more chiral (Greek *cheir* — hand) carbon atoms. In many compounds the carbon atom has the shape of a tetrahedron, in which the carbon nucleus sits in the centre of the tetrahedron and the four covalent bonds extend out to the corners of the tetrahedron:

When four different groups are attached to these bonds, the central carbon atom is said to be chiral. The four different groups may be arranged in different ways such that different compounds are formed, that is they cannot be superimposed. One compound is related to the other as a right hand is related to a left. Such chiral molecules are said to possess 'handedness' and are mirror images of each other. These mirror-image isomers constitute an enantiomeric pair. Almost all the properties of an enantiomeric pair are identical (e.g. boiling point, melting point, solubilities). Both exhibit optical activity, but in this they differ. One will rotate a plane of polarized light in a clockwise direction ((+) *dextro*-rotatory), the other in a counter-clockwise direction ((−) *levo*-rotatory).

With the existence of a large number of optical isomers in carbohydrates, it is necessary to have a reference compound; this compound is the triose glyceraldehyde. The isomer with the formula containing the hydroxyl group on the right, with the aldehyde group at the top was assigned D(+)-glyceraldehyde:

$$\begin{array}{c} \text{CHO} \\ | \\ \text{H—C—OH} \\ | \\ \text{CH}_2\text{OH} \end{array}$$

This serves as an important reference compound for carbohydrates. The D+L system has been particularly useful in relating groups of carbohydrates (such as the naturally occurring D-sugars). Carbohydrates are classified either D or L, according to their configuration in terms of D-glyceraldehyde; this classification has nothing to do with *dextro*- or *levo*-rotation.

Consider the four common hexoses; D(+)-glucose, D(+)-mannose, D(+)-galactose and D(−)-fructose (*Table 2*). D-Fructose is a structural isomer of the other three hexoses, although it is a ketose rather than an aldose. The three aldohexoses are optical isomers. However, because none are enantiomers, they are related as diastereomers, with different melting and boiling points and solubilities.

Finally, it is possible to show that two forms of D(+)-glucose exist. When D-glucose is dissolved in water and allowed to crystallize out by evaporation, a form designated as α-D-glucose is obtained. If glucose is crystallized out from acetic acid, another form, β-D-glucose is obtained. These two forms of glucose show the phenomenon of mutarotation, with different specific rotations. The explanation for the existence of two forms of D-glucose is found in the fact that the aldohexoses (and other sugars) react internally to form cyclic hemiacetals (*Figure 1*).

Figure 1. Formation of the hemiacetal forms of D-glucose. An equilibrium exists between the α and β forms and the open-chain forms.

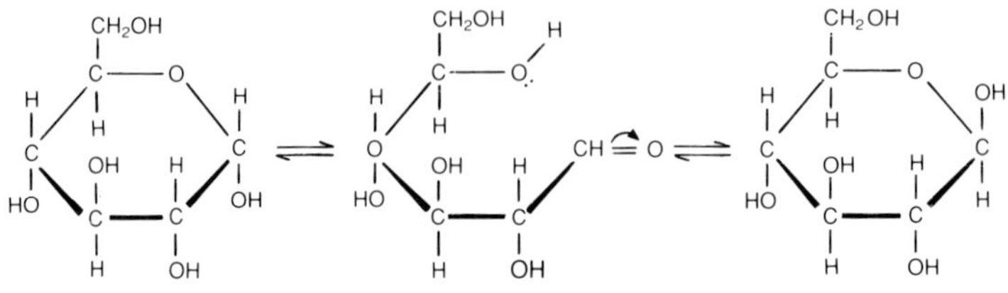

α-D-Glucose   Open D-Glucose   β-D-Glucose

3. OLIGOSACCHARIDES

3.1. Classification of oligosaccharides

Oligosaccharides are polyacetals which, as their names suggest (from the Greek *oligos* — few), contain a low (polymerization 2–10) degree of polymerization. They are composed of a number of monosaccharides linked via acetal oxygen bridges which upon hydrolysis yield one or more types of monosaccharide. Oligosaccharides are further grouped into simple (true) oligosaccharides, which are oligomers of monosaccharides that yield only monosaccharides on complete hydrolysis, and conjugate oligosaccharides, which are oligomers of monosaccharides linked to a non-saccharide such as a peptide or lipid (*Table 3*). Oligosaccharides may also be classified according to the degree of polymerization (*Table 3*) and whether or not the oligomer has an aldehyde or keto group at one end. These terminal groups are readily converted into carboxylic acids by oxidation reactions. Oligosaccharides possessing these groups are referred to as reducing oligosaccharides, while those not undergoing this oxidation reaction are referred to as non-reducing oligosaccharides (*Table 3*). Oligosaccharides, because of their sweetness, are often referred to as sugars. However, the sweetness of an oligosaccharide reduces with increasing polymerization. Beyond a degree of polymerization of four the oligosaccharide becomes tasteless.

3.2. Distribution and properties of some important oligosaccharides

The oligosaccharides most frequently encountered in nature are the disaccharides. Among the disaccharides encountered is the homodisaccharide reducing sugar, maltose (*Figure 2*), produced as an intermediate in the hydrolysis of starch by amylases. Maltose is also found free in nature in soya beans. In maltose, one molecule of glucose is linked through the hydroxyl group on the C-1 carbon atom in a glycosidic bond to the hydroxyl group on the C-4 of a second molecule of glucose (*Figure 2*). Because the configuration of the hemiacetal carbon atom involved in the bonding is of the α form, and because it is linked to the 4 position on the second glucose unit, the linkage is designated as α-1,4. The second glucose molecule possesses a free hydroxyl group, which can exist in the α or β configuration.

The homodisaccharide, cellobiose, is identical to maltose except that it has a β-1,4 glycosidic linkage (*Figure 2*). Cellobiose is a reducing sugar formed during the hydrolysis of cellulose. Lactose is a heterodisaccharide found in varying amounts in mammalian milk. Upon hydrolysis, lactose yields one molecule each of D-galactose and D-glucose. It possesses a

Table 3. Classification of simple oligosaccharides, including the names and composition of the most common oligosaccharides

Number of monosaccharides	Homo-oligosaccharides	Hetero-oligosaccharides
Reducing oligosaccharides		
2 — Disaccharides	Maltose	Lactose
	4-α-D-Glc$p$	4-β-D-Gal$p$-D-Glc
	Cellobiose	Lactulose
	4-α-D-Glc$p$-D-Glc	4-β-D-Gal$p$-D-Fru
	Isomaltose	Melibiose
	6-α-D-Glc$p$-D-Glc	6-α-D-Gal$p$-D-Glc
	Gentibiose	Turanose
	6-β-D-Glc$p$-D-Glc	3-α-D-Glc$p$-D-Fru
3 — Trisaccharides	Maltotriose	Manninotriose
	4-α-D-Glc$p$-maltose	6-α-D-Gal$p$-melibiose
4 — Tetrasaccharides	Maltotetraose	
	4-α-D-Glc$p$-maltotriose	
Non-reducing oligosaccharides		
2 — Disaccharides	Trehalose	Sucrose
	α-D-Glc-α-D-Glc$p$	β-D-Fru$f$-α-D-Glc$p$
	Isotrehalose	Isosucrose
	β-D-Glc$p$-b-D-Glc$p$	α-D-Fru$f$-β-D-Glc$p$
3 — Trisaccharides	Raffinose	
	6-α-D-Gal$p$-sucrose	
4 — Tetrasaccharides	Stachyose	
	6-α-D-Gal$p$-raffinose	

Abbreviations: Glc, glucose; Gal, galactose; Fru, fructose; f, furanose; p, pyranose.

Figure 2. Structures of some common disaccharides.

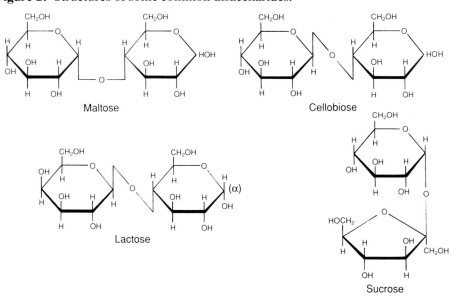

β-1,4 linkage and is a reducing sugar (*Figure 2*). Sucrose, a non-reducing heterodisaccharide is the main plant sugar, commercially extracted from beet and cane. Upon hydrolysis, sucrose yields one molecule each of D-glucose and D-fructose. The fact that it is not a reducing sugar indicates that the reducing groups in both monosaccharides must be involved in linkage between the two sugar units, that is the C-1 and C-2 carbons of glucose and fructose respectively must participate in glycoside formation. The configuration of the fructose is β, while that of glucose is α.

4. STRUCTURES AND CHARACTERISTICS OF POLYSACCHARIDES

4.1. Classification of polysaccharides

Polysaccharides are polymeric in nature and, like oligosaccharides, are polyacetals with oxygen bridges linking the monosaccharide units; however, polysaccharides may have a degree of polymerization from as few as 11 to several thousand monosaccharide units. Because solubility decreases and viscosity increases as the degree of polymerization increases, higher polysaccharides such as cellulose are insoluble in water while other polysaccharides are viscous enough to gel in an aqueous solution. All polysaccharides can be hydrolyzed with acid or enzymes to yield monosaccharides and/or monosaccharide derivatives.

There are a number of ways to classify polysaccharides (*Table 4*). Chemical classification distinguishes between simple (true) polysaccharides, which only produce monosaccharides or oligosaccharides or their derivatives (esters or ethers) upon depolymerization; or conjugate polymers, made up of a polysaccharide linked to another polymer, such as a protein (forming a glycoprotein). Polysaccharides are further subdivided into two classes: homopolysaccharides, which are simple polymers having a regular repeating unit (monomer) consisting of a single monosaccharide; and heteropolysaccharides which are made up of more than one monosaccharide (*Table 4*). The shape of polymers also significantly affects the physical properties of the polysaccharide and therefore each polymer type is further divided into linear and branched polysaccharides.

Table 4. Classes of common polysaccharides

Homopolysaccharides		Heteropolysaccharides	
Linear	**Branched**	**Linear**	**Branched**
Simple polysaccharides			
Amylose (α-D-glucan)	Amylopectin	Mannans	Gums
Cellulose (β-D-glucan)	Glycogen	Xylans	Mucilages
Chitin (D-glucosamine)	Pectins		
	Agar		
Conjugate polysaccharides			
			Lectins
			Glycoproteins
			Peptidoglycans

4.2. Plant polysaccharides

Polysaccharides have a very wide distribution in nature, where they may serve a structural function or play an important role as a stored form of energy. Cellulose forms the main constituent of the cell walls of plants. It is a homopolysaccharide composed of β-1,4-linked glucose units (*Figure 3*), composed of between 2000 and 3000 monomers depending upon the source and method of extraction. The higher plant relies on its cell wall to impart strength to the cell. Starch is the carbohydrate reserve material of most plants, and constitutes the main carbohydrate source for animals, including man. Like cellulose, starch is a homopolysaccharide composed of repeating glucose units. However, starch differs in that it is generally a mixture of two polysaccharides, amylose and amylopectin. Amylose is a linear glucan with α-1,4-linked D-glucopyranose units with a chain length of 1000–4000 monomer units. In contrast, amylopectin is a branched glucan, linked via α-1,4 or α-1,6 links (*Figure 3*).

The hemicelluloses are found in association with cellulose in plant cell walls. The commonest members of this group are the xylans, which are basically β-1,4-linked xylopyranose units, but also contain other residues such as L-arabinose and D-glucuronic acid as part of branched groups. Other classes of hemicelluloses are mannans, composed of β-1,4-linked mannopyranose units, and the glucomannans, which occur widely in conifers and are made up of both mannopyranose and glucopyranose units. Gums are complex polysaccharides exuded by plants. Gum arabic is one of the most important commercial gums and is used extensively for stiffening fabrics, for thickening medicines and cordials and is also widely used in the confectionary trade. Gum arabic is composed of a central core of 1,3-linked D-galactopyranose residues with side-chains containing L-arabinofuranose, L-rhamnopyranose and D-glucuronic acid. Other important gums include mesquite and tragacanth. Gums are generally highly branched polysaccharides composed of a mixture of monomers such as D-galactose, L-arabinose, D-xylose and D-glucuronic acid.

4.3. Bacterial polysaccharides

The polysaccharides of the bacterial cell wall are extremely complex and are usually covalently bound to polypeptide chains. Gram-positive bacteria contain teichoic acids, which are polymers of ribitol or glycerol phosphate to which are attached D-alanine units. Another bacterial component is a regularly repeating co-polymer of β-1,4-linked *N*-acetylmuramic acid and *N*-acetylglucosamine (*Figure 3*). These linear polymers are cross-linked and provide the strength of the cell wall.

Dextrans are polysaccharides produced by *Leuconostoc* sp. and are α-1,6-linked glucans. They have an important use as blood plasma substitutes.

4.4. Animal polysaccharides

Chitin is an aminopolysaccharide which forms the structural tissue of some lower animals, such as crab, shrimp and lobster shells. Chitin is composed of a chain of β-1,4-linked D-glucosamine. Evidence suggests that chitin has a very similar structure to cellulose (*Figure 3*).

Glycogen is the common carbohydrate reserve substance of animals. It is a glucan of similar structure to amylopectin, but the degree of branching is much greater.

4.5. DNA, RNA, nucleosides and nucleotides

Unlike oligo- and polysaccharides, DNA and RNA are polyesters with the sugar residues linked by phosphate bridges. DNA is the largest known polymer, with a degree of polymerization in human genes of 10^{12}. The monomers of DNA and RNA are made up of phosphorylated 2-deoxy-D-erythro-pentofuranosyl- and D-ribofuranosyl-purine and -pyrimidine bases, termed nucleotides. Nucleotides may undergo hydrolysis of their phosphoric ester

Figure 3. Examples of common polysaccharides.

Linear homopolysaccharides

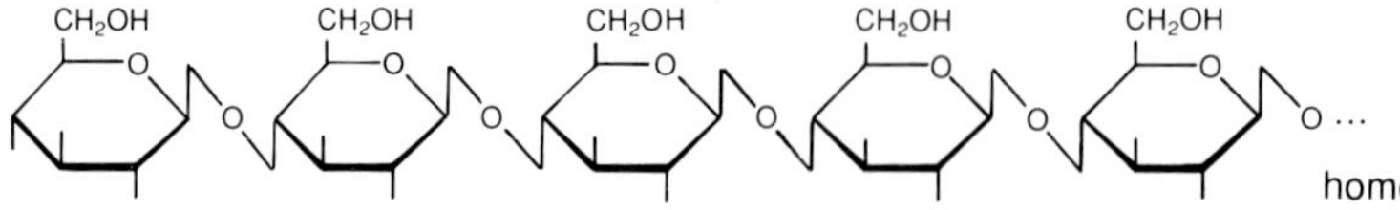

Cellulose

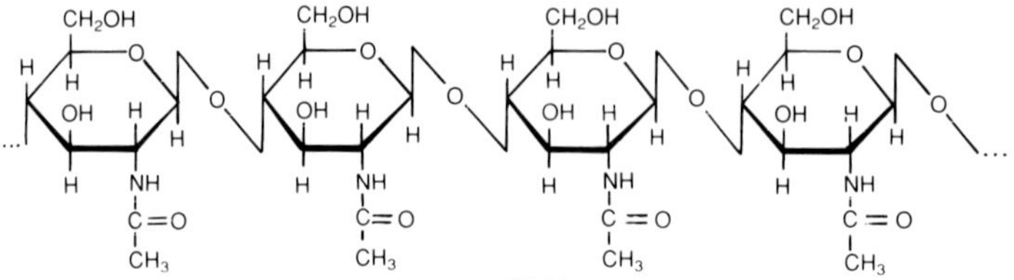

Chitin

Branched homopolysaccharide

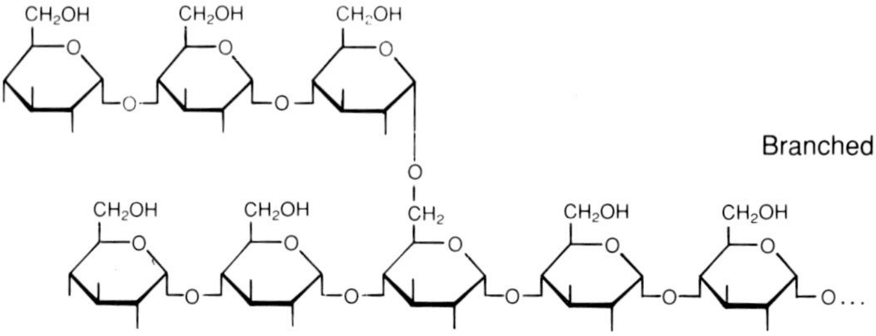

Amylopectin

Heteropolysaccharide

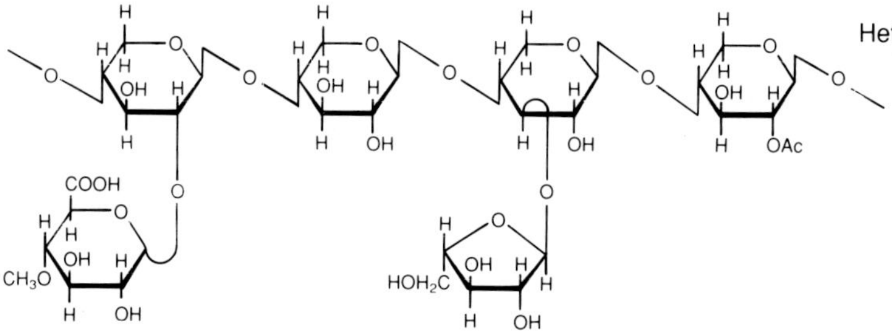

Xylan

N-acetyl glucosamine   N-acetyl muramic acid

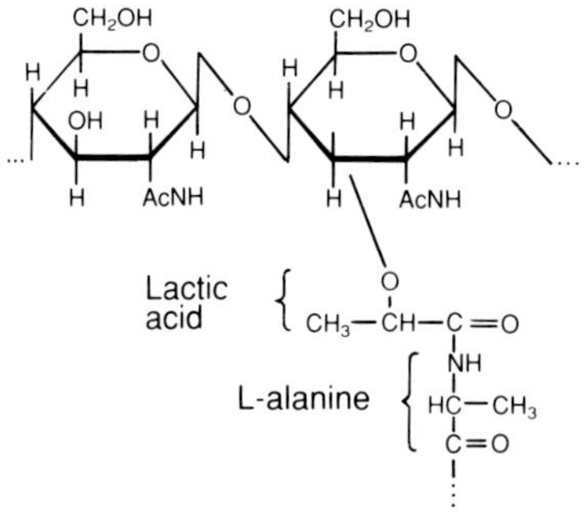

Conjugate polysaccharide

Lactic acid

L-alanine

Repeating unit of peptidoglycan

groups, forming simpler monomers — nucleosides. This group may also be divided according to their degree of depolymerization into monomers (nucleosides and nucleotides) and polymers (DNA and RNA). For more detailed information see *Chapter 11*.

4.6. Other saccharide derivatives

Carbohydrate-containing antibiotics are amongst a number of important carbohydrate derivatives which do not conform to conventional classification and are best studied individually. Despite containing both monosaccharide and oligosaccharide derivatives, carbohydrate-containing antibiotics may be closely related.

5. FURTHER READING

1. Instruction to authors (1984) *Carbohydr. Res.*, **132**, 185.

2. Angyal, S.J. (1972) in *The Carbohydrates* (W. Pigman and D. Horton eds). Academic Press, New York, Vol. IA, p. 195.

3. Angyal, S.J. (1984) *Adv. Carbohydr. Chem. Biochem.*, **42**, 15.

4. Dekker, C.A. and Goodman, L. (1970) in *The Carbohydrates* (W. Pigman and D. Horton eds). Academic Press, New York, Vol. IIA, p. 1.

5. Hanessian, S. (1983) *Total Synthesis of Natural Products*. Pergamon, New York.

6. Hanessian, S. and Haskell, T.H. (1970) in *The Carbohydrates* (W. Pigman and D. Horton eds). Academic Press, New York, Vol. IIA, p. 139.

7. Khan, R. (1981) *Adv. Carbohydr. Chem. Biochem.*, **39**, 213.

8. Pigman, W. and Horton, D. (eds) (1970, 1972, 1980) *The Carbohydrates*. Academic Press, New York, Vols IA, IB, IIA, IIB.

9. Pigman W. and Horton, D. (1972) in *The Carbohydrates* (W. Pigman and D. Horton eds). Academic Press, New York, Vol. 1A, p. 1.

10. Pigman W. (1972) in *The Carbohydrates* (W. Pigman and D. Horton eds). Academic Press, New York, Vol. IA, p. 165.

11. Stoddart, J.F. (1971) *Stereochemistry of Carbohydrates*. Wiley (Interscience), New York.

12. Tipson, R.S. and Horton, D. (eds) (1974–87) *Advances in Carbohydrate Chemistry and Biochemistry*. Academic Press, New York, Vols 29–45.

13. Umezawa, S. (1974) *Adv. Carbohydr. Chem. Biochem.*, **30**, 111.

14. Whistler, R.L. and BeMiller, J.N. (eds) (1972, 1973, 1977) *Methods in Carbohydrate Chemistry*. Academic Press, New York, Vols I, II, VI.

15. Williams, N.R. and Wander, J.D. (1980) in *The Carbohydrates* (W. Pigman and D. Horton eds). Academic Press, New York, Vol. IB, p. 761.

CHAPTER 14
SAFETY AND THE DISPOSAL OF TOXIC AND INFECTIOUS MATERIALS

D. Rickwood and T. Brown

1. PROTECTION FROM CHEMICAL HAZARDS

1.1. Risk and safety classification systems

All reputable suppliers of laboratory chemicals provide risk and safety information with those products that present a potential hazard. The label will carry a symbol indicating the hazard category (*Table 1*) and usually this will be backed up by code numbers indicating the specific risks presented by the chemical and the safety precautions that should be observed. The most frequently used code systems are the UN Hazard Classification (*Table 2*) and the European Commission Risk and Safety Phrases (EC system) (*Table 3*). The EC system has been adopted by a number of international suppliers (e.g. BDH, Serva and Fluka) and is familiar in most countries of the world, including the USA. The EC phrases are given in full as they are used elsewhere in *Biochemistry Labfax* to describe risk and safety information for various chemicals and reagents. The EC system is not, however, in universal use and several companies have code systems of their own, a notable example being Sigma-Aldrich (1). Several suppliers now publish detailed risk and safety data for their products (e.g. refs 2 and 3). General information concerning chemical hazards is also available (4–15).

Table 1. Hazard symbols

Symbol	Definition	Precaution
	Explosive	Avoid shock, friction, sparks and heat
	Oxidizing	Keep away from combustible material
	Toxic	Avoid contact
	Irritating	Do not breathe vapours and avoid contact with skin and eyes
	Harmful	Avoid contact, including inhalation

Table 1. Continued

Symbol	Definition	Precaution
	Highly flammable	Various depending on sub-category[a]
	Corrosive	Do not breathe vapours and avoid contact
	Radioactive	Avoid contact, use suitable shielding

[a] Typical sub-categories: spontaneously flammable substances — avoid contact with air; highly flammable gases — avoid contact with air, keep away from sources of ignition; substances sensitive to moisture — avoid contact with water; flammable liquids (flash point <55°C) — keep away from sources of ignition.
Reproduced from *Molecular Biology Labfax*.

Table 2. UN Chemical Hazard Classification

Hazard class	Description
Class 1	Explosive
Class 2	Gases
Class 3.1	Flammable liquids: flash point below −18°C
Class 3.2	Flammable liquids: flash point −18°C to +23°C
Class 3.3	Flammable liquids: flash point +23°C to +61°C
Class 4.1	Flammable solids
Class 4.2	Spontaneously combustible
Class 4.3	Dangerous when wet
Class 5.1	Oxidizing agent
Class 5.2	Organic peroxides
Class 6.1	Poisonous
Class 7	Radioactive
Class 8	Corrosive
Class 9	Miscellaneous dangerous substances
NR	Non-regulated

Reproduced from *Molecular Biology Labfax*.

Table 3. European Commission Risk and Safety Phrases[a]

Phrase number	Description
Risk phrases	
R1	Explosive when dry
R2	Risk of explosion by shock, friction, fire or other sources of ignition
R3	Extreme risk of explosion by shock, friction, fire or other sources of ignition
R4	Forms very sensitive explosive metallic compounds
R5	Heating may cause an explosion
R6	Explosive with or without contact with air
R7	May cause fire
R8	Contact with combustible material may cause fire
R9	Explosive when mixed with combustible material
R10	Flammable
R11	Highly flammable
R12	Extremely flammable
R13	Extremely flammable liquefied gas
R14	Reacts violently with water
R15	Contact with water liberates highly flammable gases
R16	Explosive when mixed with oxidizing substances
R17	Spontaneously flammable in air
R18	In use may form flammable/explosive vapour–air mixture
R19	May form explosive peroxides
R20	Harmful by inhalation
R21	Harmful in contact with skin
R22	Harmful if swallowed
R23	Toxic by inhalation
R24	Toxic in contact with skin
R25	Toxic if swallowed
R26	Very toxic by inhalation
R27	Very toxic in contact with skin
R28	Very toxic if swallowed
R29	Contact with water liberates toxic gas
R30	Can become highly flammable in use
R31	Contact with acid liberates toxic gas
R32	Contact with acid liberates very toxic gas
R33	Danger of cumulative effects
R34	Causes burns
R35	Causes severe burns
R36	Irritating to eyes
R37	Irritating to respiratory system
R38	Irritating to skin
R39	Danger of very serious irreversible effects
R40	Possible risk of irreversible effects
R41	Risk of serious irreversible effects
R42	May cause sensitization by inhalation
R43	May cause sensitization by skin contact
R44	Risk of explosion if heated under confinement
R45	May cause cancer

Table 3. Continued

Phrase number	Description
R46	May cause heritable genetic changes
R47	May cause birth effects
R48	Danger of serious damage to health by prolonged exposure
R14/15	Reacts violently with water, liberating highly flammable gases
R15/29	Contact with water liberates toxic, highly flammable gas
R20/21	Harmful by inhalation and in contact with skin
R20/21/22	Harmful by inhalation, in contact with skin and if swallowed
R20/22	Harmful by inhalation and if swallowed
R21/22	Harmful in contact with skin and if swallowed
R23/24	Toxic by inhalation and in contact with skin
R24/25	Toxic in contact with skin and if swallowed
R26/27	Very toxic by inhalation and in contact with skin
R26/27/28	Very toxic by inhalation, in contact with skin and if swallowed
R26/28	Very toxic by inhalation and if swallowed
R27/28	Very toxic in contact with skin and if swallowed
R36/37	Irritating to eyes and respiratory system
R36/37/38	Irritating to eyes, respiratory system and skin
R36/38	Irritating to skin and eyes
R37/38	Irritating to respiratory system and skin
R42/43	May cause sensitization by inhalation and skin contact

Safety phrases

S1	Keep locked up
S2	Keep out of reach of children
S3	Keep in a cool place
S4	Keep away from living quarters
S5	Keep contents under . . . (appropriate liquid to be specified by the manufacturer)
S6	Keep under . . . (inert gas to be specified by the manufacturer)
S7	Keep container tightly closed
S8	Keep container dry
S9	Keep container in a well ventilated place
S12	Do not keep the container sealed
S13	Keep away from food, drink and animal feeding stuffs
S14	Keep away from . . . (incompatible materials to be indicated by the manufacturer)
S15	Keep away from heat
S16	Keep away from sources of ignition — No Smoking
S17	Keep away from combustible material
S18	Handle and open container with care
S20	When using do not eat or drink
S21	When using do not smoke
S22	Do not breathe dust
S23	Do not breathe gas/fumes/vapour/spray (appropriate wording to be specified by the manufacturer
S24	Avoid contact with skin

Table 3. Continued

Phrase number	Description
S25	Avoid contact with eyes
S26	In case of contact with eyes, rinse immediately with plenty of water and seek medical advice
S27	Take off immediately all contaminated clothing
S28	After contact with skin, wash immediately with plenty of . . . (to be specified by the manufacturer)
S29	Do not empty into drains
S30	Never add water to this product
S33	Take precautionary measures against static discharges
S34	Avoid shock and friction
S35	This material and its container must be disposed of in a safe way
S36	Wear suitable protective clothing
S37	Wear suitable gloves
S38	In case of insufficient ventilation, wear suitable respiratory equipment
S39	Wear eye/face protection
S40	To clean the floor and all objects contaminated by this material use . . . (to be specified by the manufacturer)
S41	In case of fire and/or explosion do not breathe fumes
S42	During fumigation/spraying wear suitable respiratory equipment (appropriate wording to be specified)
S43	In case of fire, use . . . (indicate in the space the precise type of fire-fighting equipment. If water increases the risk, add — Never use water)
S44	If you feel unwell, seek medical advice (show the label where possible)
S45	In case of accident or if you feel unwell, seek medical advice immediately (show the label where possible)
S46	If swallowed seek medical advice immediately and show this container or label
S47	Keep at temperature not exceeding . . . °C (to be specified by the manufacturer)
S48	Keep wetted with . . . (appropriate material to be specified by the manufacturer)
S49	Keep only in the original container
S50	Do not mix with . . . (to be specified by the manufacturer)
S51	Use only in well ventilated areas
S52	Not recommended for interior use on large surface area
S53	Avoid exposure — obtain special instructions before use
S1/2	Keep locked up and out of reach of children
S3/7/9	Keep container tightly closed in a well ventilated place
S3/9	Keep in a cool, well ventilated place
S3/9/14	Keep in a cool, well ventilated place away from . . . (incompatible materials to be indicated by the manufacturer)
S3/9/14/49	Keep only in the original container in a cool, well ventilated place away from . . . (incompatible materials to be indicated by the manufacturer)
S3/9/49	Keep only in the original container in a cool, well ventilated place
S3/14	Keep in a cool place away from . . . (incompatible materials to be indicated by the manufacturer)

Table 3. Continued

Phrase number	Description
S7/8	Keep container tightly closed and dry
S7/9	Keep container tightly closed and in a well ventilated place
S20/21	When using do not eat, drink or smoke
S24/25	Avoid contact with skin and eyes
S36/37	Wear suitable protective clothing and gloves
S36/37/38	Wear suitable protective clothing, gloves and eye/face protection
S36/39	Wear suitable protective clothing and eye/face protection
S37/39	Wear suitable gloves and eye/face protection
S47/49	Keep only in the original container at temperature not exceeding . . . °C (to be specified by the manufacturer)

[a] EC Statutory Instrument 1984 No 1244: Information approved for the Classification, Packaging and Labelling of Dangerous Substances for Supply and Conveyance by Road, Part IV (risk phrases), Part V (safety phrases).
Reproduced from *Molecular Biology Labfax*.

2. DISPOSAL OF TOXIC AND INFECTIOUS MATERIALS

2.1. Disposal of toxic materials

Many of the chemicals and reagents used in biochemistry are toxic or biohazardous in various ways. Where non-toxic alternatives are available they should be used. If toxic compounds must be used, it is important to be aware of the hazards and take the necessary precautions during all operations. In addition, it is important to dispose of all toxic waste correctly. In order to fulfil these requirements it is advisable to carry out a full assessment of hazards associated with all stages of experiments.

The disposal methods can be classified into a few defined procedures:
1. Dissolve the material in a suitable solvent and burn in an approved type of incinerator with an afterburner. The temperature of operation is of paramount importance to ensure complete destruction of the substances.
2. Disposal down an approved drain with a large excess of water.
3. Oxidation with concentrated chromic acid. During this operation care is necessary to avoid splashing. Then dispose of down an approved drain (as in Method 2).
4. Oxidation with sodium hypochlorite solution (domestic bleach) for at least 24 hours at room temperature prior to disposal down an approved drain (as in Method 2).
5. Treatment with 1 M KOH or NaOH solution prior to disposal down an approved drain (as in Method 2).
6. Adsorption onto charcoal followed by incineration as described in Method 1.
7. Neutralization with 6 M sulfuric acid or hydrochloric acid followed by disposal down an approved drain (as in Method 2).
8. Protein toxins can be denatured using heat, 70% ethanol or dilute phenol or formaldehyde.
9. Bury in an approved landfill site.

Table 4. Selected list of toxic materials, their effects and recommended methods for their disposal (derived from ref. 16)

Compound name	RTECS No.	Car	Ter	HT	T	Solvent	Disposal method(s)[a]
A23187	DM 4676000				+	acetone	1, 3, 6
S-Acetylthiocholine iodide	FZ 9865000			+		water	1, 3, 4
Alamethicin	AY 1900000			+		water	1, 4
Actinomycin D	AU 1575000	+	+	+		water	1, 3, 4, 6
Actinomycin D, 7-amino	AU 1579000	+	+	+		acetone	1, 3, 4, 6
Aflatoxin B_1[b]	GY 1925000	+	+	+		chloroform	4, 5, 6
Agglutinin, castor bean	VJ 2625000	+	+	+		water	3, 4
α-Amanitin	BD 6195000	+		+		water	1, 3, 4
β-Amanitin	BD 6195100	+		+		water	1, 3, 4
Aminopterin	MA 1050000	+	+	+		water	1, 3, 4
Amphotericin B	BU 2660000			+		water, pH 2	1, 3, 4
Angiotensin I	BW 2275000		+		+	water	3, 4, 6
Angiotensin II	BW 2275000		+		+	water	3, 4, 6
Angiotensin II antagonist					+	water	3, 4, 6
Antimycin A	CD 0350000	+		+		acetone	1, 3, 4
9-β-D-Arabinofuranosyl-adenine	AU 6200000	+	+	+		1 M HCl	1, 3
1-β-D-Arabinofuranosyl-cytosine	HA 5425000	+	+		+	1 M HCl, water	1, 3
L-Arterenol	DN 6750000		+	+		water	1, 3, 4
Arsenazo III						water	9
Azaserine	VT 9625000	+	+		+	water	3, 4
Botulinus toxin, A–F	ED 9300000			+		water	1, 3, 4
Bradykinin	EE 1530000	+	+			1 M HCl	4
Brevetoxin PbTx-1	EE 4554800			+		acetone	1, 3
Brevetoxin PbTx-2	EE 4554800			+		acetone	1, 3
Brevetoxin PbTx-3	EE 4554800			+		acetone	1, 3
Brevetoxin PbTx-7	EE 4554800			+		acetone	1, 3
Brevetoxin PbTx-9	EE 4554800			+		acetone	1, 3

Table 4. Continued

Compound name	RTECS No.	Car	Ter	HT	T	Solvent	Disposal method(s)[a]
Bromelain	EF 8575000				+	water	1, 3, 4
α-Bungarotoxin	EI 6201400			+		water	1, 4, 8
β-Bungarotoxin	EI 6201000			+		water	1, 4, 8
neuronal-Bungarotoxin				+		water	1, 3, 4
Cacodylic acid	CH 7700000				+	water	9
Carbachol	GA 0875000			+		water	1, 3
Carbonyl cyanide	FG 5600000				+	DMSO	1, 4
Cardiotoxin, NNK				+		water	1, 4, 8
Cardiotoxin, NNA	FH 7482450			+		water	1, 4, 8
Cardiotoxin, NNG				+		water	1, 4, 8
Cetyltrimethylammonium bromide	BQ 7875000				+	water	1, 9
Charybdotoxin				+		water	1, 4, 8
Cholera toxin	KA 7330000					acetic acid	1, 3
Cholesterol	FZ 8400000					chloroform	4, 5
Chromomycin A3	GB 7875000	+	+	+		water	1, 3, 4
Clonidine hydrochloride	NU 2490000	+	+		+	water	1
Cobra venom factor	GG 3850000				+	water	1, 4, 8
α-Cobratoxin, NNK	GG 3850000			+		water	1, 4, 8
α-Cobratoxin, NNA	GG 4220000			+		water	1, 4, 8
ω-Conotoxin GVIA				+		water	1, 3, 4
Colcemid	GH 0800000		+		+	dilute HCl	1, 3, 6
Colchicine	GH 0700000	+	+	+		chloroform	1, 3, 6
Crotamine	GD 9046000				+	water	1, 3, 4
Crotoxin	GQ 7530000				+	water	1, 3
Cycloheximide	MA 4375000	+	+		+	ethanol	1, 3
Cytochalasin B	RO 0205000	+	+			acetone	1, 3
Cytochalasin D	GZ 4850000	+	+	+		acetone	1, 3

Table 4. Continued

Compound name	RTECS No.	Car	Ter	HT	T	Solvent	Disposal method(s)[a]
Dendrotoxin						water	1, 4, 8
Dexamethasone	TU 398000	+	+	+		water	1
1,4-Diacetoxy-2,3-dicyanobenzene			+		+	DMSO	1, 4
Diazotized sulfanilate				+	+	water	4
Diethylpyrocarbonate[c]	LQ 9350000	+	+			5% KOH	4, 5
Digitonin	IH 2050050	+	+			methanol	3, 4
Diisopropylfluorophosphate (DFP)	TE 5075000		+	+	+	water	1, 4, 5
DIIPC	FF 2175000				+	water	1, 3
Diitiazem	DL 0310000		+		+	water	1, 3
Diphtheria toxin unnicked	XW 5807200		+	+		water	1, 4, 8
17 β-Estradiol	KG 2975000	+	+			acetone	1, 3
Ethidium bromide	SF 7950000	+				water	6
N-Ethylmaleimide	UX 9625000	+			+	acetone	1, 3
Glucagon	LZ 3980000	+	+			1 M HCl	4
Hydroxyurea	YT 4900000	+	+			water	1, 3
Hygromycin B	WK 2130000	+			+	water	1, 3, 4
Indomethacin	NL 3500000	+	+			ethanol	1, 6
Iodoacetic acid	AI 3675000	+	+			water	1
Ionomycin	NO 0600000				+	acetone	1, 3
IPTG		+				water	1, 4
Isoproterenol	DO 1925000	+	+			water	1, 3, 4
Liposaccharide	OS 0895400		+			water	1, 4
Methotrexate	MA 1225000	+	+			5% KOH	1, 3, 6
N-Methyl-4-phenyl-1,2,3,6-tetrahydropyridine				+		water	1, 3

Table 4. Continued

Compound name	RTECS No.	Car	Ter	HT	T	Solvent	Disposal method(s)[a]
Mezerein	HB 5425500	+			+	DMSO	5
Microcystin-LR (MCYST-LR)	XW 5862000			+		ethanol	1, 4
Microcystin-RR (MCYST-RR)	XW 5862000			+		ethanol	1, 4
Mitomycin C	CN 0700000	+		+		water	1, 4, 8
Mojave toxin					+	water	4
Monensin	JH 2830000				+	acetone	1, 3
Monensin, methyl ester	QA 5960000			+		acetone	1, 3
Mycostatin	RF 5950000		+	+		methanol	1, 3
Myotoxin A					+	water	1, 4
Myxothiazol	QH 7580000			+		water	1, 4
Nodularin				+		methanol	1, 4, 8
Notexin	RD 3800000	+		+		water	1, 4, 8
Okadaic acid	RK 3325000			+		DMF	1, 4
Oligomycin	RN 3675000			+		acetone	1, 3
Ouabain				+		ethanol	1, 3
Palytoxin	RT 6475000			+		DMF	1, 4
Paradoxin	RV 0200000			+		water	1, 4, 8
Pertussis toxin	XW 5883750			+		water	1, 4, 8
Phalloidin	SE 9800000			+		1 M HCl	1, 3
Phenol	SJ 3325000	+				water	1
Phenylmethylsulfonyl fluoride (PMSF)	XT 8040000		+	+		ethanol	1, 5
Phorbol	GZ 0600000	+			+	DMF	4, 5
Phorbol, 12,13-diacetate	GZ 0615100	+				DMF	4, 5
Phorbol, 12,13-didecanoate	GZ 0653000	+				DMF	4, 5
4α-Phorbol, 12,13-didecanoate	GZ 0653000	+				DMF	4, 5
Phorbol, 12-myristate	GZ 0653000	+				DMF	4, 5

Table 4. Continued

Compound name	RTECS No.	Car	Ter	HT	T	Solvent	Disposal method(s)[a]
Phorbol, 12-myristate, 13-acetate	GZ 0630000	+	+			DMF	4, 5
Phorbol, 4-O-methyl, 12-myristate, 13-acetate		+				DMF	4, 5
Phorbol, 20-oxo-20-deoxy, 12,13-dibutyrate		+	+			DMF	4, 5
4α-Phorbol			+			DMF	4, 5
Phorbol, 12,13-dibutyrate						DMF	4, 5
Polymixin B	TR 1150000	+			+	Dilute HCl	1, 3
Polyoxin D	UV 7717100				+	water	1, 3, 4
Pokeweed antiviral protein I				+		water	1, 4, 8
Pokeweed antiviral protein II				+		water	1, 4, 8
Pokeweed antiviral protein S				+		water	1, 4, 8
Propidium iodide	SF 7949600	+			+	water	1, 3, 6
Protamine sulfate	VO 9470000				+	water	3, 4
Pyrazofurin	UQ 6360000	+			+	water	1, 3
Resiniferatoxin		+		+		DMSO	4
Ricin A chain	VJ 2625000	+	+	+		water	1, 4, 8
Ryanodine	VM 4025000			+		water	3, 4
Salinomycin	VO 8620000				+	water	3, 4
Sapintoxin-D	VQ 1575000				+	water	3, 4
Saxitoxin	UY 8708600			+		water	1, 3, 5
Selenomethionine	ES 7100000	+				dilute HCl	1, 3
Short neurotoxin α				+		water	4
Short neurotoxin α'				+		water	4
Spermine, tetrahydrochloride	EJ 7175000				+	water	1, 3
Streptozotocin[d]	LZ 5775000	+	+			water	3, 4
Taipoxin	WW 2200000					water	3, 4
Tetrahydroaminoacridine	AR 9532200			+	+	water	1, 3, 6

Table 4. Continued

Compound name	RTECS No.	Car	Ter	HT	T	Solvent	Disposal method(s)[a]
Tetrodotoxin	IO 1450000			+		water	1, 3, 4, 5
Tetrodotoxin, citrate free	IO 1450000			+		acetic acid	1, 3, 4, 5
Textilotoxin	YX 3962600			+		water	1, 4, 8
α-Toxin					+	water	1, 4, 8
T-2 Toxin	YD 0050000			+		1 M HCl	3, 4
Toxin II (ATX II)	CK 4438200			+		1 M HCl	3, 4
Toxin II (ATX II-Ile)				+		1 M HCl	3, 4
Toxin II (ATX II-Val)				+		water	1, 4, 8
D-Tubocurarine chloride	YO 4900000			+		water	1, 3, 4
Tunicamycin	YO 7980200	+	+			ethanol	1, 3, 4
Valinomycin	YV 9468000			+		acetone	1, 4
Verapamil	YV 8320000				+	methanol	1, 4
Vinblastin sulfate	YY 8050000	+	+	+		water	1, 3, 4
Vincristine sulfate	OH 6340000	+	+	+		water	1, 3, 4
Viroisin	ZA 5300000					water	1, 3, 4
Vitamin B$_{12}$	GG 3750000	+				ethanol	4, 6

[a] Disposal methods given in Section 2.1.
[b] Aflatoxins are fluorescent under long-wavelength ultraviolet light and their presence can be detected by this means.
[c] Diethylpyrocarbonate is of low order acute toxicity but owes its hazard primarily to its decomposition in water and, in the presence of ammonia, the formation of the carcinogen urethane. Diethylpyrocarbonate rapidly decomposes in pure water (pH 7) to ethanol and carbon dioxide, and thus can readily be disposed of by placing in aqueous solution (ammonia-free) overnight before washing down the drain. Because the compound can react in the body with endogenous ammonia and because it is highly irritant to the eyes, mucous membranes and skin, diethylpyrocarbonate must be handled with great care.
[d] Streptozotocin is a glucosidic form of the carcinogenic nitrosamide, N-nitrosomethylurea; it is unstable at high pH. There is no known reliable, inexpensive method for monitoring the environment for streptozotocin. Glassware, hoods, spills, etc., can be decontaminated with 5% KOH. Clothing can be cleaned with an alkaline detergent.

Abbreviations: Car, carcinogenic; DMF, dimethylformamide; DMSO, dimethylsufoxide; HT, highly toxic; T, toxic; Ter, teratogenic.

A list of selected biohazardous compounds and their recommended disposal procedures is given in *Table 4*: for details of disposal methods, refer to the above list.

2.2. Disposal of biohazardous materials

Often there may be a risk of infection from experimental material. If at all possible, avoid using any material for which the infectious risk is uncertain, especially where there is a risk of viral infection. It is important to sterilize all infectious or potentially infectious material before it is disposed of through the normal waste disposal routes.

One of the most frequent ways of sterilizing infected material is by autoclaving. *Table 5* lists the pressure and temperatures used for this process.

Typically 20 p.s.i. (1.3 bar) for 20 min will sterilize all but the largest volumes and containers. It is important to ensure that the autoclave is correctly serviced and maintained at all times.

If autoclaving is unsuitable or unavailable, then it is possible to use chemical disinfection. The choice of disinfection agent depends on the infectious hazard. *Table 6* lists the various methods that can be used and their effectiveness against different infectious agents.

Table 5. Pressure–temperature relationships for autoclaves

Pressure[a]			Temperature		Pressure[a]			Temperature	
p.s.i.	bar	kPa	°C	°F	p.s.i.	bar	kPa	°C	°F
0	0	0	100	212	30	2.0	207	135	274
5	0.3	34	108	227	35	2.3	241	138	281
10	0.7	69	115	239	40	2.7	276	142	287
15	1.0	103	121	250	50	3.3	345	148	298
20	1.3	138	126	259	60	4.0	414	153	307
25	1.7	172	130	267	80	5.3	552	162	331

[a] Pressure above atmospheric.

Table 6. The effectiveness of disinfectants against different types of infectious agents

Type of disinfectant	Bacteria		Viruses		Fungi
	Vegetative	Spores	Lipid	Non-lipid	
5% Hypochlorite	+ + +	+ + +	+ + +	+ + +	+
Phenolics	+ + +	−	+ + +	+	+ + +
90% Ethanol	+ + +	−	+ + +	+	−
2% Glutaraldehyde	+ + +	+ + +	+ + +	+ + +	+ + +
37% Formaldehyde	+ + +	+ + +	+ + +	+ + +	+ + +

− Ineffective, do not use.
+ Effective in some cases.
+ + + Effective, recommended method.

3. REFERENCES

1. *Sigma-Aldrich Safety Data.* (1990) Aldrich Chemical Company Ltd, Gillingham.

2. *BDH Hazard Data Sheets.* (1989) BDH, Poole.

3. Lenga, R.E. (ed.) (1988) *The Sigma-Aldrich Library of Chemical Safety Data.* Sigma-Aldrich Corp, Milwaukee.

4. Sax, N.I. and Lewis, R.J. (1988) *Dangerous Properties of Industrial Materials. (7th edn).* Van Nostrand Reinhold, New York, Vols 1–3.

5. Bretherick, L. (ed.) (1986) *Hazards in the Chemical Laboratory. (4th edn).* Royal Society of Chemistry, London.

6. National Research Council (1981) *Prudent Practices for Handling Hazardous Chemicals in Laboratories.* National Academy Press, Washington DC.

7. National Research Council (1981) *Prudent Practices for Disposal of Chemicals from Laboratories.* National Academy Press, Washington DC.

8. Sax, N.I. and Lewis, R.J. (1986) *Rapid Guide to Hazardous Chemicals in the Workplace.* Van Nostrand Reinhold, New York.

9. Steere, N.V. (ed.) (1971) *CRC Handbook of Laboratory Safety. (2nd edn).* CRC Press, Boca Raton.

10. Lefevre, M.J. (1980) *First Aid Manual for Chemical Accidents.* Dowden, Hutchinson and Ross, Stroudsburg.

11. Pitt, M.J. and Pitt, E. (1985) *Handbook of Laboratory Waste Disposal.* Wiley, New York.

12. Young, J.A. (ed.) (1987) *Improving Safety in the Chemical Laboratory.* Wiley, New York.

13. Pipitone, D.A. (1984) *Safe Storage of Laboratory Chemicals.* Wiley, New York.

14. Collings, A.J. and Luxon, S.G. (eds) (1982) *Safe Use of Solvents.* Academic Press, Orlando.

15. Anon (1975) *Toxic and Hazardous Industrial Chemicals Safety Manual.* International Technical Information Institute, Tokyo.

16. Snow, J.T. (1982) Handling of carcinogens and hazardous compounds. Calbiochem Corporation, La Jolla, CA.

CHAPTER 15
SOURCES OF FURTHER BIOCHEMICAL DATA
J. A. A. Chambers

The information presented in this book is intended to be of general use to those planning biochemical experiments. A truly comprehensive coverage of biochemistry is beyond the scope of this, or any other such book. This section includes a bibliography of other information sources for more specialized coverage of aspects of biochemistry.

1. REFERENCE BOOKS

The Biology Data Book. FASEB, Bethesda, MD, USA.

Fasman, G.D. ed. *CRC Handbook of Biochemistry and Molecular Biology*, CRC Press, Boca Raton, Florida (13 vols including index).

Dawson *et al. Data for Biochemical Research (3rd edn).* Oxford University Press.

Scott, T. A and Eagleson, M. eds. *Encyclopedia of Biochemistry.* de Gruyter, Berlin (a translation of the German *Lexicon der Biochemie*).

2. DATABASES

The number of publications in biochemistry and related fields is too large to keep up with unaided. A number of databases now allow searching of the literature for relevant papers in minutes rather than days. The following commercial databases include information of use to the biochemist. All of them are available through two commercial vendors, Dialog and STN, and may also be available through other networks. When substantially similar databases are available from both vendors, the descriptions have been fused and differences indicated. Some of these databases also have an instructional file associated with them.

Beilstein/Beilstein Online
Available through: STN (Beilstein), Dialog (Beilstein Online)
Producer: Beilstein Institute for Organic Chemistry
Coverage: 1830–1979
File size: 3.8 million records
Updates: Updated periodically
File type: Structure, numeric
Content: Organic chemistry
Language: English with some fields in German

BEILSTEIN contains organic chemical structures, preparation and reaction information, and numeric property information. The source for BEILSTEIN is the *Beilstein Handbook of Organic Chemistry* (main volume and supplements 1–5) for the period 1830–1979. Records are structure searchable and all text fields are searchable. The Dialog File starts in 1779 and the database allows definition of reactants and products. Chemical structures, chemical names, physical properties such as boiling point and relative density, chemical properties, preparation data and virtually every data field can be searched.

BIOQUIP: the biotechnology equipment supplier data bank

Available through:	STN
Producer:	DECHEMA Deutsches Gesellschaft für Chemisches Apparatewesen, Chemische Technik und Biotechnologie e.V.
Coverage:	Current data
File size:	1000 product groups and 400 companies
Updates:	Reloaded periodically
File type:	Directory
Content:	Biotechnology equipment
Language:	English, German, French

This is a product file containing information on 400 manufacturers of biotechnology equipment world-wide. The source for this file is the 1988 ACHEMA yearbook. Company names, addresses and a full text product description are searchable in English. Product names, descriptions and classification information are searchable in English, French and German.

BIOSIS Previews/RN

Available through:	STN, an almost identical database: Biosis Previews is available from Dialog
Producer:	BIOSIS
Coverage:	1969 to date
File size:	7.96 million records
Updates:	Weekly
File type:	Bibliographic
Content:	Biological sciences
Language:	English

The world's largest and most comprehensive life science database, BIOSIS Previews/RN covers original research reports, reviews and selected US patents in biological and biomedical areas, with subject coverage ranging from aerospace biology to zoology. Sources include periodicals, journals, conference proceedings, reviews, reports, patents and short communications. Bibliographic information, supplementary terms, abstracts and CAS Registry Numbers are all searchable.

CA

Available through:	STN, the same information is available through Dialog in the CA Search, Chemname and Chemsearch files
Producer:	Chemical Abstracts Service
Coverage:	1967 to date
File size:	10.2 million records
Updates:	Biweekly
File type:	Bibliographic
Content:	Chemistry
Language:	English

The Chemical Abstracts database covers all areas of chemistry, biochemistry and chemical engineering. CA contains records for documents reported in the printed *Chemical Abstracts* (CA) from 1967 to the present. Sources include journals, patents, technical reports, books, conference proceedings and dissertations from all areas of the field world-wide. Bibliographic terms, indexing terms, CAS Registry Numbers, substance names and structures are searchable. Over 87% of the records also contain CA abstracts, the text of which is searchable. Protein and nucleic acid sequences are also searchable. CAOLD is a similar but limited file covering the period prior to 1967 in which CAS Registry Numbers are searchable. The

CA file is accompanied by the Registry File, which includes structures, formulae, chemical and trade names for approximately 12 million substances. This file can be searched for all of these properties.

CABA/CAB Abstracts

Available through:	STN (CABA), Dialog (CAB Abstracts)
Producer:	CAB International
Coverage:	1979 to date
File size:	1.87 million records
Updates:	Monthly
File type:	Bibliographic
Content:	Agriculture
Language:	English

This database from the Commonwealth Agricultural Bureaux covers world-wide literature from all areas of agriculture and related sciences, including biotechnology, forestry and veterinary medicine. Sources include journals, books, reports, published theses, conference proceedings and patents. Bibliographic information, indexing terms and abstracts are searchable. Coverage in the Dialog file begins with 1972.

CANCERLIT

Available through:	Dialog
Producer:	US National Cancer Institute
Coverage:	1963 to the present
File size:	823 000 records
Updates:	Monthly
Content:	Cancer research
Language:	English

This is a comprehensive database of published literature on cancer research selected from more than 3500 biomedical journals, meeting proceedings, books, reports, etc. Topics include experimental and clinical therapy, chemical and viral carcinogenesis, biochemistry, immunology and physiology of cancer. *Medical Subject Heading* controlled vocabulary is used in indexing.

CASREACT

Available through:	STN
Producer:	Chemical Abstracts Service
Coverage:	1985 to date
File size:	Over 50 000 records and 740 000 single-step and 1.5 million multi-step reactions
Updates:	Biweekly
File type:	Bibliographic, structure
Content:	Chemical reactions
Language:	English

CASREACT contains information on reactions of organic substances, including organometallics and biomolecules, selected from papers reported in 104 core journals. CASREACT contains single and multi-step reaction information for all reactants, products, reagents, solvents and catalysts. CAS Registry Numbers, textual reaction information and yields are searchable. Bibliographic information, in-depth substance and subject indexing, and the associated CA abstract are displayable but not searchable.

CEBA

Available through:	STN
Producer:	Deutsche Gesellschaft für Chemisches Apparatewesen Chemische Technik und Biotechnologie e.V. and the Royal Society of Chemistry
Coverage:	1975 to date
File size:	134 000 records
Updates:	Monthly
File type:	Bibliographic
Content:	Chemical engineering and biotechnology
Language:	German with titles in English and German

The CEBA Chemical Engineering and Biotechnology Abstracts database contains references to chemical equipment and biotechnology publications. Sources for CEBA are approximately 950 German and foreign-language journals, reports, conference papers, books and dissertations. Bibliographic information, indexing terms and abstracts are all searchable. This file combines the older DECHEMA file and CEA (Chemical Engineering Abstracts) file produced by RSC.

Chapman and Hall chemical database (formerly HEILBRONN)

Available through:	Dialog
Producer:	Chapman & Hall Ltd
Coverage:	Current
File size:	96 976 records with information on 200 000 substances
Updates:	Semiannually

This database represents the complete text of the Chapman & Hall chemical dictionaries *Dictionary of Organic Compounds* and *Dictionary of Organometallic Compounds*. Other sources include *Carbohydrates, Amino Acids and Peptides, The Dictionary of Antibiotics and Related Compounds* and *The Dictionary of Organophosphorus Compounds*. Useful as a resource in chemical and chemical hazard identification, this database allows searching by name, synonym, Registry Numbers, molecular formula and weight, sources and physical properties.

CHEMLIST

Available through:	STN
Producer:	American Petroleum Institute
Coverage:	1979 to date
File size:	83 000 records
Updates:	Weekly
File type:	Directory
Content:	Regulated chemicals
Language:	English

The Regulated Chemicals Listing file contains information about chemical substances listed on the EPA TSCA (Environmental Protection Agency Toxic Substances Control Act) Inventory or those subject to regulations under TSCA or similar legislation. Chemical names, regulatory information and CAS Registry Numbers are searchable.

CJACS

Available through:	STN
Producer:	American Chemical Society
Coverage:	1982 to date (Langmuir — 1985 to date)
File size:	117 463 records
Updates:	Biweekly
File type:	Full-text
Content:	Chemical journals
Language:	English

Available on STN only, CJACS contains the complete and searchable text of 18 primary journals. These include: *Analytical Chemistry, Biochemistry, Environmental Science & Technology, Journal of Agricultural & Food Chemistry, Journal of Medicinal Chemistry, The Journal of Organic Chemistry, Macromolecules* and *Journal of the American Chemical Society*. All information is searchable except graphics and some chemical or mathematical equations. CAS Registry Numbers are also searchable.

CJAOAC

Available through:	STN
Producer:	Association of Official Analytical Chemists
Coverage:	1987 to date
File size:	1106 records
Updates:	Bimonthly
File type:	Full text
Content:	Analytical chemical journals
Language:	English

CJAOAC contains research papers in analytical chemistry. The source is the full text of the *Journal of the Association of Official Analytical Chemists*. All information from the journal is searchable except graphics and some chemical or mathematical equations.

CJELSEVIER

Available through:	STN
Producer:	Elsevier Science Publishers, B.V.
Coverage:	January 1990 to date
File size:	4573 records
Updates:	Weekly
File type:	Full-text
Content:	Chemical journals
Language:	English, French and German

CJELSEVIER contains research papers in chemistry from *Analytica Chimica Acta, Applied Catalysis, Carbohydrate Research* and the *Journal of Organometallic Chemistry*. All information from the journals is searchable except graphics and some chemical and mathematical equations.

CJRSC

Available through:	STN
Producer:	The Royal Society of Chemistry
Coverage:	1987 to date
File size:	17 788 records
Updates:	Biweekly
File type:	Full-text
Content:	Chemical journals
Language:	English

CJRSC contains full texts of research papers and communications on physical and analytical chemistry from 10 peer-reviewed journals of the Royal Society of Chemistry. All information in the journals is searchable except graphics and some chemical or mathematical equations.

CJVCH

Available through:	STN
Producer:	VCH Verlagsgesellschaft
Coverage:	1988 to date
File size:	1589 records
Updates:	Monthly
File type:	Full-text
Content:	Chemical journals
Language:	English

CJVCH contains research papers in chemistry with emphasis on inorganic and organic chemistry. Source for CJVCH is the international English edition of *Angewandte Chemie*. All information from the journals is searchable except graphics and some chemical or mathematical equations.

CSCHEM

Available through:	STN
Producer:	Directories Publishing Company, Inc.
Coverage:	Current data
File size:	133 584 records
Updates:	Reloaded annually
File type:	Directory
Content:	Chemical manufacturers
Language:	English

The CHEM SOURCES database contains information on chemical products and their suppliers. The CSCHEM file is equivalent to the Chemical and Trade Names sections of the printed directories CHEM-SOURCES-USA and CHEM SOURCES-INTERNATIONAL. Chemical names, names of suppliers, CAS Registry Numbers, and classification of trade name products are all searchable.

CSCORP

Available through:	STN
Producer:	Directories Publishing Company, Inc.
Coverage:	Current data
File size:	1600 records
Updates:	Reloaded annually
File type:	Directory
Content:	Chemical manufacturers
Language:	English

CSCORP contains directory information about companies that supply chemicals and chemical products, including the addresses of chemical product suppliers. The file is equivalent to the Company Directory and Trade Name sections of the printed directories CHEMSOURCES-USA and CHEM SOURCES-INTERNATIONAL. Company names and codes, addresses of the main office, divisions, branches, telephone numbers and classifications of chemical products are all searchable.

CSNB

Available through:	STN
Producer:	The Royal Society of Chemistry
Coverage:	1981 to date
File size:	22 800 records
Updates:	Monthly
File type:	Bibliographic
Content:	Health & safety
Language:	English

The Chemical Safety NewsBase provides access to chemical information related to fire and explosions, storage and transport, toxic substances, studies on laboratory animals, waste removal and other subjects related to health and safety in chemistry. Sources for CSNB include journals, books, conference papers, legislation and press releases. Bibliographic information, indexing terms, abstracts and CAS Registry Numbers are all searchable.

DEQUIP

Available through:	STN
Producer:	Deutsche Gesellschaft für Chemisches Apparatewesen, Chemische Technik und Biotechnologie e.V.
Coverage:	Current data
File size:	8000 product groups and 3500 companies
Updates:	Reloaded annually
File type:	Directory
Content:	Chemical engineering and biotechnology equipment
Language:	English, French and German

The DECHEMA Equipment Suppliers Data Bank supplies information on manufacturers of apparatus and technical equipment for chemical engineering and biotechnology. Companies from all parts of the world and their products (including protected trade names) are reviewed. This information corresponds to the *ACHEMA Yearbook* 1988. Company names, addresses and a full-text product description are searchable in English. Product names, descriptions and classification information are searchable in English, French and German.

Derwent Biotechnology Abstracts

Available through:	Dialog
Provider:	Derwent Inc.
Coverage:	1982 to date
File size:	123 000 records
Updates:	Monthly

This database has a comprehensive coverage of publications describing research in all fields of biotechnology including genetic engineering, biochemical engineering, fermentation, cell culture and waste disposal. Coverage includes journals and patents and each entry has an abstract of up to 200 words and complete bibliographic information.

EMBASE

Available through:	STN
Producer:	Elsevier Science Publishers B.V.
Coverage:	From 1980 to date
File size:	Approximately 3.3 million records
Updates:	Weekly
File type:	Bibliographic
Content:	Medicine, pharmacology, bioscience
Language:	English

The EMBASE file contains 44 printed abstract bulletins (*Excerpta Medica*) and two printed drug literature bibliographies (*Adverse Reaction Titles* and *Drug Literature Index*) and other information not included in these publications. Worldwide literature on biomedical topics is included in the database. The records in the file contain bibliographic and indexing information, including MALIMET/EMTREE drug and medical descriptors, EMTREE codes, EMTAGS, chemical names (drug trade names), corporate names (manufacturers) and CAS Registry Numbers. All of this information is searchable.

FSTA/Food Science and Technology Abstracts

Available through:	STN (FSTA), Dialog (Food Science and Technology Abstracts)
Producer:	International Food Information Service
Coverage:	From 1969 (Dialog) and 1981 (STN)
File size:	201 636 records
Updates:	Monthly
File type:	Bibliographic
Content:	Food science
Language:	English

FSTA provides world-wide coverage of all scientific and technological aspects of the processing and manufacturer of human food products. Coverage includes basic food sciences, biotechnology, hygiene and toxicology, engineering, packaging and all individual foods and food products. Sources for FSTA include 1800 journals, books, reviews, conference proceedings, patents, standards and legislation. Bibliographic information, indexing terms and abstracts are all searchable. The Dialog file begins coverage with 1969.

HODOC

Available through:	STN
Producer:	CRC Press
Coverage:	*CRC Handbook of Data on Organic Compounds*
File size:	25 500 compounds
Updates:	Reloaded annually
File type:	Numeric
Content:	Chemical and physical property data
Language:	English

HODOC covers the most frequently used physical and chemical data of organic compounds and is an extensive source of spectral data. When available, physical data in this database include boiling point, melting point, density, refractive index, optical rotary power, solubility, crystal property and molecular weight calculated according to the International Atomic Weight values. Source for HODOC is the *CRC Handbook of Data on Organic Compounds* (HODOC II). Substance information, property information, element terms and CAS Registry Numbers are all searchable.

IFICDB

Available through:	STN
Producer:	IFI/Plenum Data Corporation
Coverage:	1950 to date
File size:	2.08 million records
Updates:	Weekly
File type:	Bibliographic
Content:	US patents
Language:	English

The IFI Comprehensive file contains records for US utility patents, reissue patents and defensive publications. IFICDB includes patents related to chemistry from 1950 to the present; design patents from 1980 to the present; and mechanical and electrical patents from 1963 to the present; and indexing by Uniterms (see IFIREF below) for chemical patents. Source for IFICDB is the US Patent and Trademark Office *Official Gazette*. Bibliographic information, indexing terms, abstracts and CAS Registry Numbers are all searchable.

IFIPAT

Available through:	STN
Producer:	IFI/Plenum Data Corporation
Coverage:	1950 to date
File size:	2.22 million records
Updates:	Weekly
File type:	Bibliographic
Content:	US patents
Language:	English

The IFI Patent file contains records for all granted US utility patents, reissue patents and defensive publications. IFIPAT includes patents related to chemistry from 1950 to the present; design patents from 1980 to the present; and mechanical and electrical patents from 1963 to the present. Source for IFIPAT is the US Patent and Trademark Office *Official Gazette*. Bibliographic information, abstracts and CAS Registry Numbers are all searchable.

IFIREF

Available through:	STN
Producer:	IFI/Plenum Data Corporation
Coverage:	Current
File size:	110 000 and 50 000 Uniterms
Updates:	Reloaded quarterly
File type:	Bibliographic
Content:	US patents
Language:	English

IFI Uniterm and US Class Reference file is a special reference file for the other IFI files. IFIREF contains US Patent Office classification codes and titles, plus authority data for general, fragment and compound Uniterms (five-digit codes or text terms used as search shortcuts). Sources for IFIREF include the US Patent and Trademark Office *Manual of Classification, Uniterm Authority Lists* and *IFI Compound Term Vocabulary*. Classification codes, indexing terms, molecular formula information and Uniterm codes are all searchable.

IFIRXA

Available through:	STN
Producer:	IFI/Plenum Data Corporation
Coverage:	1975 to date
File size:	32 000 records
Updates:	Bimonthly
File type:	Bibliographic
Content:	US patents
Language:	English

IFI Reassignment and Reexamination file contains records, since 1975, for those US patents that have been reassigned, because ownership of the patent has been transferred, or reexamined, since 1981, because of patentability of questions. Source for IFIRXA is the US Patent and Trademark Office. Bibliographic information, former and new assignments and reassignments or reexamination dates are all searchable.

IFIUDB

Available though:	STN
Producer:	IFI/Plenum Data Corporation
Coverage:	1950 to date
File size:	2.22 million records
Updates:	Weekly
File type:	Bibliographic
Content:	US patents
Language:	English

IFIUDB contains records for granted US utility patents, reissue patents and defensive publications. The IFIUDB database contains in-depth indexing by Uniterms (and all IFIPAT information) for all chemical patients. Source for IFIUDB is the US Patent and Trademark Office *Official Gazette*. Bibliographic information, indexing terms, abstracts and CAS Registry Numbers are all searchable.

INPADOC

Available through:	STN
Producer:	International Patent Document Center
Coverage:	1968 to date
File size:	17.4 million patent records, 20 million legal records
Updates:	Weekly
File type:	Bibliographic
Content:	International patents
Language:	Legal status in English, rest of text in language of the original patent

The International Patent Documentation file contains bibliographic and family data for patent documents and utility models of 56 patent issuing organizations, including the European Patent Organization and the World Intellectual Property Organization (WIPO). Bibliographic information, patent family information and classification codes are all searchable.

INPAMONITOR

Available through:	STN
Producer:	International Patent Document Center
Coverage:	Current records (4 weeks)
File size:	80 000 patent records, 250 000 legal status records
Updates:	Weekly
File type:	Bibliographic
Content:	Current international patents
Language:	English

INPAMONITOR contains the most current bibliographic citations on patent and utility models published during the past 4 weeks. Included are the patent and utility models published by 56 patent organizations as well as the legal status data published by 11 patent organizations.

JICST-E

Available through:	STN
Producer:	Japan Information Center of Science and Technology
Coverage:	1985 to date
File size:	1 357 655 records
Updates:	Biweekly
File type:	Bibliographic
Content:	Multidisciplinary
Language:	English

JICST-E contains abstracts of Japanese literature on chemistry and the chemical industry, engineering, pharmacology, the life sciences and medical science. Sources include journals, serials, conference proceedings and technical reports. Bibliographic information, indexing terms and abstracts are all searchable. An on-line thesaurus is available for the controlled terms (/CT) field.

Life Sciences Collection

Available through:	Dialog
Producer:	Cambridge Scientific Abstracts
Coverage:	1978 to date
File size:	1 400 000 records
Updates:	Monthly

Life Sciences Collection contains abstracts of literature in the fields of animal behavior, biochemistry, ecology, endocrinology, entomology, genetics, immunology, microbiology, oncology, neuroscience, toxicology and virology, among others. This file corresponds to a series of 17 abstracting journals including *Marine Biotechnology Abstracts, Oncogenes and Growth Factor Abstracts* and *Human Genome Abstracts: Basic Research and Clinical Applications*. The coverage includes journal articles, conference proceedings and reports.

MEDLINE

Available through:	STN, Dialog
Producer:	US National Library of Medicine
Coverage:	1972 to date
File size:	5 867 148 million records
Updates:	Biweekly
File type:	Bibliographic
Content:	Medicine
Language:	English

MEDLINE contains information on every area of medicine. The MEDLINE database corresponds to the *Index Medicus*, the *Index to Dental Literature* and the *International Nursing Index*, Bibliographic information, indexing terms, abstracts and CAS Registry Numbers are all searchable. An on-line thesaurus is available for the Medical Subject Headings (/MN). Controlled Terms (/CT) and Chemical Name (/CN) fields. The Dialog file starts coverage with 1966.

PHYTOMED

Producer:	Biologische Bundesanstalt für Land- and Forstwirtschaft
Coverage:	1965 to date
File size:	362 000 records
Updates:	Quarterly
File type:	Bibliographic
Content:	Plant science
Language:	English and German

PHYTOMED contains citations of literature in the field of plant science (phytomedicine), toxicology and ecology. Sources for PHYTOMED include journals, books, conference proceedings, reprints and non-conventional literature. Bibliographic information, indexing terms and abstracts are all searchable in German. Titles appear in the original language of publication. Most are available in English (all titles have appeared in English since 1984). Abstracts and bibliographic information appear in either German or English.

3. NUCLEIC ACID SEQUENCE DATABASES

GenBank

GenBank is a nucleic acid sequence database of all published, and many unpublished, DNA and RNA sequences, and currently contains some 45 000 entries of 56 000 nucleotides. The database is run by IntelliGenetics, Inc. on behalf of the United States Department of Energy.

NCBI-GenBank

Producer:	National Center for Biotechnology Information, National Library of Medicine, National Institutes of Health
Coverage:	1981 to date
File size:	86 626 sequences, 110 208 764 bases
Updates:	Bi-monthly updates for CD-ROM; daily updates by FTP* to 'ncbi.nlm.nih.gov.'
File type:	ASCII flat file
Content:	Nucleic acid sequences
Language:	English

NCBI-Genbank is a nucleic acid sequence database that contains sequences directly submitted by authors. Searchable fields include: locus, definition, accession, keywords, source, organism, bibliographic information, comments, features, base count and origin.

Contact address:	NCBI GenBank National Library of Medicine Bldg. 38A, Rm. 8N-809 8600 Rockeville Pike Bethesda, MD 20894
Electronic mail:	info@ncbi.nlm.nih.gov

*File transfer protocol (FTP): the transfer of files from a donor to a recipient at the request of the recipient over a network such as INTERNET.

4. CONTACT ADDRESSES

STN

United States:
STN International
c/o Chemical Abstracts Service
PO Box 3012
Columbus
OH 43210-0012

Phone: 800-753-4227 (North America) or 614-447-3600 (world-wide)
Fax: 614-447-3713

Europe:
STN International
c/o FIZ Karlsruhe
Postfach 2465
W-7500 Karlsruhe 1
Germany

Phone: (+49) 7247/808-555
Fax: (+49) 7247/808-666

Japan:
STN International
c/o The Japan Information Center of Science and Technology (JICST)
5-2, Nagato 2-chome
Chiyoda-ku, Tokyo 100
Japan

Phone: (+81) 03-3581-6411
Fax: (+81) 3-3581-6446

Dialog

United States:
Dialog Information Services Inc.
3460 Hillview Avenue
Palo Alto
CA 94304

Phone: 800-334-3564 or 415-858-3785
Fax: 415-858-7069

Europe:
Dialog Information Retrieval Service
PO Box 188
Oxford
OX1 5AX
UK

Phone: (+44)(0)865 326226
Fax: (+44)(0)865 736354

CHAPTER 16
ATOMIC WEIGHTS AND MATHEMATICAL FORMULAE
D. Rickwood

1. ATOMIC WEIGHTS
Table 1. Atomic weights

Name	Symbol	Atomic number	Atomic weight
Aluminum	Al	13	26.98
Antimony	Sb	51	121.75
Argon	Ar	18	39.95
Arsenic	As	33	74.92
Barium	Ba	56	137.34
Beryllium	Be	4	9.01
Bismuth	Bi	83	208.98
Boron	B	5	10.81
Bromine	Br	35	79.91
Cadmium	Cd	48	112.40
Calcium	Ca	20	40.08
Carbon	C	6	12.01
Cerium	Ce	58	140.12
Cesium	Cs	55	132.91
Chlorine	Cl	17	35.45
Chromium	Cr	24	51.99
Cobalt	Co	27	58.93
Copper	Cu	29	63.54
Dysprosium	Dy	66	162.50
Erbium	Er	68	167.26
Europium	Eu	63	151.96
Fluorine	F	9	18.99
Gadolinium	Gd	64	157.25
Gallium	Ga	31	69.72
Germanium	Ge	32	72.59
Gold	Au	79	196.90
Hafnium	Hf	72	178.49
Helium	He	2	4.00
Holmium	Ho	67	164.93
Hydrogen	H	1	1.01
Indium	In	49	114.82
Iodine	I	53	126.90
Iridium	Ir	77	192.20
Iron	Fe	26	55.85
Krypton	Kr	36	83.80
Lanthanum	La	57	138.91
Lead	Pb	82	207.19

Table 1. Continued

Name	Symbol	Atomic number	Atomic weight
Lithium	Li	3	6.94
Lutetium	Lu	71	174.97
Magnesium	Mg	12	24.31
Manganese	Mn	25	54.94
Mercury	Hg	80	200.59
Molybdenum	Mo	42	95.94
Neodymium	Nd	60	144.24
Neon	Ne	10	20.18
Nickel	Ni	28	58.71
Niobium	Nb	41	92.91
Nitrogen	N	7	14.01
Osmium	Os	76	190.20
Oxygen	O	8	16.00
Palladium	Pd	46	106.40
Phosphorus	P	15	30.97
Platinum	Pt	78	195.09
Potassium	K	19	39.10
Praseodymium	Pr	59	140.91
Rhenium	Re	75	186.20
Rhodium	Rh	45	102.91
Rubidium	Rb	37	85.47
Ruthenium	Ru	44	101.07
Samarium	Sm	62	150.35
Scandium	Sc	21	44.96
Selenium	Se	34	78.96
Silicon	Si	14	28.09
Silver	Ag	47	107.87
Sodium	Na	11	22.99
Strontium	Sr	38	87.62
Sulfur	S	16	32.06
Tantalum	Ta	73	180.95
Tellurium	Te	52	127.60
Terbium	Tb	65	158.92
Thallium	Tl	81	204.37
Thorium	Th	90	232.04
Thulium	Tm	69	168.93
Tin	Sn	50	118.69
Titanium	Ti	22	47.90
Tungsten	W	74	183.85
Uranium	U	92	238.03
Vanadium	V	23	50.94
Xenon	Xe	54	131.30
Ytterbium	Yb	70	173.04
Yttrium	Y	39	88.91
Zinc	Zn	30	65.37
Zirconium	Zr	40	91.22

2. MATHEMATICAL FORMULAE

2.1. Lengths, areas and volumes in some common geometric figures

Planar figures

Triangle. If p_a, p_b and p_c are the perpendiculars from A, B and C to a, b and c respectively,

$$\text{area} = \tfrac{1}{2}ap_a = \tfrac{1}{2}bp_b = \tfrac{1}{2}cp_c = \sqrt{\{s(s-a)(s-b)(s-c)\}} \quad \left(s = \frac{a+b+c}{2}\right).$$

Rectangle. If the sides are a and b,

$$\text{area} = ab.$$

Parallelogram. If the sides are a and b, perpendicular height p, and angle between a and b is $\theta°$,

$$\text{area} = ap = ab \sin \theta.$$

Rhombus. If the diagonals are c and d,

$$\text{area} = \tfrac{1}{2}cd.$$

Trapezium. If the parallel sides are a and b, and perpendicular height p,

$$\text{area} = \tfrac{1}{2}p(a+b).$$

Quadrilateral. If the diagonals are c and d, and the angle between them $\theta°$,

$$\text{area} = \tfrac{1}{2}cd \sin \theta.$$

Regular polygon. If there are n sides, each of length l,

$$\text{area} = \tfrac{1}{4}nl^2 \cot \frac{180}{n};$$

$$\text{radius of inscribed circle} = \frac{l}{2} \cot \frac{180}{n};$$

$$\text{radius of circumscribed circle} = \frac{l}{2} \operatorname{cosec} \frac{180}{n}.$$

Circle (radius r). Circumference $= 2\pi r$.

Length of arc subtended by radii at an angle of $\theta° = \dfrac{\pi r \theta}{180}$.

Length of chord subtended by radii at an angle of $\theta° = 2r \sin \dfrac{\theta}{2}$.

Area of circle $= \pi r^2$.

Area of sector, with the radii at an angle of $\theta°$, and length of arc $s = \dfrac{\pi r^2 \theta}{360}$,

$$= \dfrac{sr}{2}.$$

Area of segment, with the radii at an angle of $\theta° = \dfrac{\pi r^2 \theta}{360} - \dfrac{r^2 \sin \theta}{2}$.

Area of ring between two circles, not necessarily concentric, but with one enclosing the other, and having radii r_1 and $r_2 = \pi(r_1 + r_2)(r_1 - r_2)$.

Ellipse. If the semi-axes are a and b

$$\text{Circumference} = 2\pi \sqrt{\left(\dfrac{a^2 + b^2}{2}\right)} \text{ (approx.).}$$

$$\text{Area} = \pi ab.$$

Solids

Pyramid (regular). If the slant height $= l$, length of one side of the base $= a$ and the number of sides n,

$$\text{lateral area} = \tfrac{1}{2} nal.$$

In any pyramid,

$$\text{volume} = \tfrac{1}{3} \text{ area of base} \times \text{height}.$$

Sphere (radius r).

$$\text{Surface area} = 4\pi r^2.$$
$$\text{Volume} = \tfrac{4}{3} \pi r^3.$$

Area of curved surface of spherical segment of height $h = 2\pi rh$.

Volume of spherical segment of height $h = \tfrac{1}{3}\pi h^2(3r - h) = \tfrac{1}{6}\pi h(h^2 + 3a^2)$,

where a is the radius of the base of segment.

Spheroid. Major and minor axes a and b respectively, and

$$\varepsilon = \text{eccentricity} = 1 - \left(\frac{b}{a}\right)^2.$$

Oblate spheroid. Formed by rotating an ellipse about its minor axis.

$$\text{Surface area} = 2\pi a^2 + \pi \frac{b^2}{\varepsilon} \log_e \frac{1+\varepsilon}{1-\varepsilon}.$$

$$\text{Volume} = \tfrac{4}{3}\pi a^2 b.$$

Prolate spheroid. Formed by rotating an ellipse about its major axis.

$$\text{Surface area} = 2\pi b^2 + 2\pi \frac{ab}{\varepsilon} \sin^{-1} \varepsilon.$$

$$\text{Volume} = \tfrac{4}{3}\pi ab^2.$$

Cylinder (right). If the radius of the base is r, and l the length,

$$\text{area of curved surface} = 2\pi rh;$$

$$\text{volume} = \pi r^2 h.$$

Cone (right). If the radius of the base is r, and h the height,

$$\text{area of curved surface} = \pi r \sqrt{(r^2 + h^2)};$$

$$\text{volume} = \frac{\pi}{3} r^2 h.$$

Truncated cone (right). If the radii of the larger and smaller circles are r_1 and r_2 respectively, and the distance between them h,

$$\text{area of curved surface} = \pi(r_1 + r_2)\sqrt{\{h^2 + (r_1^2 - r_2^2)\}};$$

$$\text{volume} = \pi \frac{h}{3}(r_1^2 + r_1 r_2 + r_2^2).$$

2.2. Inter-relations of sides and angles in a plane triangle

In any triangle with angles A, B and C, and opposite sides, a, b and c, respectively, if R and r are the radii of the circumscribed and inscribed circles respectively, and $s = \tfrac{1}{2}(a+b+c)$, then

$$\frac{a}{\sin A} = \frac{b}{\sin B} = \frac{c}{\sin C} = \frac{abc}{2\sqrt{\{s(s-a)(s-b)(s-c)\}}} = 2R.$$

$$a^2 = b^2 + c^2 - 2bc.\cos A.$$

$$a = b \cos C + c \cos B.$$

$$\tan\tfrac{1}{2}(A - B) = \frac{a - b}{a + b} \cot\tfrac{1}{2}C.$$

$$\frac{a + b}{a - b} = \frac{\sin A + \sin B}{\sin A - \sin B} = \frac{\tan\tfrac{1}{2}(A + B)}{\tan\tfrac{1}{2}(A - B)} = \frac{\cot\tfrac{1}{2}C}{\tan\tfrac{1}{2}(A - B)}.$$

$$\sin A = \frac{2}{bc} \sqrt{\{s(s - a)(s - b)(s - c)\}}.$$

$$r = \sqrt{\left\{\frac{(s - a)(s - b)(s - c)}{s}\right\}}.$$

$$\sin\tfrac{1}{2}A = \sqrt{\left\{\frac{(s - b)(s - c)}{bc}\right\}}.$$

$$\cos\tfrac{1}{2}A = \sqrt{\left\{\frac{s(s - a)}{bc}\right\}}.$$

$$\tan\tfrac{1}{2}A = \sqrt{\left\{\frac{(s - b)(s - c)}{s(s - a)}\right\}} = \frac{r}{s - a}.$$

2.3. Trigonometrical data

Values of functions of some angles

	0°	30°	45°	60°	90°
sin	0	$\tfrac{1}{2}$	$\tfrac{\sqrt{2}}{2}$	$\tfrac{\sqrt{3}}{2}$	1
cos	1	$\tfrac{\sqrt{3}}{2}$	$\tfrac{\sqrt{2}}{2}$	$\tfrac{1}{2}$	0
tan	0	$\tfrac{\sqrt{3}}{3}$	1	$\sqrt{3}$	∞

Values for angles greater than 90°

$$\sin x = \sin(180 - x) = -\sin(180 + x) = -\sin(360 - x).$$

$$\cos x = -\cos(180 - x) = -\cos(180 + x) = \cos(360 - x).$$

$$\tan x = -\tan(180 - x) = \tan(180 + x) = -\tan(360 - x).$$

Relations of functions

$$\sin x = \cos(90 - x) = \frac{1}{\operatorname{cosec} x}.$$

$$\cos x = \sin(90 - x) = \frac{1}{\sec x}.$$

$$\tan x = \cot(90 - x) = \frac{1}{\cot x} = \frac{\sin x}{\cos x}.$$

$$\sin^2 x + \cos^2 x = 1.$$

$$1 + \tan^2 x = \sec^2 x.$$

$$1 + \cot^2 x = \operatorname{cosec}^2 x.$$

$$\operatorname{cosec} x = \cot x/2 - \cot x.$$

$$\sin(x \pm y) = \sin x \cos y \pm \cos x \sin y.$$

$$\cos(x \pm y) = \cos x \cos y \mp \sin x \sin y.$$

$$\tan(x \pm y) = \frac{\tan x \pm \tan y}{1 \mp \tan x \tan y}$$

$$\sin 2x = 2 \sin x \cos x.$$

$$\cos 2x = \cos^2 x - \sin^2 x = 2\cos^2 x - 1 = 1 - 2\sin^2 x.$$

$$\tan 2x = \frac{2 \tan x}{1 - \tan^2 x}.$$

$$\sin \frac{x}{2} = \pm \sqrt{\left(\frac{1 - \cos x}{2}\right)}.$$

$$\cos \frac{x}{2} = \pm \sqrt{\left(\frac{1 + \cos x}{2}\right)}.$$

$$\tan \frac{x}{2} = \pm \sqrt{\left(\frac{1-\cos x}{1+\cos x}\right)} = \frac{1-\cos x}{\sin x} = \frac{\sin x}{1+\cos x}.$$

$$\sin x \pm \sin y = 2 \sin \tfrac{1}{2}(x \pm y) \cos \tfrac{1}{2}(x \mp y).$$

$$\cos x + \cos y = 2 \cos \tfrac{1}{2}(x+y) \cos \tfrac{1}{2}(x-y).$$

$$\cos x - \cos y = -2 \sin \tfrac{1}{2}(x+y) \sin \tfrac{1}{2}(x-y).$$

$$\tan x \pm \tan y = \frac{\sin(x \pm y)}{\cos x \cos y}.$$

2.4. Mathematical series

$$e^x = 1 + x + \frac{x^2}{2!} + \frac{x^3}{3!} + \dots.$$

$$\log_e(1+x) = x - \tfrac{1}{2}x^2 + \tfrac{1}{3}x^3 - \tfrac{1}{4}x^4 + \dots$$

when $-1 < x < +1$.

$$\log_e x = 2(p + \tfrac{1}{3}p^3 + \tfrac{1}{5}p^5 + \dots)$$

where $p = \dfrac{x-1}{x+1}$ and $x > 0$.

$$\sin x = x - \frac{x^3}{3!} + \frac{x^5}{5!} - \dots.$$

$$\cos x = 1 - \frac{x^2}{2!} + \frac{x^4}{4!} - \dots.$$

$$\tan x = x + \frac{x^3}{3} + \frac{2x^5}{15} + \frac{17x^7}{315} + \frac{62x^9}{2835} + \dots.$$

Arithmetic progression
First term = a, last term = l, common difference = d, number of terms = n, sum of n terms = s.

$$l = a + (n-1)d$$

$$s = \frac{n}{2}(a+l) = \frac{n}{2}\{2a + (n-1)d\}.$$

Geometric progression
First term = a, last term = l, common ratio = r, number of terms = n, sum of n terms = s.

$$l = ar^{n-1}.$$

$$s = a\frac{1-r^n}{1-r} = \frac{lr-a}{r-1}.$$

$$1 + 2 + 3 + \ldots + n = \frac{n(n+1)}{2}.$$

$$1^2 + 2^2 + 3^2 + \ldots + n^2 = \frac{n(n+1)(2n+1)}{6}.$$

$$1^3 + 2^3 + 3^3 + \ldots + n^3 = \frac{n^2(n+1)^2}{4}.$$

$$n! = 1 \times 2 \times 3 \times \ldots \times n = e^{-n}n^n\sqrt{2\pi n} \text{ (approx.)}.$$

2.5. Mathematical constants

$e = 2.718282$	$\pi = 3.141593$
$e^2 = 7.389056$	$\pi^2 = 9.869604$
$e^{-1} = 0.367897$	$\pi^{-1} = 0.318310$
$e^{1/2} = 1.648721$	$\pi^{1/2} = 1.772454$

2.6. Standard equations

Straight line
If m is the slope (= tangent of angle of inclination), and c the intercept on the Y axis,

$$y = mx + c.$$

If the intercepts on the X and Y axes are a and b respectively,

$$\frac{x}{a} + \frac{y}{b} = 1.$$

If the length of the perpendicular from the origin is p, and its angle of inclination θ,

$$x \cos\theta + y \sin\theta = p.$$

Circle
If the centre is at a, b, and its radius is c,

$$(x-a)^2 + (y+b)^2 = c^2.$$

If the centre is at the origin,

$$x^2 + y^2 = c^2.$$

Parabola
If the origin is at the vertex, and f is the distance from the focus to the vertex,

$$y^2 = 4fx$$

$$2f = p = \text{semi-latus rectum}$$

$$y^2 = 2px.$$

If the pole is at the focus,

$$r = \frac{p}{1 - \cos\theta}.$$

If the pole is at the vertex,

$$r = \frac{2p \cos\theta}{\sin^2\theta}.$$

Ellipse
If the centre is at the origin, the semi-axes a and b,

$$\frac{x^2}{a^2} + \frac{y^2}{b^2} = 1.$$

Where the pole is at the centre,

$$r^2 = \frac{a^2 b^2}{a^2 \sin\theta + b^2 \cos^2\theta}.$$

Hyperbola
If the centre is at the origin, and semi-axes a and b,

$$\frac{x^2}{a^2} - \frac{y^2}{b^2} = 1.$$

Where the pole is at the centre,

$$r^2 = \frac{a^2 b^2}{a^2 \sin^2\theta - b^2 \cos^2\theta}.$$

Quadratic

$$y = ax^2 + bx + c.$$

When

$$y = 0, x = \frac{-b \pm \sqrt{(b^2 - 4ac)}}{2a}.$$

Exponential

$$y = ae^{bx}.$$

When

$$y = 1, x = -\frac{\log_e a}{b}.$$

2.7. Differentials and integrals

Differentials

$$d \cdot ax = a \cdot dx.$$

$$d \cdot uv = u \cdot dv + v \cdot du.$$

$$d \cdot \frac{u}{v} = \frac{v \cdot du + u \cdot dv}{v^2}.$$

$$d \cdot x^n = nx^{n-1} \cdot dx.$$

$$d \cdot x^y = yx^{y-1} dx + x^y \log_e x \cdot dy.$$

$$d \cdot e^{ax} = ae^{ax} \cdot dx.$$

$$da^x = a^x \log_e a \cdot dx.$$

$$d \log_a x = x^{-1} \log_a e \cdot dx.$$

$$d \sin x = \cos x \cdot dx.$$

$$d \cos x = -\sin x \cdot dx.$$

$$d \tan x = \sec^2 x \cdot dx.$$

Integrals

$$\int a \cdot dx = ax.$$

$$\int a \cdot f(x)\, dx = a\int f(x)\cdot dx.$$

$$\int (u+v)\, dx = \int u \cdot dx + \int v \cdot dx$$

u and v being any functions of x.

$$\int u \cdot dv = uv - \int v \cdot du.$$

$$\int x^n \cdot dx = \frac{x^{n+1}}{n+1}.$$

$$\int x^{-1} \cdot dx = \log \pm x.$$

$$\int e^{ax} \cdot dx = e^{ax}/a.$$

$$\int b^{ax} \cdot dx = b^{ax}/a \log b.$$

$$\int \log x \cdot dx = x \log x - x.$$

$$\int \sin x \cdot dx = -\cos x.$$

$$\int \cos x \cdot dx = \sin x.$$

$$\int \tan x \cdot dx = \log \cos x \text{ or } \log \sec x.$$

INDEX

acetylation   135, 221, 233–238
acids (selected)   22
acid–base indicators   29
actin-binding proteins   167
actinins   167
actinogelins   167
actinomycin   126, 148
actin   148, 167
acylation   135, 216
acylglycerols   271
adhesins   167
ADP-ribosylation   234, 236
adseverins   168
affinity chromatography   67
agarose gels   74
albumins   168
alkylating agents   148
α-amanitin   126
Amberlite resins   59, 61
amino acids   22, 31, 33
    abbreviations   261
    modifications of   216, 233
aminopeptidases   154
ammonium sulfate precipitation   33–35
Ampholines — see ampholytes
ampholytes   72
anchorins — see ankyrin
animal
    DNA   257–259
    glycoproteins   196
    lipids   268, 269, 271
    polysaccharides   313
ankyrins   168
annexins   168
antibiotics   255
apolipoproteins   169
arachidonic acid   291
atomic weights   345
autoclaves   329
autoradiography   39
    intensifier screens for   39
axonins   168

bacteria
    chlorophylls   279
    disinfection of   329
    DNA   256
    membranes   193
    polysaccharides   313
bacteriorhodopsins — see opsins
balanced salt solutions   10
base pairing of DNA   250
bases (selected)   22
bases of nucleic acids   248, 250
Becquerel   37
bile salts   300
Biogel   51
blood coagulation factors   170
bradykinin — see kinins
brevins   170
buffers
    acetate   9
    compatibility   2
    Good's   1, 3, 8
    pH ranges   2
    phosphate buffers   9
    reactivity   2
    temperature effects   1
    zwitterionic   1, 3, 8

cadherins   170
calbindins   170
calcimedins — see annexins
caldesmon   170
calelectrins — see annexins
calmodulin   171
calpactins — see annexins
calponins   171
carboxypeptidases   155
carbohydrates
    hydrolytic enzymes for   161
    linkage to proteins   198
    monosaccharides   305
    oligosaccharides   310

stereoisomerism   309
transporters of   205
carcinogens   323-328
carotenoids   279
carrier ampholytes — see ampholytes
cascade mechanisms   215
caseins   171
CD antigens   171
centrifugation
   calculation of RCF   85
   rotors
      applications   85
      derating   86
      k-factors   85
   tubes and bottles
      chemical resistance   89
      disinfection   96
      properties of   87, 88
      sterilization   96
Cerenkov counting   41
ceruloplasmins   171
chaotropic agents   13
chaperonins   172
chelating agents   11-13
chemotherapy agents   255
chlorophylls   172, 279
chromatography media
   affinity chromatography   67
   agarose   54
   cellulose media   50, 58
   controlled pore glass   57
   ion-exchange   58-63
   polymer based   56, 59, 63
   reversed phase   64
   silica media   50, 55, 62, 64
   size exclusion   52-57
   suppliers   66
chromochelatins — see metallothioneins
chromogranins   172
cingulins   172
clathrins   172
CMC — see critical micellar concentration
cobamide   137
coenzyme A   135, 253
coenzymes   130, 132-138, 252-254
cofilins   172
collagens   172
conalbumins   172
connexins   173
controlled-pore glass   57
critical micellar concentration   18
crystallins   173

Curies   37
cytochromes   134, 173

DAG — see diacylglycerol
databases   331
   contact addresses   343
denaturing agents
   nucleic acids   80
   proteins   13-15, 69
density from refractive index   98-99
deoxyribonucleases — see nucleases
deoxyribonucleosides   250
deoxyribonucleotides   251
desmins   174
desmoplakins   174
destrins   174
detergents
   anionic   13
   cationic   19
   non-ionic detergents   20
   zwitterionic detergents   19
diacylglycerol   294
disaccharides — see oligosaccharides
disinfection procedures   96, 329
dithiothreitol   17
DNA
   amounts per cell   256-259
   assays   262
   base pairing   250
   bases   248, 250
   synthesis inhibitors   255
DNase — see nucleases
Dowex resins   60
DTT — see dithiothreitol
dyneins   174
dystrophin   174

eicosanoids   288
EGF — see epidermal growth factor
elastins   174
elastonectins   174
electrophoresis
   buffers   71, 80
   denaturing (SDS-PAGE)   69
   isoelectric focusing   69, 73
   molecular weight markers   75, 81-83
   non-denaturing gels   69
   nucleic acids   74
   nucleoproteins   79
   proteins   69
   staining recipes   74, 84
   two-dimensional   71

endoproteases 151
entactins 175
enzymes
    allosteric 107, 115
    assays 130, 140
    Bi Bi reaction kinetics 111, 112
    coenzymes 130, 132–138, 252–254
    direct linear plot 107
    Eadie-Hofstee plot 106
    Hane-Woolf plot 105
    Hill plot 109
    inhibitors 115–129, 145
        competitive 117
        irreversible 115
        mixed inhibition 121
        non-competitive 120
        uncompetitive 118
    kinetics 101–125
    Lineweaver-Burke 105
    Michaelis-Menten equation 101, 103
    pH effects 130, 131
    progress curves 102
    storage 141
    substrate inhibition 124
epidermal growth factor 175
erythrocruorins 175
extensins 175

fatty acids
    monoenoic 268
    polyunsaturated 269
    saturated 268
ferredoxins 175
ferritins 175
fibrinogens 175
fibrins 175
fibronectins 176
filaggrins 176
filamins 176
fimbrins 176
flagellins 176
flavin coenzymes 137, 253
fluorography 39

G proteins 183
gamma counting 43
gangliosides 278
gel electrophoresis — see electrophoresis
gelsolins 176
genetic code 261
glyceroglycolipids 275

glycerophospholipids 272
glycocalyx of animals 196
glycosylated proteins 194
    modification of 205
    functions of 213
glycosyltransferases 207
Good's buffers 1, 3, 8
granins 177
growth factors 175

hazards
    biohazard disposals 329
    classification of 318
    symbols for 317
    toxic chemical disposal 322
heat-shock proteins 177
heme proteins 173, 175, 177, 182
hemocyanin 177
hemoglobins 177
hemopexins 177
high mobility group proteins 178
Hill plot 109
histones 178
HLA antigens 178
HMG proteins 178
HPLC media 50
hydrolytic enzymes 145–165
hydroxyeicosatetraenoic acids 290

immunoglobulins 178–180
infectious material, disposal of 329
inhibitors
    analogs 255
    disposal of 323
    enzymes 115–129, 236
    irreversible 115
    nucleases 148
    proteases 156–160
    protein synthesis 126 ff.
    respiratory 128
    reversible 115
    transcription 126
integrins 178
interferons 179
interleukins 179
involucrins 179
ion exchange media
    cellulose 58
    polymer 59–61
    silica 62–63
isoelectric focusing 69, 73
isoprenylation 233–238

k-factors of rotors   85
kallikrein   152
keratins   181
kinesins   181
kininogens   181
kinins   181

laminins   181
lamins   181
legumins   181
leukotrienes   290
lipid Chemical Abstract registry
   numbers   301

MAP — see microtubule-associated
   proteins
mathematical data
   area calculation   347
   constants   353
   differentials and integrals   355
   series expressions   352
   trigonometrical data   350
   volume calculation   347
membrane
   glycerophospholipids   273
   lipids   282
   proteins   197
metalloproteins   175, 177
metallothioneins   182
methylation   136, 233–238
microtubule-associated proteins   182
monosaccharides
   distribution of   306
   lipid linked   375
   nucleic acids sugars   248
   properties of   306
   stereoisomerism of   309
   types   305

mucins   182
muscle proteins   167, 183, 188
myoglobins   182
myosins   183

nicotinamide coenzymes   132, 142, 254
non-zwitterionic buffers   1, 5
nucleases
   inhibitors of   148
   properties of   146
nucleic acids
   base pairing   250
   bases   248, 250
   databases   342

   optical properties   250
   precipitants   14
nucleosides   250
nucleotides   250
   analogs   255
   derived compounds   252

oligosaccharides
   distribution of   310
   linkage to proteins   200
   lipid linked   275
   properties of   310
opsins   183
osteocalcins   183
osteonectins   183
osteopontins   184

patatins   184
PBS — see phosphate buffered saline
PCNA — see proliferating cell nuclear
   antigen
Pharmalyte   72
phosphate-buffered saline   10
phosphorylation   233–238
ping-pong kinetics   113
plakoglobins   184
plant
   DNA   260
   lipids   269, 273, 284, 286
   polysaccharides   313
platelet activating factor   275, 294
plectins   184
polyacrylamide gels   70, 72, 73, 78
polyADP-ribosylation — see ADP-
   ribosylation
polysaccharides   312
polyribosomes — see ribosomes
polysomes — see ribosomes
porins   184
precipitating agents   13–15
prolamins   185
proliferating cell antigen   185
prostaglandins   288
prostanoids   288
protamines   185
proteases
   aminopeptidases   154
   broad specificity   150
   carboxypeptidases   155
   endoproteases   151
   inhibitors   156–160
   types of   150–155
protein precipitants   14
protein synthesis factors   186
proteinases — see proteases
proteins
   glycosylated   194
   modification
      chemical   216–220

    reagent specificity   221
    reagents   216
    enzymic   234
        function of   235
        inhibitors of   237
        reversible   238
proteoglycans   197
protoplasting enzymes   162

radioactivity units   37
radioisotopes
    assays   141
    decontamination methods   46
    detection methods   41, 43
    half-life   39, 40
    properties of   38
    shielding for   46
    tracers   43
redox proteins   188, 173, 175
reference books   331
refractive index measurement of
    density   98–99
reverse phase chromatography   64–66
ribonucleases — *see* nucleases
ribonucleosides   250
ribonucleotides   250
ribosomal proteins   185
ribosomes
    electrophoresis of   74, 79
    proteins of   185
    RNA of   81
RNase — *see* nucleases
rotors for centrifuges
    applications   85
    chemical resistance of   89
    disinfection and sterilization of   97
    derating   86
    types   85

safety — *see* hazards
salts (selected)   22
scintillants   42
scintillation counting fluids   41
SDS   14, 15, 19, 69
Sephacryl   50, 53
Sephadex   50, 52
Sepharose   50, 54
Servalyte   72
size exclusion chromatography   52–57
SLS — *see* SDS
sodium dodecyl sulfate — *see* SDS
solvents (selected)   22

spectrins   185
sphingolipids   277
sterilization methods   96, 329
steroids
    analogs   296
    effects of   298
    hormones   295
sterols   279
sucrose
    distribution of   310
    properties of solutions   99
sugars — *see* monosaccharides and
    oligosaccharides
Superose   54
synapsins   187

talins   187
tau proteins   187
tenascins   188
teratogens   323
terpenoids   278
toxic chemical disposal   322
transcription inhibitors   126, 255
transferrins   188
trigonometrical formulae   350
Triton detergents   20
tropoelastins   188
tropomyosins   188
troponins   188
tubulins   188
two-dimensional gel electrophoresis   71

ubiquitins   189
Ultrogel   51

vicilins   189
vimentins   189
vinculins   189
virus
    DNA   260
    glycosylated proteins   194
    disinfection   329
vitamins   130
vitronectins   189

waxes   270

X-ray film for autoradiography   39

zwitterionic buffers   1, 3, 8
zwitterionic detergents   19